Communications
in Computer and Information Science   **511**

Commenced Publication in 2007
Founding and Former Series Editors:
Alfredo Cuzzocrea, Dominik Ślęzak, and Xiaokang Yang

More information about this series at http://www.springer.com/series/7899

Guy Plantier · Tanja Schultz
Ana Fred · Hugo Gamboa (Eds.)

Biomedical Engineering Systems and Technologies

7th International Joint Conference, BIOSTEC 2014
Angers, France, March 3–6, 2014
Revised Selected Papers

 Springer

Editors
Guy Plantier
ESEO
Angers Cedex 02
France

Tanja Schultz
Cognitive Systems Lab.
Karlsruhe Institute of Technology
Karlsruhe, Baden-Württemberg
Germany

Ana Fred
Technical University of Lisbon
Lisbon
Portugal

Hugo Gamboa
New University of Lisbon
Lisboa
Portugal

ISSN 1865-0929   ISSN 1865-0937 (electronic)
Communications in Computer and Information Science
ISBN 978-3-319-26128-7   ISBN 978-3-319-26129-4 (eBook)
DOI 10.1007/978-3-319-26129-4

Library of Congress Control Number: 2015956367

Printed on acid-free paper

This Springer imprint is published by SpringerNature
The registered company is Springer International Publishing AG Switzerland

Preface

The present book includes extended and revised versions of a set of selected papers from the 7th International Joint Conference on Biomedical Engineering Systems and Technologies (BIOSTEC 2014), held in Angers, Loire Valley, France, during March 3–6, 2014.

Co-organized by the ESEO Group, BIOSTEC was sponsored by the Institute for Systems and Technologies of Information, Control and Communication (INSTICC), in cooperation with the Special Interest Group on Artificial Intelligence (ACM Sigart), the Special Interest Group on Bioinformatics, Computational Biology, and Biomedical Informatics (ACM SIGBio), the Association for the Advancement of Artificial Intelligence (AAAI), EUROMICRO, and the International Society for Telemedicine & eHealth (ISfTeH). BIODEVICES was also technically co-sponsored by the European Society for Engineering and Medicine (ESEM), the Biomedical Engineering Society (BMES), and IEEE Engineering in Medicine and Biology Society (IEEE EMBS).

The main objective of the International Joint Conference on Biomedical Engineering Systems and Technologies is to provide a point of contact for researchers and practitioners interested in both theoretical advances and applications of information systems, artificial intelligence, signal processing, electronics, and other engineering tools in knowledge areas related to biology and medicine.

BIOSTEC is composed of five complementary and co-located conferences, each specialized in at least one of the aforementioned main knowledge areas, namely:

- International Conference on Biomedical Electronics and Devices – BIODEVICES
- International Conference on Bioimaging - BIOIMAGING
- International Conference on Bioinformatics Models, Methods, and Algorithms – BIOINFORMATICS
- International Conference on Bio-inspired Systems and Signal Processing – BIOSIGNALS
- International Conference on Health Informatics – HEALTHINF

The purpose of the International Conference on Biomedical Electronics and Devices (BIODEVICES) is to bring together professionals from electronics and mechanical engineering interested in studying and using models, equipment, and materials inspired from biological systems and/or addressing biological requirements. Monitoring devices, instrumentation sensors and systems, biorobotics, micro-nanotechnologies, and biomaterials are some of the technologies addressed at this conference.

The International Conference on Bioimaging (BIOIMAGING) encourages authors to submit papers to one of the main conference topics, describing original work, including methods, techniques, advanced prototypes, applications, systems, tools, or survey papers, reporting research results and/or indicating future directions.

The International Conference on Bioinformatics Models, Methods, and Algorithms (BIOINFORMATICS) intends to provide a forum for discussion to researchers and practitioners interested in the application of computational systems and information technologies to the field of molecular biology, including, for example, the use of statistics and algorithms to understand biological processes and systems, with a focus on new developments in genome bioinformatics and computational biology. Areas of interest for this community include sequence analysis, biostatistics, image analysis, scientific data management and data mining, machine learning, pattern recognition, computational evolutionary biology, computational genomics, and other related fields.

The goal of the International Conference on Bio-inspired Systems and Signal Processing (BIOSIGNALS) is to bring together researchers and practitioners from multiple areas of knowledge, including biology, medicine, engineering, and other physical sciences, interested in studying and using models and techniques inspired from or applied to biological systems. A diversity of signal types can be found in this area, including image, audio, and other biological sources of information. The analysis and use of these signals is a multidisciplinary area including signal processing, pattern recognition, and computational intelligence techniques, among others.

The International Conference on Health Informatics (HEALTHINF) aims to be a major meeting point for those interested in understanding the human and social implications of technology, not only in health-care systems but in other aspects of human–machine interaction such as accessibility issues and the specialized support to persons with special needs.

The joint conference, BIOSTEC, received 362 paper submissions from 59 countries, which demonstrates the success and global dimension of this conference. From these, 51 papers were published as full papers, 86 were accepted for short presentation, and another 93 for poster presentation. These numbers, leading to a "full-paper" acceptance ratio of about 14 % and an oral paper acceptance ratio close to 38 %, show the intention of preserving a high-quality forum for the next editions of this conference.

We would like to thank all participants. First of all the authors, whose quality work is the essence of the conference, and the members of the Program Committee, who helped us with their expertise and diligence in reviewing the conference papers. Last but not least, we would like to express our gratitude to the INSTICC staff for their excellent organizational support. Without their help, this conference would not happen.

April 2015

Alberto Cliquet Jr.
Mário Forjaz Secca
Jan Schier
Oscar Pastor
Christine Sinoquet
Harald Loose
Marta Bienkiewicz
Christine Verdier
Guy Plantier
Tanja Schultz
Ana Fred
Hugo Gamboa

Organization

Conference Co-chairs

Guy Plantier	ESEO, GSII, France
Tanja Schultz	Karlsruhe Institute of Technology, Germany
Ana Fred	Instituto de Telecomunicações, IST, University of Lisbon, Portugal
Hugo Gamboa	CEFITEC/FCT - New University of Lisbon, Portugal

Program Co-chairs

BIODEVICES

Alberto Cliquet Jr.	University of São Paulo and University of Campinas, Brazil

BIOIMAGING

Mário Forjaz Secca	CEFITEC, FCT/UNL, Portugal
Jan Schier	Institute of Information Theory and Automation of the ASCR, Czech Republic

BIOINFORMATICS

Oscar Pastor	Universidad Politécnica de Valencia, Spain
Christine Sinoquet	University of Nantes, France

BIOSIGNALS

Harald Loose	Brandenburg University of Applied Sciences, Germany

HEALTHINF

Marta Bienkiewicz	TUM, Technische Universität München, Germany
Christine Verdier	LIG - Joseph Fourier University of Grenoble, France

Organizing Committee

Marina Carvalho	INSTICC, Portugal
Helder Coelhas	INSTICC, Portugal
Vera Coelho	INSTICC, Portugal
Bruno Encarnação	INSTICC, Portugal
Ana Guerreiro	INSTICC, Portugal
André Lista	INSTICC, Portugal

Andreia Moita	INSTICC, Portugal
Raquel Pedrosa	INSTICC, Portugal
Vitor Pedrosa	INSTICC, Portugal
Cláudia Pinto	INSTICC, Portugal
Susana Ribeiro	INSTICC, Portugal
Sara Santiago	INSTICC, Portugal
Mara Silva	INSTICC, Portugal
José Varela	INSTICC, Portugal
Pedro Varela	INSTICC, Portugal

BIODEVICES Program Committee

Dharma Agrawal	University of Cincinnati, USA
Sameer Antani	National Library of Medicine, USA
Nizametin Aydin	Yildiz Technical University, Turkey
Dinesh Bhatia	Deenbandhu Chhotu Ram University of Science and Technology, India
Egon L. van den Broek	University of Twente/Radboud UMC Nijmegen, The Netherlands
Tom Chen	Colorado State University, USA
Wenxi Chen	The University of Aizu, Japan
Mireya Fernández Chimeno	Universitat Politècnica de Catalunya, Spain
James M. Conrad	University of North Carolina at Charlotte, USA
Carlos Correia	University of Coimbra, Portugal
Vladimir Devyatkov	Bauman Moscow State Technical University, Russian Federation
Maeve Duffy	NUI Galway, Ireland
G.S. Dulikravich	Florida International University, USA
Elisabetta Farella	University of Bologna, USA
Paddy French	Delft University of Technology, The Netherlands
Anselmo Frizera	UFES, Brazil
Joaquim Gabriel	Instituto de Engenharia Mecânica, University of Porto, Portugal
Hugo Gamboa	CEFITEC/FCT - New University of Lisbon, Portugal
Juan Carlos Garcia Garcia	Universidad de Alcala, Spain
Javier Garcia-Casado	Universitat Politècnica de València, Spain
Miguel Angel García Gonzalez	Universitat Politècnica de Catalunya, Spain
Xiyi Hang	California State University, USA
Hans L. Hartnagel	Technische Universität Darmstadt, Germany
Toshiyuki Horiuchi	Tokyo Denki University, Japan
Leonid Hrebien	Drexel University, USA
Jiri Jan	University of Technology Brno, Czech Republic
Sandeep K. Jha	Indian Institute of Technology Delhi, India
Alberto Cliquet Jr.	University of São Paulo and University of Campinas, Brazil

Bozena Kaminska	Simon Fraser University, Canada
Michael Kraft	Fraunhofer-Institute for Microelectronic Circuits and Systems, Germany
Ondrej Krejcar	University of Hradec Kralove, Czech Republic
Hongen Liao	Tsinghua University, China
Chwee Teck Lim	National University of Singapore, Singapore
Chin-Teng Lin	National Chiao Tung University, Taiwan
Mai S. Mabrouk	Misr University for Science and Technology, Egypt
Jarmo Malinen	Aalto University, Finland
Dan Mandru	Technical University of Cluj Napoca, Romania
Cheryl D. Metcalf	University of Southampton, UK
Vojkan Mihajlovic	Holst Centre/imec the Netherlands, The Netherlands
Saibal Mitra	Missouri State University, USA
Joseph Mizrahi	Technion, Israel Institute of Technology, Israel
Jorge E. Monzon	Universidad Nacional del Nordeste, Argentina
Raimes Moraes	Universidade Federal de Santa Catarina, Brazil
Raul Morais	UTAD - Escola de Ciencias e Tecnologia, Portugal
Umberto Morbiducci	Politecnico di Torino, Italy
Alexandru Morega	University Politehnica of Bucharest, Romania
Michael R. Neuman	Michigan Technological University, USA
Robert W. Newcomb	University of Maryland, USA
Mónica Oliveira	University of Strathclyde, UK
Abraham Otero	Universidad San Pablo CEU, Spain
Gonzalo Pajares	Universidad Complutense de Madrid, Spain
Danilo Pani	University of Cagliari, Italy
Sofia Panteliou	University of Patras, Greece
Kwang Suk Park	College of Medicine, Seoul National University, Republic of Korea
Jürgen Popp	Friedrich-Schiller-Universität Jena, Germany
Mark A. Reed	Yale University, USA
Dmitry Rogatkin	Moscow Regional Research and Clinical Institute "MONIKI", Russian Federation
Seonghan Ryu	Hannam University, Republic of Korea
Chutham Sawigun	Mahanakorn University of Technology, Thailand
Michael J. Schöning	FH Aachen, Germany
Fernando di Sciascio	Institute of Automatics National University of San Juan, Argentina
Rahamim Seliktar	Drexel University, USA
Mauro Serpelloni	University of Brescia, Italy
Alcimar Barbosa Soares	Universidade Federal de Uberlândia, Brazil
Sameer Sonkusale	Tufts University, USA
Paul Sotiriadis	National Technical University of Athens, Greece
Anita Lloyd Spetz	Linköpings Universitet, Sweden
João Paulo Teixeira	Polytechnic Institute of Bragança, Portugal
Bruno Wacogne	FEMTO-ST UMR CNRS 6174, France

Tim Wark	Commonwealth Scientific and Industrial Research Organisation, Australia
Huikai Xie	University of Florida, USA
Sen Xu	Bristol-Myers Squibb, Co., USA
Wuqiang Yang	University of Mancheter, UK
Hakan Yavuz	Çukurova Üniversity, Turkey
Stefan Zappe	Carnegie Mellon University, USA
Aladin Zayegh	Victoria University, Australia

BIODEVICES Additional Reviewer

W. Andrew Berger	University of Scranton, USA

BIOIMAGING Program Committee

Jesús B. Alonso	Universidad de Las Palmas de Gran Canaria, Spain
Fernando Alonso-Fernandez	Halmstad University, Sweden
Peter Balazs	University of Szeged, Hungary
José Bioucas-Dias	Instituto Superior Técnico, Lisboa, Portugal
Egon L. van den Broek	University of Twente/Radboud UMC Nijmegen, The Netherlands
Tom Brown	University of St. Andrews, UK
Katja Buehler	Vrvis Research Center, Austria
Begoña Calvo	University of Zaragoza, Spain
M. Emre Celebi	Louisiana State University in Shreveport, USA
Phoebe Chen	La Trobe University, Australia
László Czúni	University of Pannonia, Hungary
Dimitrios Fotiadis	University of Ioannina, Greece
Jing Hua	Wayne State University, USA
Patrice Koehl	University of California, USA
Kristina Lidayova	Uppsala University, Sweden
Xiongbiao Luo	Nagoya University, Japan
Bjoern Menze	ETH Zurich, MIT, Switzerland
Kunal Mitra	Florida Institute of Technology, USA
Joanna Olszewska	University of Gloucestershire, UK
Kalman Palagyi	University of Szeged, Hungary
George Panoutsos	The University of Sheffield, UK
Joao Papa	UNESP - Universidade Estadual Paulista, Brazil
Ales Prochazka	Institute of Chemical Technology, Czech Republic
Miroslav Radojevic	Erasmus MC, Biomedical Imaging Group Rotterdam, Rotterdam, The Netherlands
Carlo Sansone	University of Naples, Italy
Jan Schier	Institute of Information Theory and Automation of the ASCR, Czech Republic
Vaclav Smidl	UTIA, AV CR, Czech Republic

Michal Sorel	Institute of Information Theory and Automation of the ASCR, Czech Republic
Jindrich Soukup	Institute of Information Theory and Automation of the ASCR, Czech Republic
Filip Sroubek	Institute of Information Theory and Automation of the ASCR, Czech Republic
Carlos M. Travieso	University of Las Palmas de Gran Canaria, Spain
Vladimír Ulman	Masaryk University, Czech Republic
Sandra Rua Ventura	School of Allied Health Science, Portugal
Yuanyuan Wang	Fudan University, China
Pew-Thian Yap	University of North Carolina at Chapel Hill, USA
Zeyun Yu	University of Wisconsin at Milwaukee, USA
Li Zhuo	Beijing University of Technology, China

BIOIMAGING Additional Reviewers

| Thung Kim-Han | University of Malaya, Malaysia |
| Pei Zhang | The University of North Carolina at Chapel Hill, USA |

BIOINFORMATICS Program Committee

Mohamed Abouelhoda	Nile University, Egypt
Rini Akmeliawati	International Islamic University Malaysia, Malaysia
Tatsuya Akutsu	Kyoto University, Japan
Mar Albà	UPF/IMIM, Spain
Hesham Ali	University of Nebraska at Omaha, USA
Jens Allmer	Izmir Institute of Technology, Turkey
Paul Anderson	College of Charleston, USA
Péter Antal	Budapest University of Technology and Economics, Hungary
Joel Arrais	Universidade de Coimbra, Portugal
Charles Auffray	CNRS Institute of Biological Sciences, France
Rolf Backofen	Albert-Ludwigs-Universität, Germany
Davide Bau	Centre Nacional d'Analisi Genomica (CNAG), Spain
Tim Beissbarth	University of Göttingen, Germany
Shifra Ben-Dor	Weizmann Institute of Science, Israel
Thomas Holberg Blicher	University of Copenhagen, Denmark
Ulrich Bodenhofer	Johannes Kepler University Linz, Austria
Vincent Breton	CNRS, France
Carlos Brizuela	Centro de Investigación Científica y de Educación Superior de Ensenada, Baja California, Mexico
Egon L. van den Broek	University of Twente/Radboud UMC Nijmegen, The Netherlands
Conrad Burden	Australian National University, Australia
Heorhiy Byelas	University Medical Center Groningen, The Netherlands
Joaquín Cañizares	Universitat Politecnica de Valencia, Spain

Claudia Consuelo Rubiano Castellanos	Universidad Nacional de Colombia - Bogota, Colombia
Wai-Ki Ching	The University of Hong Kong, Hong Kong, SAR China
Francisco Couto	Universidade de Lisboa, Portugal
Antoine Danchin	AMAbiotics SAS, France
Thomas Dandekar	University of Würzburg, Germany
Sérgio Deusdado	Instituto Politecnico de Bragança, Portugal
Eytan Domany	Weizmann Institute of Science, Israel
Guozhu Dong	Wright State University, USA
Richard Edwards	University of Southampton, UK
André Falcão	Universidade de Lisboa, Portugal
Pedro Fernandes	Instituto Gulbenkian de Ciência, Portugal
Fabrizio Ferre	University of Rome Tor Vergata, Italy
António Ferreira	Universidade de Lisboa, Portugal
Liliana Florea	Johns Hopkins University, USA
Gianluigi Folino	Institute for High Performance Computing and Networking, National Research Council, Italy
Andrew French	University of Nottingham, UK
Holger Fröhlich	University of Bonn, Germany
Max H. Garzon	The University of Memphis, USA
Julian Gough	University of Bristol, UK
Reinhard Guthke	Hans Knoell Institute, Germany
Joerg Hakenberg	Medical School at Mount Sinai, USA
Emiliano Barreto Hernandez	Universidad Nacional de Colombia, Colombia
Hailiang Huang	Massachusetts General Hospital, USA
Yongsheng Huang	University of Michigan, USA
Daisuke Ikeda	Kyushu University, Japan
Seiya Imoto	University of Tokyo, Japan
Sohei Ito	National Fisheries University, Japan
Bo Jin	Sigma-aldrich, USA
Giuseppe Jurman	Fondazione Bruno Kessler, Italy
Tamer Kahveci	University of Florida, USA
Sami Khuri	San José State University, USA
Inyoung Kim	Virginia Tech, USA
Jirí Kléma	Czech Technical University in Prague, Czech Republic
Sophia Kossida	Academy of Athens, Greece
Malgorzata Kotulska	Wroclaw University of Technology, Poland
Ivan Kulakovskiy	VIGG RAS, EIMB RAS, Russian Federation
Lukasz Kurgan	University of Alberta, Canada
Yinglei Lai	George Washington University, USA
Ana Levin	GEM Biosoft, Spain
Matej Lexa	Masaryk University, Czech Republic
Xiaoli Li	Nanyang Technological University, Singapore
Leo Liberti	Ecole Polytechnique, France
Michal Linial	The Hebrew University of Jerusalem, Israel

Antonios Lontos	Frederick University, Cyprus
Shuangge Ma	Yale University, USA
Ketil Malde	Institute of Marine Research, Norway
Xizeng Mao	University of Georgia, Athens, USA
Elena Marchiori	Radboud University, The Netherlands
Majid Masso	George Mason University, USA
Pavel Matula	Masaryk University, Czech Republic
Petr Matula	Masaryk University, Czech Republic
Nuno Mendes	Instituto de Biologia Experimental e Tecnológica, Portugal
Nikolaos Michailidis	Aristoteles University of Thessaloniki, Greece
Satoru Miyano	The University of Tokyo, Japan
Saad Mneimneh	Hunter College CUNY, USA
Pedro Tiago Monteiro	INESC-ID, Portugal
Bernard Moret	École Polytechnique Fédérale de Lausanne, Switzerland
Burkhard Morgenstern	University of Göttingen, Germany
Shinichi Morishita	University of Tokyo, Japan
Vincent Moulton	University of East Anglia, UK
Chad Myers	University of Minnesota, USA
Radhakrishnan Nagarajan	University of Kentucky, USA
Arcadi Navarro	Universitat Pompeu Fabra, Spain
Jean-Christophe Nebel	Kingston University, UK
José Luis Oliveira	Universidade de Aveiro, Portugal
Hakan S. Orer	Hacettepe University, Turkey
Allan Orozco	University of Costa Rica, Costa Rica
Gustavo Parisi	Universidad Nacional de Quilmes, Argentina
Oscar Pastor	Universidad Politécnica de Valencia, Spain
Florencio Pazos	National Centre for Biotechnology, Spain
Marco Pellegrini	Consiglio Nazionale delle Ricerche, Italy
Matteo Pellegrini	University of California, Los Angeles, USA
Horacio Pérez-Sánchez	Catholic University of Murcia, Spain
Guy Perrière	Université Claude Bernard - Lyon 1, France
Michael R. Peterson	University of Hawaii at Hilo, USA
Francisco Pinto	Universidade de Lisboa, Portugal
Olivier Poch	Université de Strasbourg, France
Giuseppe Profiti	University of Bologna, Italy
Marylyn Ritchie	Pennsylvania State University, USA
Miguel Rocha	University of Minho, Portugal
David Rocke	University of California, Davis, USA
Simona E. Rombo	Università degli Studi di Palermo, Italy
Chiara Romualdi	Università degli Studi di Padova, Italy
Eric Rouchka	University of Louisville, USA
Carolina Ruiz	WPI, USA
J. Cristian Salgado	University of Chile, Chile
Armindo Salvador	Universidade de Coimbra, Portugal

Joao C. Setubal	Universidade de São Paulo, Brazil
Ugur Sezerman	Sabanci University, Turkey
Hamid Reza Shahbazkia	Universidade do Algarve, Portugal
Christine Sinoquet	University of Nantes, France
Pavel Smrz	Brno University of Technology, Czech Republic
Gordon Smyth	Walter and Eliza Hall Institute of Medical Research, Australia
Sérgio Sousa	REQUIMTE, Universidade do Porto, Portugal
Peter F. Stadler	Universität Leipzig - IZBI, Germany
Yanni Sun	Michigan State University, USA
David Svoboda	Masaryk University, Czech Republic
Peter Sykacek	BOKU: University of Natural Resources and Life Sciences, Austria
Sandor Szedmak	University of Innsbruck, Austria
Gerhard Thallinger	Graz University of Technology, Austria
Silvio C.E. Tosatto	Università di Padova, Italy
Jyh-Jong Tsay	National Chung Cheng University, Taiwan
Alexander Tsouknidas	Aristotle University of Thessaloniki, Greece
Massimo Vergassola	University of California San Diego (UCSD), USA
Allegra Via	Università di Roma La Sapienza, Italy
Juris Viksna	University of Latvia, Latvia
Thomas Werner	University of Michigan, Germany
Yanbin Yin	Northern Illinois University, USA
Jingkai Yu	Institute of Process Engineering, Chinese Academy of Sciences, China
Qingfeng Yu	Stowers Institute for Medical Research, USA
Erliang Zeng	University of Notre Dame, USA
Leming Zhou	University of Pittsburgh, USA

BIOINFORMATICS Additional Reviewers

Zhe Ji	Harvard Medical School, Broad Institute, USA
Artem Kasianov	VIGG, Russian Federation
Qin Ma	Computational Systems Biology Lab, USA
Min Zhao	Vanderbilt University, USA

BIOSIGNALS Program Committee

Jean-Marie Aerts	Katholieke Universitëit Leuven, Belgium
Jesús B. Alonso	Universidad de Las Palmas de Gran Canaria, Spain
Fernando Alonso-Fernandez	Halmstad University, Sweden
Sergio Alvarez	Boston College, USA
Julián David Arias	Universidad de Antioquia, Colombia
Luis Azevedo	Anditec, Portugal
Richard Bayford	Middlesex University, UK
Eberhard Beck	Brandenburg University of Applied Sciences, Germany

Peter Bentley	UCL, UK
Jovan Brankov	Illinois Institute of Technology, USA
Egon L. van den Broek	University of Twente/Radboud UMC Nijmegen, The Netherlands
Tolga Can	Middle East Technical University, Turkey
Maria Claudia F. Castro	Centro Universitário da FEI, Brazil
M. Emre Celebi	Louisiana State University in Shreveport, USA
Sergio Cerutti	Polytechnic University of Milan, Italy
Bor-Sen Chen	Tsing Hua University, Taiwan
Liang-Gee Chen	National Taiwan University Taiwan
Yang Quan Chen	University of California at Merced, USA
Joselito Chua	Monash University, Australia
Adam Czajka	Research and Academic Computer Network, Warsaw University of Technology, Poland
Justin Dauwels	NTU, Singapore
Gordana Jovanovic Dolecek	Institute INAOE, Mexico
Pedro Encarnação	Experiência, Portugal
Luca Faes	Università degli Studi di Trento, Italy
Dimitrios Fotiadis	University of Ioannina, Greece
Esteve Gallego-Jutglà	University of Vic, Spain
Hugo Gamboa	CEFITEC/FCT - New University of Lisbon, Portugal
Arfan Ghani	University of Bolton Greater Manchester, UK
Verena Hafner	Humboldt-Universität zu Berlin, Germany
Md. Kamrul Hasan	Bangladesh University of Engineering and Technology (BUET), Bangladesh
Thomas Hinze	Friedrich Schiller University Jena, Germany
Donna Hudson	University of California, San Francisco, USA
Nuri Firat Ince	University of Houston, USA
Christopher James	University of Warwick, UK
Jiri Jan	University of Technology Brno, Czech Republic
Bart Jansen	Vrije Universiteit Brussel, Belgium
Ashoka Jayawardena	University of New England, Australia
Visakan Kadirkamanathan	The University of Sheffield, UK
Yasemin Kahya	Bogazici University, Turkey
Shohei Kato	Nagoya Institute of Technology, Japan
Natalya Kizilova	Kharkov National University, Ukraine
Dagmar Krefting	Hochschule für Technik und Wirtschaft Berlin - University of Applied Sciences, Germany
KiYoung Lee	AJOU University School of Medicine, Republic of Korea
Lenka Lhotska	Czech Technical University in Prague, Czech Republic
Chin-Teng Lin	National Chiao Tung University, Taiwan
Ana Rita Londral	Universidade de Lisboa, Portugal
Harald Loose	Brandenburg University of Applied Sciences, Germany
Wenlian Lu	The University of Warwick, UK
Hari Krishna Maganti	Samsung, UK

Andreas Voss	University of Applied Sciences Jena, Germany
Yuanyuan Wang	Fudan University, China
Quan Wen	University of Electronic Science and Technology of China, China
Kerstin Witte	Otto von Guericke University Magdeburg, Germany
Didier Wolf	Research Centre for Automatic Control - CRAN CNRS UMR 7039, France
Dongrui Wu	GE Global Research, USA
Huikai Xie	University of Florida, USA
Pew-Thian Yap	University of North Carolina at Chapel Hill, USA
Chia-Hung Yeh	National Sun Yat-sen University, Taiwan
Nicolas Younan	Mississippi State University, USA
Rafal Zdunek	Wroclaw University of Technology, Poland
Huiru Zheng	University of Ulster, UK
Li Zhuo	Beijing University of Technology, China

BIOSIGNALS Additional Reviewer

| Soo Yeol Lee | Kyung Hee University, Republic of Korea |

HEALTHINF Program Committee

Nabil Alrajeh	King Saud University, Saudi Arabia
Sergio Alvarez	Boston College, USA
Flora Amato	Università degli Studi di Napoli Federico II, Italy
Francois Andry	Philips HealthCare, USA
Turgay Ayer	Georgia Tech, USA
Philip Azariadis	University of the Aegean, Greece
Adrian Barb	Penn State University, USA
Rémi Bastide	Jean-Francois Champollion University, France
Bert-Jan van Beijnum	University of Twente, The Netherlands
Marta Bienkiewicz	TUM, Technische Universität München, Germany
Egon L. van den Broek	University of Twente/Radboud UMC Nijmegen, The Netherlands
Federico Cabitza	Università degli Studi di Milano-Bicocca, Italy
Eric Campo	LAAS CNRS, France
Guillermo Lopez Campos	The University of Melbourne, Australia
Mamede de Carvalho	University of Lisbon, Portugal
Philip K. Chan	Florida Institute of Technology, USA
James Cimino	NIH Clinical Center, USA
Krzysztof Cios	Virginia Commonwealth University, USA
Miguel Coimbra	Universidade do Porto, Portugal
Emmanuel Conchon	University of Toulouse, IRIT/ISIS, France
Carlos Costa	Universidade de Aveiro, Portugal
Ricardo João Cruz-Correia	Universidade do Porto, Portugal

Chrysanne Di Marco	University of Waterloo, Canada
Yichuan (Daniel) Ding	University of British Columbia, Canada
Liliana Dobrica	University Politehnica of Bucharest, Romania
Stephan Dreiseitl	Upper Austria University of Applied Sciences at Hagenberg, Austria
Derek Flood	Dundalk Institute of Technology, Ireland
José Fonseca	UNINOVA, Portugal
Daniel Ford	IBM Research, USA
Christoph M. Friedrich	University of Applied Science and Arts Dortmund, Germany
Ioannis Fudos	University of Ioannina, Greece
Angelo Gargantini	University of Bergamo, Italy
Laura Giarré	Università degli Studi di Palermo, Italy
David Greenhalgh	University of Strathclyde, UK
Murat Gunal	Turkish Naval Academy, Turkey
Andrew Hamilton-Wright	Mount Allison University, Canada
Jesse Hoey	University of Waterloo, Canada
Chun-Hsi Huang	University of Connecticut, USA
Vojtech Huser	National Institutes of Health (NIH), USA
Ivan Evgeniev Ivanov	Technical University Sofia, Bulgaria
Stavros Karkanis	Technological Educational Institute of Central Greece, Greece
Anastasia Kastania	Athens University of Economics and Business, Greece
Helen Kelley	University of Lethbridge, Canada
Andreas Kerren	Linnaeus University, Sweden
Waldemar W. Koczkodaj	Laurentian University, Canada
Nan Kong	Purdue University, USA
Dimitri Konstantas	University of Geneva, Switzerland
Eduan Kotze	University of the Free State, South Africa
Giuseppe Liotta	University of Perugia, Italy
Daniel Lizotte	University of Waterloo, Canada
Nicolas Loménie	Université Paris Descartes, France
Guillaume Lopez	Aoyama Gakuin University, Japan
Martin Lopez-Nores	University of Vigo, Spain
Vanda Luengo	LIG Laboratory, France
Michele Luglio	University of Rome Tor Vergata, Italy
Mia Markey	The University of Texas at Austin, USA
Alda Marques	University of Aveiro, Portugal
José Luis Martínez	Universidade Carlos III de Madrid, Spain
Paloma Martínez	Universidad Carlos III de Madrid, Spain
Maria di Mascolo	Gscop Laboratory, INPG, France
Sally Mcclean	University of Ulster, UK
Marilyn McGee-Lennon	University of Glasgow, UK
Gianluigi Me	Università degli Studi di Roma Tor Vergata, Italy
Gerrit Meixner	Heilbronn University, Germany

Mohyuddin Mohyuddin   King Abdullah International Medical Research Center
                         (KAIMRC), Saudi Arabia
Christo El Morr   York University, Canada
Roman Moucek   University of West Bohemia, Czech Republic
Radhakrishnan Nagarajan University of Kentucky, USA
Ebrahim Nageba   Université Claude Bernard Lyon, France
Hammadi Nait-Charif   Bournemouth University, UK
Tadashi Nakano   Osaka University, Japan
Goran Nenadic   University of Manchester, UK
José Luis Oliveira   Universidade de Aveiro, Portugal
Rui Pedro Paiva   University of Coimbra, Portugal
Chaoyi Pang   The Australian e-Health Research Centre, CSIRO,
                         Australia
Danilo Pani   University of Cagliari, Italy
José J. Pazos-Arias   University of Vigo, Spain
Rosario Pugliese   Università di Firenze, Italy
Juha Puustjärvi   University of Helsinki, Finland
Arkalgud Ramaprasad   University of Miami, USA
Zbigniew W. Ras   University of North Carolina at Charlotte, USA
Reinhard Riedl   Bern University of Applied Sciences, Switzerland
Marcos Rodrigues   Sheffield Hallam University, UK
Valter Roesler   Federal University of Rio Grande do Sul, Brazil
Elisabetta Ronchieri   INFN, Italy
George Sakellaropoulos   University of Patras, Greece
Ovidio Salvetti   National Research Council of Italy - CNR, Italy
Akio Sashima   AIST, Japan
Jacob Scharcanski   UFRGS – Universidade Federal do Rio Grande do Sul,
                         Brazil
Bettina Schnor   Potsdam University, Germany
Luca Dan Serbanati   Politehnica University of Bucharest, Romania
Alejandro Pazos Sierra   University of A Coruña, Spain
Carla Simone   Università degli studi di Milano-Bicocca, Italy
Irena Spasic   University of Cardiff, UK
Jan Stage   Aalborg University, Denmark
Francesco Tiezzi   IMT - Institute for Advanced Studies Lucca, Italy
Ioannis G. Tollis   University of Crete, Greece
Vicente Traver   ITACA, Universidad Politécnica de Valencia, Spain
Alexey Tsymbal   Siemens AG, Germany
Gary Ushaw   Newcastle University, UK
Aristides Vagelatos   CTI, Greece
Christine Verdier   LIG - Joseph Fourier University of Grenoble, France
Francisco Veredas   Universidad de Málaga, Spain
Justin Wan   University of Waterloo, Canada
Rafal Wcislo   AGH - University of Science and Technology
                         in Cracow, Poland
Janusz Wojtusiak   George Mason University, USA

Serhan Ziya   University of North Carolina, USA
André Zúquete   IEETA/IT/Universidade de Aveiro, Portugal

HEALTHINF Additional Reviewers

Giovanni Bottazzi   University of Rome "Tor Vergata", Italy
Alessia Dessì   University of Cagliari, Italy
Giancarlo Facoetti   University of Bergamo, Italy
Iade Gesso   Università degli Studi di Milano-Bicocca, Italy
Atif Khan   University of Waterloo, Canada
Elyes Lamine   Ecole d'ingénieurs ISIS, France
Chengbo Li   University of Waterloo, Canada
Rhiannon Rose   University of Waterloo, Canada
Thomas Scheffler   Beuth University of Applied Sciences, Berlin, Germany
Andrea Vitali   Università degli studi di Bergamo, Italy

Invited Speakers

Mário Oliveira   Santa Marta Hospital, Portugal
Patrick Flandrin   ENS de Lyon, France
Cesare Stefanini   Scuola Superiore Sant'Anna, Italy
Arcadi Navarro   Universitat Pompeu Fabra, Spain
Boudewijn Lelieveldt   Leiden University Medical Center, The Netherlands

Contents

Invited Paper

Managing Systems in Cardiac Remote Monitoring: A Complex Challenge Turned into an Important Clinical Tool

Mario Oliveira[✉], Pedro S. Cunha, and Nogueira da Silva

Laboratory of Pacing and Electrophysiology, Cardiology Departement,
Santa Marta Hospital, Lisbon, Portugal
m.martinsoliveira@gmail.com

1 Introduction

With the increasing awareness of indications for cardiac devices, especially related to the clinical benefits of implantable cardioverter-defibrillators (ICD) and cardiac resynchronization therapy (CRT), the number of patients with implanted electronic devices (CIED) has been growing steadily [1–3]. Recent data from the country members of the European Heart Rhythm Association (EHRA) have shown a general trend to increase in the number of centers implanting CIED, and, despite great differences in implanting rates among EHRA members, a steadily increase in the number of implants in almost all countries [4]. Consequently, more patients require regular follow-up to ascertain technical integrity. In 2007, over 1.6 million CIED were implanted in the United States and Europe, which would translate to over 5.5 million patient-encounters per year [5]. In view of these numbers, facilitated methods for follow-up are required. This growth, primarily resulting from more complex devices, in a population that most of the times present heart failure (HF) symptoms with significant underlying comorbidities, has led to difficulties in providing the specialized follow-up. Current suggestions regarding in-office visits for patients with an ICD or a CRT recommend a minimum frequency of 2 to 4 scheduled appointments per year [6, 7]. In fact, considerable human and logistical resources are needed to provide appropriate care, particularly for regular interrogation of the technical parameters of different CIED, detection and resolution of problems, identification and treatment of arrhythmias via the ICD, ensuring biventricular stimulation to optimize CRT, and specialized clinical care. These services can only be provided by hospital teams that are trained and able to perform tasks that are complex and challenging. Therefore, remoe monitoring (RM) has gained significant attention regarding the potential impact on CIED management, particularly in order to perform face-to-face visits less frequently, while maintaining safety and effectiveness, allowing health team to check the device between follow-ups in a more comprehensive care for heart health.

© Springer International Publishing Switzerland 2015
G. Plantier et al. (Eds.): BIOSTEC 2014, CCIS 511, pp. 3–15, 2015.
DOI: 10.1007/978-3-319-26129-4_1

The advent of RM systems for CIED, which provides an access to complete information on device performance, as an alternative to the traditional outpatient visits, offers many options and, at the same time, raises many questions with regard to its implementation, organization of the obtained wealth of data, safety, legal issues and reimbursement [8].

In the last ten years, telemedicine systems for RM of these devices have become a reality and are increasingly used in clinical practice, enabling changes in the specialized follow-up of this population, with well documented benefits and levels of safety [9–11]. Today, the broad application of RM systems supports the capability to improve the care of CIED recipients, contributing to the optimization of healthcare resources. The ability of the device follow-up clinic personnel to review data has expanded and has challenged the traditional model of patient care after CIED implantation. Undoubtedly, this technology is becoming an integral part of the future treatment for many of these patients.

2   Remote Monitoring of Cardiac Electronic Implantable Devices

CIED require long-term regular follow-up interrogation in dedicated clinics. However, due to the increasing population referred for implantation and the resources required, routine in-clinic follow-up contribute with a significant burden to the already overstrained electrophysiology teams and hospital services. Also, time spent in patient's travelling to the hospital and waiting time in the outpatient clinic may be an important issue in optimal care.

RM is expanding rapidly for chronic follow-up. Therefore, issues ranging from clinical and technological aspects (particularly concerning the long-term performance of the devices) to implementation, management and organization, legal questions, data protection, and funding still a matter of debate. All major CIED manufacturers have developed systems to allow patients to have their devices interrogated remotely, using standard phone lines but also wireless cellular technology to extend telemetry links into the patient's location.

How does RM work? RM allows you to send comprehensive device information over a standard phone line or a mobile phone (GSM) network to a central computer (server), which can be reviewed by your hospital team on a secure website, allowing routine device follow-up or a special situation to be reviewed quickly and efficiently. It may represent a safe and effective alternative to conventional follow-up programmes, and contribute to cost savings in health care. This technology has been proven to be reliable, allowing early identification of device malfunction and minimizing the risk of underreporting. All RM systems are slightly different and use slightly different technology. The different cardiac implantable devices currently available are listed in Table 1 and Fig. 1.

The device can be interrogated manually using a wand linked to the monitor in the patient's home (usually by the patient's bedside), or automatically using wireless systems, in which data are sent regularly without the patient's involvement at intervals set by the hospital team. All systems allow data to be sent when scheduled and according to clinical circumstances, as agreed between the patient and the team.

Data are transmitted to a central (internet-based) data repository. The information is formatted and transmitted to a central information service, through the internet fixed telephone lines at standard call cost in the CareLink™, Latitude™ and Merlin™ systems, and through the GSM cell network for the Home Monitoring™ and Smart-View™ systems (Table 1). Each center has access limited to its patient's using a password code in a web page in order to analyze the dynamic parameters of various

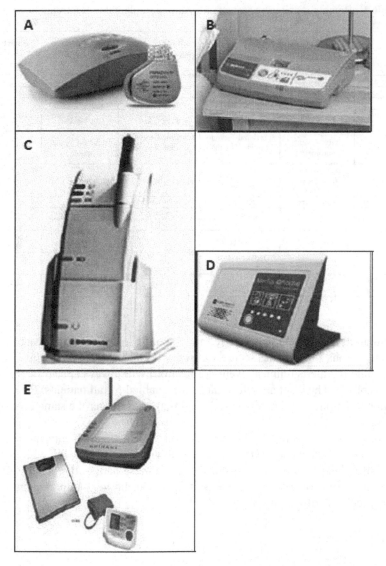

Fig. 1. Remote Monitoring Systems for follow-up of cardiac implantable devices. A – SmartView (Sorin, Italy); B – CareLink (Medtronic, USA); C – Home Monitoring (Biotronik, Germany); D – Merlin (S. Jude Medical, USA); E – Latitude (Boston Scientifgic, USA).

devices, including stored and measured information about the lead(s), sensor(s), battery and the implanted pulse generator function, as well as data collected about the patient's heart rhythm, detection of arrhythmia episodes recorded on intracavitary electrograms, therapies delivered by the device, the percentage of different pacing modes and tachyarrhythmias treated. These parameters can be configured individually to define alert levels according to the potential clinical impact of the alterations detected.

Table 1. Remote monitoring systems for cardiac implantable electronic devices.

	Home Monitoring™ (Biotronik, Germany)	CareLink Network™ (Medtronic Inc., USA)	Latitude System™ (Boston Scientific, USA)	Merlin.net™ (St. Jude Medical, USA)	Smartview Solution™ (Sorin, Italy)
FDA approval	2001	2005	2006	2007	2013
Characteristics	portable	stationary	stationary	stationary	stationary
Telemetry	wireless	wireless/antenna	wireless/antenna	wireless	wireless
Communication to server	mobile/analog line	analog line	analog line	analog line	mobile/analog line
Transmission	daily + events (automatic)	scheduled + events (initiated by the patient)	scheduled + events (initiated by the patient)	initiated by the patient	
Active notification of events	fax, internet, e-mail, text message	e-mail, text message	fax, phone	fax, internet, EMR	fax, internet, e-mail, text message
Events detection	<24 hours	<24 hours	<24 hours	-	<24 hours
Data storage	long-term	long-term	long-term	long-term	long-term
EMR interphase	HL7	HL7	HL7	HL7	PDF, XML
Holter transmission	>45 sec	10 sec	10 sec	30 sec	real-time + episodes
Sensor	IC monitor	Optivol, Cardiac compass	weight, blood pressure	-	SonR
Repercussion on battery longevity	low	high	high	-	very low

Yellow alerts may be remotely selected and re-configured according to each patient's clinical indications through the secure website without bringing the patient into the hospital, while red alerts (clinical event notification) will always appear on the secure website (Table 2). Thus, not all yellow alerts are enabled for all patients. Patients may trigger more than one alert simultaneously and may also trigger the same type of alert several times.

Who deal with the telemonitoring information? The patient's healthcare team, that has access to the clinical and device status. The device managing team, while working in a multidisciplinary approach, may communicate these data to the health following team in a collaborative patient management. This can be very important in clinical conditions, such as arrhythmic events or HF (Fig. 2).

Table 2. Remote yellow alert configuration (A) and red alert notifications (B) A - Yellow alert configuration B -Red alert notifications

A - Yellow alert configuration

- **Arrhythmias**
  - Shock therapy delivered to convert arrhythmia
  - Accelerated arrhythmia episode (ventricular)
  - Atrial Arrhythmia Burden
  - Patient-triggered event stored

- **Weight change**
  - At least .91 kg average over two days or at least 2.27 kg in a week

- **Battery**
  - Voltage was too low for projected remaining capacity
  - Explant indicator reached

- **Cardiac Resynchronization Therapy Pacing**
- **Right Ventricular Pacing**
- **Therapy history corruption detected**

- **Ventricular pacing leads**
  - Low right ventricular intrinsic amplitude
  - Low left ventricular intrinsic amplitude
  - Low left ventricular pacing lead impedance
  - High left ventricular pacing lead impedance

- **Atrial pacing leads**
  - Low atrial intrinsic amplitude
  - Low atrial pacing lead impedance
  - High atrial pacing lead impedance

B -Red alert notifications

- Remote monitoring disabled due to limited battery capacity
- High or low shock lead impedance
- High or low shock lead impedance detected when attempting to deliver a shock
- High or low right ventricular pacing lead impedance
- High voltage detected on shock lead during charge
- Tachy mode set to value other than Monitor + Therapy
- PG has detected a possible device malfunction
- Device parameter error

Receive all the diagnostic data remotely that we would normally get with the patient in front of us has a role in the surveillance of the device function, but has also the potential to improve quality of care and improve patient outcome, and, in addition, to reduce the number of out-patient clinic visits and healthcare costs [11–13].

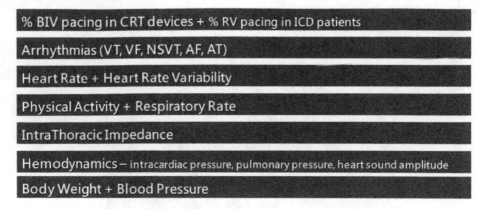

Fig. 2. Device based monitoring features in heart failure. BIV = biventricular pacing; CRT = cardiac resynchronization therapy; RV = right ventricle; ICD = implantable cardioverter-defibrillator; VT = ventricular tachycardia; VF = ventricular fibrillation; NSVT = non-sustained ventricular tachycardia; AF = atrial fibrillation; AT = atrial tachycardia.

3   Advantages of Remote Monitoring Systems

Telemedicine, in general, is recognized by governmental and medical agencies as an innovation for improving the access to healthcare services, with a potential reduction in patient hospital stays related to cardiac events [2].

One of the main functions of systems for monitoring CIED is to detect malfunctions as early as possible. Failure to apply therapies when required and problems of lead and/ or generator malfunction may only occur between scheduled hospital visits. Therefore, a rapid detection of technical failures followed by a fast response to alerts by the cardiologist or allied professional can prevent potential harm to the device patient. Electronic malfunctions in these devices are unpredictable, with ICD leads representing the most common cause of complications, with an incidence ranging between 2 % and 15 % at five years [14]. Recalls, although uncommon, are an important factor in decisions concerning the frequency of consultations, clinical management and inappropriate detections. RM systems provide regular assessment of the function of the various components of implanted devices, as well as detection and characterization of arrhythmias, therapies applied, and even identification of factors that could indicate risk of hospitalization for decompensated HF [11].

RM capabilities are associated with patient's convenience, better device surveillance, clinic efficiency, and a potential improvement in monitoring arrhythmias and HF. Scheduled remote CIED follow-up can save the time and efforts of the cardiologist and allied professional, and of the patient and the accompanying family by avoiding in-hospital visits for CIED follow-up. It has been suggested that RM can substantially reduce the number of hospital visits, freeing up hospital staff to attend other patients and to perform other tasks [2].

If RM is applied, accessibility increases due to the networking of patient data, but also because patient-initiated interrogations (non scheduled follow-ups initiated manually by the patient as a result of a clinical event) allow information analysis avoiding unnecessary in-clinic visits. By using RM with an individualized approach, the team's decision is based on better health care, with a positive impact in quality of life, less time spent by the patient and caregivers, quicker follow-ups, with more efficient use of hospital resources contributing to costs savings [2, 15]. Finally, it is expected that RM can counteract the pending imbalance between the annually increasing load of CIED population.

4   Cardiac Implantable Device Based Monitoring in Heart Failure

Patients with chronic HF are prone to frequent exacerbation of symptoms that can substantially increase hospital admissions carrying a heavy economic burden. The current guidelines for the management of chronic HF include the use of ICDs and CRT-Ds as the standard care in selected patients [16]. A multidisciplinary approach, which requires specialized knowledge as well as frequent monitoring, may contribute to improve clinical outcome and keep HF patients in an optimized health-care program.

The use of RM has emerged as a possible way to improve the management of these patients by providing more frequent assessment of the function of the various components of CIED, as well as detection and characterization of arrhythmias, therapies applied, and even identification of factors that could indicate risk of hospitalization for decompensated HF [9]. The ability of CIED to continuously monitor variables such as mean heart rate (and heart rate variability), episodes of arrhythmias, patient's daily activity, changes in intra-thoracic impedance (for the detection of fluid accumulation), and the integrity of the CIED system appears very attractive and may provide early warning of changes in cardiac status or of safety issues, and allow adequate clinical management. It has been suggested that blood pressure, atrial fibrillation episodes, intrathoracic impedance (a surrogate marker of pulmonary congestion), heart rate variability (giving information about autonomic nervous system activity), percentage of time in biventricular pacing and appropriate ICD shocks can predict HF events and change outcomes [17, 18]. This important subject remains under active investigation. Until now, and despite the potential impact on the healthcare system of different follow-up RM strategies for HF management, published data are conflicting. In the ALTITUDE study, patients followed remotely showed higher survival rates than those followed in-clinic [19]. Also, survival outcomes were better than those observed previously in clinical trials, suggesting that closer management with RM allows to clinicians to intervene more effectively with impact on survival.

Management adapted in response to monitoring intrathoracic impedance presented low sensivity and low positive predictive values, and, in previous studies, has not been shown to improve outcomes [20–22]. However, recent findings indicate that the sensi-tivity of intrathoracic impedance monitoring is superior to daily weight monitoring for predicting worsening HF events [23]. Also, device-based RM has been shown to identify patients in increased risk of HF hospitalization and improve prognosis by allowing timely detection of HF decompensation and therapeutic intervention, while reducing the number of total clinical visits and visits for HF [17, 24, 25]. More recently, early physical activity levels, measured by an accelerometer on implanted CRT and ICD devices, showed a strong value in the prediction of outcome in patients with chronic HF [26]. Newly unpublished studies showed that at one-year, significantly more patients with an ICD or a CRT and chronic HF with reduced left ventricular ejection fraction, randomized to home monitoring scored better on a composite endpoint that included all-cause mortality and specific cardiac measures [27], and that patients with high adherence to RM, measured as weekly transmission of data at least 75 % of the time, had a 58 % reduced likelihood of mortality, compared to patients not using RM, and a 35 % reduced likelihood of mortality, compared to those with low adherence to RM [28].

These studies reflect the importance of a potential role for the use of RM device–based diagnostics in the ever-growing population of HF patients with an implanted CIED. Further data are required to validate the use of device-based algorithms in HF. Current investigation efforts are ongoing in this area in order to provide physicians with technology support to improve management of cardiac status. Meanwhile, the general applicability of this RM approach is uncertain and a guideline recommendation about the positive impact of CIED algorithms in chronic HF patients is not yet possible.

5   Cost-Benefit Ratio of Remote Monitoring

The increasing number of patients referred for CIED implantation and the resources required for routine in-clinic CIED follow-up causes a significant burden to the electrophysiology departments and hospital services contributing to increasing health care costs. In-person hospital visits usually entail transportation to the hospital, time spent waiting and loss of productivity, which represent additional costs for patients and for the health system.

RM is a safe technology, widely accepted by patients and physicians, for its convenience, reassurance, and diagnostic potential. It has been associated with a significant reduction of hospital visits while maintaining levels of safety, but also with a reduction of the number of patients lost to follow-up, and a shorter interval between detection of actionable events and a clinical decision when compared with conventional follow-up [29, 30]. In the CONNECT trial, the mean length of hospital stays were significantly shorter in the RM group, representing a potential advantage in cost-saving for both hospital care and patients [30].

Analysis of quality of life showed that 93–97 % of patients were satisfied with the convenience and feasibility of RM, with high levels of satisfaction among both ICD or CRT-D patients and physicians [31, 32].

Prospective health-economic studies are important to determine the clinical and economic benefits of systematic RM in patients with ICD and CRT-D. In the EVOLVO study, the authors aimed to measure the benefits and economic evaluation of RM in chronic HF patients with an ICD or a CRT-D. The results showed cost savings of €888.10 per patient over the 16-month follow-up period [12]. In a recent meta-analysis, regarding the economic impact of RM in patients with HF, there was a significantly lower number of hospitalizations for HF (p < 0.001) and for any cause (p = 0.003) compared to conventional follow-up [33]. The difference in costs between RM and usual care ranged from €300 to €1000, favouring a RM strategy. A recent single-center prospective randomized study showed that the time spent by the hospital staff was significantly reduced in the RM group, with costs savings for both patients and the hospital [34]. The ECOST trial, designed to compare prospectively the safety and the costs of remote ICD monitoring with standard ambulatory follow-up, showed that the direct non-hospital-related costs of RM were 26 % lower than the costs of ambulatory follow-up over a period of 27 months after ICD implantation [35]. The savings observed in this study were particularly significant in view of the greater efficacy of RM (fewer shocks delivered) with equivalent safety (no increase in major adverse events) when compared with ambulatory visits. Increasing the time interval between in-office follow-up visits may be safe if adequate RM is performed.

These cost savings, combined with a quality-adjusted life years gain, suggest that adoption of a RM program will be a progressive dominant technology over existing standard care, particularly in centers performing a large amount of CIED implants.

Overall, the cost–benefit analysis seems to be clearly in favor of RM. A RM network enabling technology and medical information, promptly providing physicians with data comparable to an in-clinic follow-up visit, and offering patients a reduction in the time spent during routine follow-up, may contribute to a closer monitoring of CIED functions, with a favourable impact in costs, irrespective of the patient location, while offering practitioners ability to proactively respond to changes in the device status.

6 Legal Aspects and Data Protection

Despite its potential to reduce the number of visits, RM cannot replace direct contact with the physician, which is important to many patients. The consensus document of the Heart Rhythm Society and the European Heart Rhythm Association on the follow-up of CIED recommends an in-person visit at least once a year and that RM should take place every 3–6 months [9]. When RM is proposed, the patient must be told clearly how this complex system works, its potential benefits and limitations, and that it cannot replace the emergency department since the data transmitted are not analyzed immediately.

Regarding medico legal implications, questions related to the use of RM have been raised. A recent review of publications on RM of CIED revealed that 38 % of the studies included legal and technical issues among the disadvantages of remote follow-up [36]. Standardization and consensus are essential concerning the requirement and ability of the follow-up team to respond to alerts, to deal with information arriving outside the hospital's normal working hours, to manage the human resources required and to allocate responsibility. Data obtained by RM should be reviewed within reasonable time and frequency during office hours and immediate action should be undertaken when problems are identified.

The informed consent must cover the authorization for transmission of data, recording, and its use for clinical and scientific purposes, respecting privacy and confidentiality. There is also the question of whether to inform all patients of the option of

RM. These and other issues must be discussed thoroughly from a multidisciplinary perspective, taking into account that active communication between the treating physicians provide comprehensive information about the findings.

The implementation of a RM program will require a reorganization of the duties of the health team, who will require access to the servers hosting the data repository and will need to manage the large quantity of data transmitted. Clinical decisions will need to be taken regarding the management of alerts, telephone contact with patients and type of information provided, requests for unscheduled visits, measures to increase monitoring if necessary, and reprogramming and maintenance of equipment.

The involvement of data protection commissions is also of considerable importance to ensure that all the components of the system respect legal requirements and confidentiality. Although there have been no reports of security breaches by software attacks to date, the servers that contain patient data are potentially vulnerable to hackers. Therefore, powerful security software must be installed and constant vigilance is required to ensure that the systems are able to resist possible intrusions.

Legal and safety aspects and cost-effectiveness are subject of much debate justifying further studies focusing on how to best allocate this new technology in clinical practice, regarding the use of RM as the new standard of care for follow-up of patients with CIED.

7   Future Directions in Remote Monitoring Technologies

As RM becomes accepted by patients, physicians, and health systems, it may play an increased role in the care of patients with CIED. Integration of the growing amount of RM information in a multidisciplinary approach, the increase of complex workload and compatibility of the relevant data with hospital electronic medical records represent a challenge in optimizing the clinical management of this population. Also, reimbursement issues still need to be addressed in several countries.

The proliferation of CIED and the expanding of HF population raise the question of HF monitoring. This complex task, combining multiple features that provide additional information to HF specialists, requires a well trained team, a detailed analysis and integration of all data received, and an adequate strategy to allow early intervention. This will certainly involve new dedicated devices to monitor cardiac rhythm and hemodynamic parameters to improve patients management. Additional studies in this area are needed to evaluate whether advances in HF monitoring will result in outcomes improvement and reduction in health care costs to justify the widespread indications for these technologies.

References

1. Stellbrink, C., Trappe, H.J.: The follow-up of cardiac devices: what to expect for the future? Eur. Heart J. Suppl. 9(Suppl. I), I113–I115 (2007)
2. Stoepel, C., Boland, J., Busca, R., Saal, G., Oliveira, M.: Usefulness of remote monitoring in cardiac implantable device follow-up. Telemed J E Health. 15(10), 1026–1030 (2009)

3. Marinskis, G., van Erven, L., Bongiorni, M.G., Lip, G.Y., Pison, L., Blomström-Lundqvist, C.: Practices of cardiac implantable electronic device follow-up: results of the European Heart Rhythm Association survey. Europace **14**(3), 423–425 (2012)
4. Arribas, F.: The EHRA White Book. Europace **14**(Suppl. 3), ii1–ii55 (2012)
5. Joglar, J.A.: Remote monitoring of cardiovascular implantable electronic devices: new questions raised. Pacing Clin. Electrophysiol. **32**(12), 1489–1491 (2010)
6. Dubner, S., Auricchio, A., Steinberg, J.S., Vardas, P., Stone, P., Brugada, J., et al.: ISHNE/EHRA expert consensus on remote monitoring of cardiovascular implantable electronic devices (CIEDs). Europace **14**, 278–293 (2012)
7. Tracy, C.M., et al.: 2012 ACCF/AHA/HRS Focused Update of the 2008 Guidelines for Device-Based Therapy of Cardiac Rhythm Abnormalities. JACC **60**(14), 1297–1313 (2012). Device-Based Therapy Guideline Focused Update October 2, 2012
8. Sticherling, C., Kühne, M., Schaer, B., Altmann, D., Osswald, S.: Remote monitoring of cardiovascular implantable electronic devices. Prerequisite or luxury? Swiss Med Wkly **139**(41–42), 596–601 (2009)
9. Wilkoff, B.L., Auricchio, A., Brugada, J., et al.: HRS/EHRA expert consensus on the monitoring of cardiovascular implantable electronic devices (CIEDs): description of techniques, indications, personnel, frequency and ethical considerations. Heart Rhythm. **5**, 907–925 (2008)
10. Raatikainen, M.J., Uusimaa, P., van Ginneken, M.M., et al.: Remote monitoring of implantable cardioverter defibrillator patients: a safe, time-saving, and cost effective means for follow-up. Europace. **10**, 1145–1151 (2008)
11. Oliveira, M., Silva, P.C., Silva, N.: Remote monitoring for follow-up of patients with implantable cardiac devices. Rev. Port. Cardiol. **32**, 185–190 (2013)
12. Landolina, M., Perego, G.B., Lunati, M., Curnis, A., Guenzati, G., Vicentini, A., Parati, G., Borghi, G., Zanaboni, P., Valsecchi, S., Marzegalli, M.: Remote monitoring reduces healthcare use and improves quality of care in heart failure patients with implantable defibrillators. the evolution of management strategies of heart failure patients with implantable defibrillators (EVOLVO) study. Circulation **125**, 2985–2992 (2012)
13. Burri, H.: Remote follow-up and continuous remote monitoring, distinguished. Europace **15**, i14–i16 (2013)
14. Kleeman, T., Becker, T., Doenges, K., Vater, M., Senges, J., Schneider, S., Saggau, W., Weisse, U., Seidl, K.: Annual rate of transvenous defibrillation lead defects in implantable cardioverter-defibrillators over a period of > 10 years. Circulation **115**, 2474–2480 (2007)
15. Burri, H., Heidbüchel, H., Jung, W., Brugada, P.: Remote monitoring: a cost or an investment? Europace **13**, ii44–ii48 (2011)
16. McMurray, J.J., Adamopoulos, S., Anker, S.D., Auricchio, A., et al.: ESC Guidelines for the diagnosis and treatment of acute and chronic heart failure 2012. Eur. Heart J. **33**, 1787–1847 (2012)
17. Whellan, D.J., Ousdigian, K.T., Al-Khatib, S.M., et al.: Combined heart failure device diagnostics identify patients at higher risk of subsequent heart failure hospitalizations: results from PARTNERS HF (Program to Access and Review Trending Information and Evaluate Correlation to Symptoms in Patients With Heart Failure) study. J. Am. Coll. Cardiol. **55**, 1803–1810 (2010)
18. Sousa, C., Leite, S., Lagido, R., Ferreira, L., Silva-Cardoso, J., Maciel, M.J.: Telemonitoring in heart failure: a state-of-the-art review. Rev. Port. Cardiol. **33**, 229–239 (2014)

19. Saxon, L.A., Hayes, D.L., Gilliam, F.R., Heidenreich, P.A., Day, J., Seth, M., Meyer, T.E., Jones, P.W., Boehmer, J.P.: Long-term outcome after ICD and CRT implantation and influence of remote device follow-up. the ALTITUDE survival study. Circulation **122**, 2359–2367 (2010)

20. Conraads, M., Tavazzi, L., Santini, M., et al.: Sensitivity and positive predictive value of implantable intrathoracic impedance monitoring as a predictor of heart failure hospitalizations: the SENSE-HF trial. Eur. Heart J. **32**, 2266–2273 (2011)

21. Adamson, P.B., Gold, M.R., Bennett, T., et al.: Continuous hemodynamic monitoring in patients with mild to moderate heart failure: results of the Reducing Decompensation Events Utilizing Intracardiac Pressures in Patients with Chronic Heart Failure (REDUCEhf) trial. Congest Heart Fail. **17**, 248–254 (2011)

22. Braunschweig, F., Ford, I., Conraads, V., et al.: Can monitoring of intrathoracic impedance reduce morbidity and mortality in patients with chronic heart failure? rationale and design of the Diagnostic Outcome Trial in Heart Failure (DOT-HF). Eur. J. Heart Fail **10**, 907–916 (2008)

23. Abraham, W.T., Compton, S., Haas, G., Foreman, B., Canby, R.C., Rishel, R., et al.: Intrathoracic impedance vs daily weight monitoring for predicting worsening heart failure events: results of the Fluid Accumulation Status Trial (FAST). Congest. Heart Fail. **17**(2), 51–55 (2011)

24. Catanzariti, D., Lunati, M., Landolina, M., et al.: Monitoring intrathoracic impedance with an implantable defibrillator reduces hospitalizations in patients with heart failure. Pacing Clin. Electrophysiol. PACE **32**, 363–370 (2009)

25. Landolina, M., Perego, G.B., Lunati, M., et al.: Remote monitoring reduces healthcare use and improves quality of care in heart failure patients with implantable defibrillators: the Evolution of Management Strategies of Heart Failure Patients With Implantable Defibrillators (EVOLVO) study. Circulation **125**, 2985–2992 (2012)

26. Conraads, V.M., Spruit, M.A., Braunschweig, F., Cowie, M.R., Tavazzi, L., Borggrefe, M., Hill, M.R.S., Jacobs, S., Gerritse, B., van Veldhuisen, D.J.: Physical activity, measured with implanted devices, predicts patient outcome in chronic heart failure. Circ. Heart Fail. **7**(2), 279–287 (2014)

27. Hindricks, G. et al.: IN-TIME: the influence of implant-based home monitoring on the clinical management of heart failure patients with an impaired left ventricular function. Eur. Heart J., **34**(suppl 1) (2013)

28. Mittal, S. et al.: Increased adherence to remote monitoring is associated with reduced mortality in both pacemaker and defibrillator patients. Heart Rhythm 2014, the Heart Rhythm Society's 35th Annual Scientific Sessions (2014)

29. Varma, N., Epstein, A.E., Irimpen, A., Schweikert, R., Love, C.: efficacy and safety of automatic remote monitoring for implantable cardioverter-defibrillator follow-up. the Lumos-T Safely Reduces Routine Office Device Follow-Up (TRUST) trial. Circulation **122**, 325–332 (2010)

30. Crossley, G., Boyle, A., Vitense, H., Chang, Y., Mead for the CONNECT Investigators, R.H.: The CONNECT (Clinical Evaluation of Remote Notification to Reduce Time to Clinical Decision) trial. the value of wireless remote monitoring with automatic clinician alerts. J. Am. Coll. Cardiol. **57**, 1181–1189 (2011)

31. Masella, C., Zanaboni, P., di Stasi, F., et al.: Assessment of a remote monitoring system for implantable cardioverter defibrillators. J. Telemed. Telecare **14**, 290–294 (2008)

32. Marzegalli, M., Lunati, M., Landolina, M., et al.: Remote monitoring of CRT–ICD: the multicenter Italian CareLink evaluation – ease of use, acceptance, and organizational implications. Pacing Clin. Electrophysiol. **31**, 1259–1264 (2008)

33. Klersy, C., De Silvestri, A., Gabutti, G., Raisaro, A., Curti, M., Regoli, F., Auricchio, A.: Economic impact of remote patient monitoring: an integrated economic model derived from a meta-analysis of randomized controlled trials in heart failure. Eur. J. Heart Fail. **13**(4), 450–459 (2011)
34. Calò, L., Gargaro, A., De, R.E., Palozzi, G., Sciarra, L., Rebecchi, M., Guarracini, F., Fagagnini, A., Piroli, E., Lioy, E., Chirico, A.: Economic impact of remote monitoring on ordinary follow-up of implantable cardioverter defibrillators as compared with conventional in-hospital visits. A single-center prospective and randomized study. J. Interv. Card. Electrophysiol. **37**(1), 69–78 (2013)
35. Guedon-Moreau, L., Lacroix, D., Sadoul, N., Clementy, J., Kouakam, C., Hermida, J.-S., Aliot, E., Kacet, S., ECOST trial Investigators.: Costs of remote monitoring vs. ambulatory follow-ups of implanted cardioverter defibrillators in the randomized ECOST study. Europace. doi:10.1093/europace/euu012
36. Costa, P.D., Rodrigues, P.P., Reis, A.H., et al.: A review on remote monitoring technology applied to implantable electronic cardiovascular devices. Telemed. J. E. Health **16**, 1042–1050 (2010)

Biomedical Electronics and Devices

Simple Fabrication Method of Micro-Fluidic Devices with Thick Resist Flow Paths Designed Arbitrarily Using Versatile Computer Aided Design Tools

Toshiyuki Horiuchi[✉], Shinpei Yoshino, and Jyo Miyanishi

Tokyo Denki University, Tokyo, Japan
horiuchi@cck.dendai.ac.jp

Abstract. A simple and easy fabrication method of micro-fluidic devices using only single lithography processes was proposed. Thick resist patterns printed using a simple and handmade contact-lithography tool were directly used as flow-path patterns. Because pattern widths were as large as 50–200 μm, low-cost film reticles were applicable. Accordingly, various flow-path shapes were designed arbitrarily using a versatile computer aided design tool, and easily obtained in very quick turn-around times at very low costs. The sidewalls of flow paths were controlled almost vertical using long-wavelength exposure light. The flow paths formed on a silicon wafer chip were easily capped by sandwiching them between a vessel and lid plates using bolts and nuts. Even when two colored waters were simultaneously injected from the two inlet ports of a swirl-type micro-mixer, they did not leak at all in spite of such simple assemblies. The new fabrication method of micro-fluidic devices using thick resist patterning will be useful for various applications.

Keywords: Micro-fluidic device · Lithography · Contact exposure · Thick resist · Blue filter · Vertical sidewall · Computer aided design

1 Introduction

Recently, according to increase of old age peoples, various medical, pharmaceutical, and biochemical devices for monitoring health or disease are frequently used. In such devices, micro-fluidic devices are typical and valuable ones [1–4]. Using small devices, it becomes possible to reduce quantities of specimen fluids, medicines, and regents. In addition, because it becomes easier to keep and change temperatures of testing samples, effective and efficient tests are executed.

However, it is necessary to fabricate devices with low-costs in short turn-around times easily. Because biochemical materials are flown, the devices should be disposable, and fabricated at low-costs. In addition, because flow path shapes should be varied matching to requirements, they should be fabricated or remade in short turn-around times flexibly and easily. For these reasons, a new simple and easy fabrication method of micro-fluidic devices is developed here.

Various methods have been proposed for fabricating micro-fluidic devices. As materials for fabricating flow paths, glass, silicon, poly-dimethyl-siloxane (PDMS),

© Springer International Publishing Switzerland 2015
G. Plantier et al. (Eds.): BIOSTEC 2014, CCIS 511, pp. 19–33, 2015.
DOI: 10.1007/978-3-319-26129-4_2

poly-methyl-methacrylate (PMMA), and others are used. To fabricate flow path grooves in glass or silicon substrates, they are masked by resist patterns, and etched using deep reactive ion etching (RIE) or wet etching [5, 6]. On the other hand, soft PDMS and PMMA flow paths are generally fabricated by replicating convex patterns of metal dies [7–9]. Hot emboss stamping, nano-imprint, injection mold, and others are used as the replication methods [10, 11]. However, it takes considerably long times to prepare necessary flow paths by these methods. In addition, if glass or silicon substrates are etched, it needs advanced and expensive lithography and etching systems. On the other hand, in the case of replicating molds, expensive molds should be prepared in advance. Accordingly, it is difficult to change designs and prepare various molds. It is the aim of this research to clear these inconveniences.

In the proposed new method [12, 13], micro-fluidic devices with arbitrarily and freely designed flow paths are easily fabricated at low costs and without preparing expensive systems or tools for short turn-around times. As a material for flow paths, negative resist SU-8 (MicroChem) is directly used [12–15], and flow paths are fabricated easily by single lithography processes. Because the main component of SU-8 is an epoxy resin, the material is inert to various body fluids such as blood and sputum [16].

2   Total Process for Fabricating Micro-Fluidic Devices

In the new method, flow paths are fabricated as groove patterns printed in thick SU-8 coated on silicon wafers. Film masks are used as original pictures with flow path patterns. Because precise alignment is not required between masks and wafers, patterns can be printed without using mask aligners, if only a short-wavelength visible or near ultra-violet light source is prepared, as shown in Fig. 1. In general, required device numbers are very small comparing with semiconductor mass-production devices. Therefore, strong light sources with high powers are not necessary. Costs of film masks are very cheap, and turn-around times are as short as 3–4 working days including checks of flow path patterns without using the express orders. Accordingly, masks with new or modified flow-path patterns are easily and quickly prepared without much caring costs and loss times.

Flow-path patterns can be designed using two dimensional versatile computer aided design (CAD) tools (Auto desk, AutoCAD 2012). Because a standard film mask has a large size corresponding to an A3 paper (420×297 mm^2), and the patterning field size is 400×277 mm^2, various similar flow paths with parametrical sizes are simultaneously delineated in the large area, if necessary. The large film mask can be cut in free-size small sheets using scissors, cutters, or knives. Although customers or researchers of micro-fluidic devices have to learn the CAD software to design flow paths, it is very easy to delineate lines, circles and rectangles. In addition, flow-path patterns are very simple and not generally complicated as mechanical design drawings. Therefore, everyone can use the software soon, and design the necessary flow paths freely. CAD data of flow paths required from film-mask maker is DXF files expressed using a file-name extension of ".dxf" that are used as de facto standards of intermediate files, and almost all the CAD software can convert the designed data and the DXF file data reciprocally.

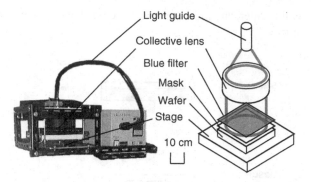

Fig. 1. Simple exposure system used for this research. (Source: BIODEVICES 2014, p. 6).

Because a lot of masks with different flow paths can be simultaneously fabricated, various experiments can be done using them. If the flow-path sizes are small, several tens working masks are simultaneously delineated on one standard film-mask sheet with drawing sizes of 277 × 400. Figure 2 shows an example of sheet design of a film mask containing 35 flow-path masks with a size of 50 mm square. Because the fabrication costs of a standard film sheet is approximately 20,000 yen (€140), costs for an individual working mask is only less than 600 yen (€4.2), for example.

Figure 3 shows the total processes for fabricating micro-fluidic devices. To print flow-path patterns on silicon wafers, negative resist SU-8 100 (MicroChem) was coated approximately in 100 μm, softly baked for 1 h at 95°C in an oven, and cooled down to room temperature, as shown in Fig. 3(a). After a film mask was put on a wafer for contacting them, they are exposed to the light, as shown in Fig. 3(b). Next, the mask

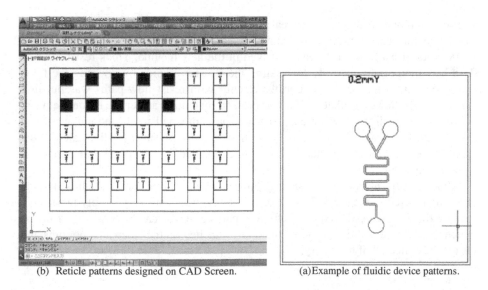

(b) Reticle patterns designed on CAD Screen.   (a)Example of fluidic device patterns.

Fig. 2. Example of film-reticle patterns designed using versatile CAD software for printing fluidic paths.

was removed from the wafer, and the wafer was developed by dipping it in a developer, as shown in Fig. 3(c).

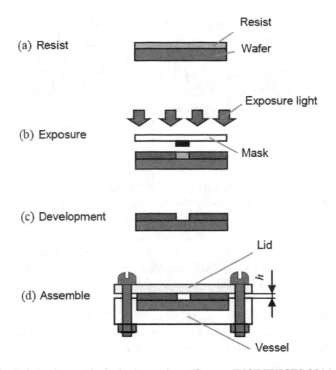

Fig. 3. Fabrication method of micro-mixer. (Source: BIODEVICES 2014, p. 9).

In conventional methods, how to seal and cap the flow paths was also a very important subject. For this reason, a very simple and easy seal measure for fabricating micro-fluidic devices with resist flow paths was contrived in the new method. Thick resist films of SU-8 had appropriate elasticity for sealing the flowing liquids by pressing the films adding relevant forces. So, a silicon wafer chip with resist flow-paths was put in a concave of acrylic vessel plate, and it was pressed by putting an acrylic lid on them and binding the lid with the vessel plate using four pairs of small screw and nut, as shown in Fig. 3(d). Using only a screw driver, the flow paths were successfully sealed, if the stick-out height h shown in the figure was controlled in an appropriate range, as explained in chapter 4.

Some ports for injecting and ejecting fluids were also fabricated on the lids and fine tubes are adhered to the ports. Thus, micro-fluidic devices are very simply fabricated. Because the lids are easily removed off, flow-paths after use can be observed or investigated in detail. In addition, residual materials attached on the sidewalls, bottoms and lids can be collected, if necessary.

3   Fabrication of Flow Paths with Vertical Sidewalls

It was considered that the sidewalls of flow paths should be vertical and perpendicular to the silicon substrates. If the cross sections of flow paths are almost rectangular, the areas of them are easily calculated, and it becomes easy to measure or guess the flow velocities and volumes. In addition, irregularity or peculiarity of the flow will not be occurred. For this reason, how to make sidewalls of thick SU-8 vertical was investigated.

In the case of contact exposure lithography, all the diffracted light from the mask patterns reach to resist films, including the 0th-order transparent light through the clear parts of the mask This is a very important point different from projection exposure using lens or mirror optics. In the case of such projection lithography, diffraction light rays with large angles from the optical axis are blocked by the lens or mirror apertures and frames. Accordingly, the light ray angles are limited corresponding to the aperture sizes or the numerical apertures (NAs).

However, pattern widths required for micro-fluidic devices are as wide as 50–200 µm. For this reason, diffraction angles from the flow-path patterns become very small. The first order diffraction angle θ from lines-and-spaces (L&S) patterns with a pitch of p is shown by Eq. (1).

$$\theta = \sin^{-1} \frac{\lambda}{p} \tag{1}$$

Here, λ is the wavelength of exposure light. Accordingly, when λ equals to 405 nm, and p equals to 100 µm, sin θ becomes 0.00405 rad, and θ becomes 0.25°.

Therefore, if the exposure light rays were controlled to illuminate the mask films in almost perpendicular to them, spreads of light rays caused by defocuses from the masks were suppressed in small ranges. Accordingly, distribution profiles of light intensity in the pattern width direction slightly degrade at the bottom of the resist layer. On the other hand, absolute light intensity largely decreases caused by the light absorption while the rays are passing through the resist. As a result, if the light intensity is almost the same and the absolute intensity is much reduced, sensitized widths differ between at the surface and bottom of the resist, as shown in Fig. 4(a). Therefore, it is necessary to suppress the light absorption in the resist as small as possible for printing patterns with the same widths at the surface and bottom, that is, with vertical sidewalls.

Figure 5 shows spectral transmittance of a SU-8 resist film with a thickness of 100 µm. Although the recommended wavelength for exposing SU-8 was 365 nm, measured transmittance at the wavelength of 365 nm was approximately 60 %, and this meant that the absorption ratio was 40 %. The wavelength of 365 nm is the one appropriate for obtaining high sensitivity, and the absorption is too high to obtain almost the same light intensity at the surface and bottom. However, it is also known from Fig. 5, in the wavelength ranges longer than 400 nm, light transmittances for 100-µm thick SU-8 are more than 90 %, that is, the absorption ratios are less than 10 %. Accordingly, light intensity curve differences between the resist surface and the bottom became small, as schematically shown in Fig. 4(b). Therefore, it was supposed that if a slice level for exposing the resist was set at the intensity shown in Fig. 4(b), width differences of space patterns between the surface and bottom became very small.

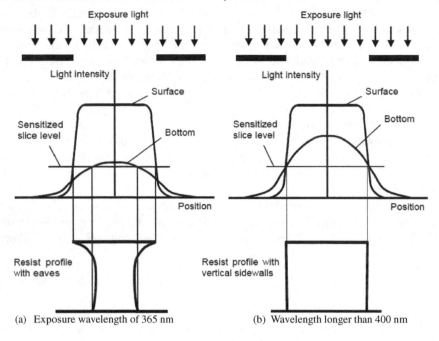

(a) Exposure wavelength of 365 nm   (b) Wavelength longer than 400 nm

Fig. 4. Light intensity curve differences between the resist surface and bottom.

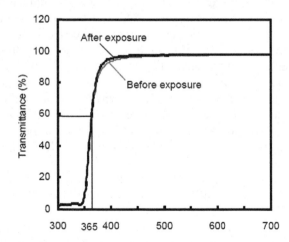

Fig. 5. Spectral transmittance of 100-μm thick SU-8. (Source: BIODEVICES 2014, p. 7).

For this reason, longer wavelengths were selected to suppress the absorption of the light in the resist using filters. Referring to Fig. 5, a blue filter cutting the light with wavelengths of shorter than 380 nm and a band-pass filter of 405 nm were individually inserted, and the effects were experimentally certified.

As an exposure source, a ultra-violet lamp with a spectral intensity shown in Fig. 6 was used. When the lamp was used as it was, a strong spectral line was observed at the wavelength of 365 nm. At first, this spectral line was removed by inserting a blue filter with a spectral transmittance shown in Fig. 7. The spectral intensity of the exposure light was improved, as shown in Fig. 8. It is known from Fig. 8 that light rays with wavelengths of less than 380 nm were almost completely removed. Profiles of flow-path patterns are compared, as shown in Fig. 9. When the exposure light was used as it was without the blue filter, although the parts near the top surface are strongly sensitized, parts near the bottom are not sufficiently sensitized. For this reason, surface parts hanged down like eaves, as shown in Fig. 9(a). On the other hand, when the blue filter was inserted, because parts near the bottom were sensitized similarly with the surface parts, patterns with almost vertical sidewalls were obtained, as shown in Fig. 9(b).

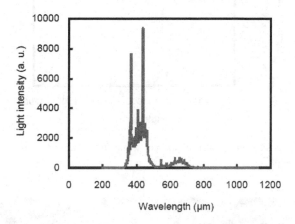

Fig. 6. Spectral light intensity of exposure source. (Source: BIODEVICES 2014, p. 6).

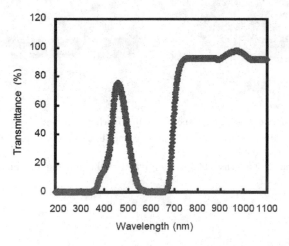

Fig. 7. Spectral transmittance of blue filter. (Source: BIODEVICES 2014, p. 7).

Furthermore, results for inserting a 405-nm band path filter were also investigated. Full width half maximum of the filter was 10 nm. Figure 10 compares the line pattern widths of L&S patterns printed using the blue and 405-nm filters. The widths were measured both at the surface and the bottom. Pattern-width differences between the surface and the bottom were very small for both filters. However, the width differences were much smaller for the blue filter. For this reason, it was decided that the blue filter should be inserted.

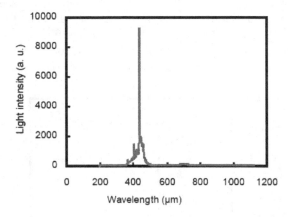

Fig. 8. Spectral intensity of exposure source when the blue filter is inserted. (Source: BIODEVICES 2014, p. 7).

(a) T-top pattern with eaves obtained by exposing without using the blue filter.

(b) Trench pattern with perpendicular sidewalls obtained by exposing with using the blue filter.

Fig. 9. Comparison of patterns obtained with and without using the blue filter. (Source: BIODEVICES 2014, p. 7).

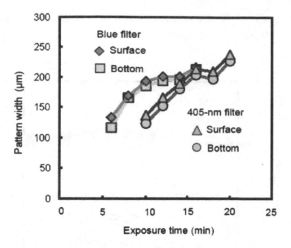

Fig. 10. Evaluation results of sidewall perpendicularity. (Source: BIODEVICES 2014, p. 8).

Thus, the method to fabricate resist flow-paths with vertical sidewalls was estab-lished. It was demonstrated that various flow paths were easily fabricated, as shown in Fig. 11. Not only simple flow paths, but also flow paths with various obstacles were successfully fabricated.

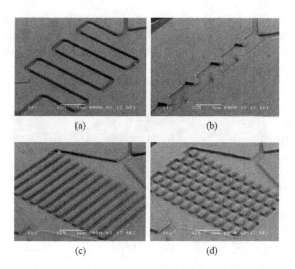

Fig. 11. Printed various flow-path patterns.

Next, pattern-width homogeneity in the 15-mm square exposure field was investi-gated using 100-μm lines-and-spaces patterns. Figure 12 shows the results. Line-width fluctuation was approximately within 106 ± 6 µm, and sufficiently small for using the exposure tool for fabricating micro-fluidic devices.

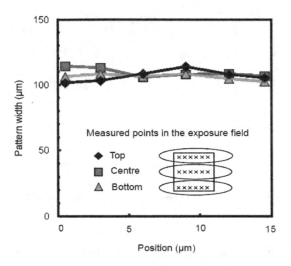

Fig. 12. Homogeneity of line pattern width in the exposure field. Line-and-space patterns of 100-µm were used for the evaluation. (Source: BIODEVICES 2014, p. 8).

Pattern-width homogeneity was also checked using a typical micro-mixer pattern shown in Fig. 13. Closely measured flow path widths were 100 ± 6 µm, and they were almost homogeneous, as shown in Fig. 14.

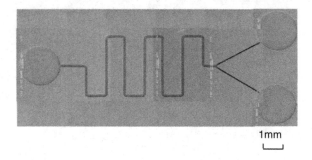

Fig. 13. Flow-path patterns used for width evaluation. (Source: BIODEVICES 2014, p. 8).

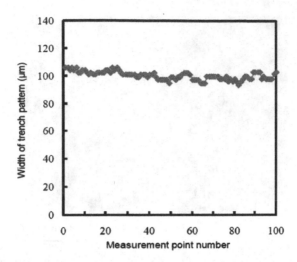

Fig. 14. Width homogeneity of flow-path patterns. (Source: BIODEVICES 2014, p. 8).

4   Assemblies of Micro-Fluidic Devices

Methods to seal the resist flow-paths by pressing the cap lids were investigated. Since the main component of the resist SU-8 is an epoxy resin, the resist has considerably large elasticity. Accordingly, it was considered that seals to prevent leaks of liquids flown in the flow paths would be capable if flat lids were appropriately pressed on the resist flow paths. For this reason, the resist film with flow-paths on a silicon wafer chip was sandwiched by an acrylic vessel and lid plates. To prevent side slips of wafer chips in the direction parallel to the wafer surface, concaves for placing the wafer chips were fabricated on the surfaces of vessel plates. If the lids were bound with the vessels using bolts and nuts, flat parts of the resist surfaces worked as sealing surfaces for capping the flow paths by the lid plates. Therefore, if the stick-out height h of the resist surface from the vessel surface, shown in Fig. 3, was appropriately controlled, flow paths were capped without leaking the liquids. Under the condition of 100-μm thick SU-8, the best stick-out height h was 30–40 μm.

The concaves of the vessel plates were fabricated using a three dimensional cutting machine. Giving three dimensional shape data to the machine, designed vessel plates were automatically machined out. Next, fine tubes were attached to the lid using adhesive. Figure 15 shows an example of assembled micro-mixer. A silicon chip with flow paths shown in Fig. 13 was enclosed between the lid and the vessel. If the lids and vessel plates were not sufficiently thick, they were curved, and the resist surfaces were not pressed uniformly. As a result, the fluids sometimes leaked out through the clearances between the resist surfaces and lids. The thicknesses of the lid and vessel were decided to 5 mm finally.

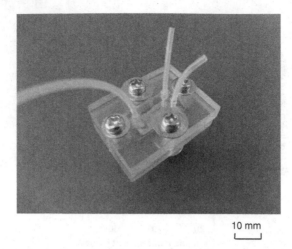

10 mm

Fig. 15. Outside View of the fabricated micro-mixer. (Source: BIODEVICES 2014, p. 9).

5   Flow Tests and Consideration

Fabricated micro-fluidic devices were tested whether liquids are flown without leaks. As a result, colored water injected from the inlets were ejected from the outlet without leaking through the only simply contacted seals between the lid and the resist film.

On the other hand, it was clarified that if two colored waters were injected from the Y-shape inlets, they flew as two parallel flows, and not mixed at all, as shown in Fig. 16. Because the injection volume rate was 10 $\mu\ell$/h, and the main flow path has a cross section of 100-μm width and 100-μm depth, the flow velocity was calculated to be 2.6 mm/s. Therefore, Reynold's number Re was calculated as

$$Re = \frac{VD}{v} = 0.2 \qquad (2)$$

This Re value means that the flow is a typical laminar flow. Accordingly, the phenomena that two colored waters were not mixed probably depended on the low Re flow conditions. However, because widths and depths of flow paths of micro-fluidic devices are generally very small, it is difficult to make the Re value large enough for changing the flow to the turbulent one using ordinary flow paths.

Therefore, it will be necessary to give some seeds of turbulence to flows for mixing fluids in such small mixers. In the past research, it was demonstrated that waters colored in red and blue using paints were successfully mixed, as shown in Fig. 17. In this case, pattern widths of Y-shape inlets were 73 μm, and widths of main swirl flow paths were 135 μm, and the depth was 380 μm. Accordingly, the aspect ratios were 5.2 and 2.8 at the inlet parts and swirl parts, respectively. These aspect ratios were considerably larger than those of the flow shown in Fig. 16. On the other hand, red and blue flows were mixed by passing through right-angle corners. For this reason, making such corners and

Yellow water   Blue water

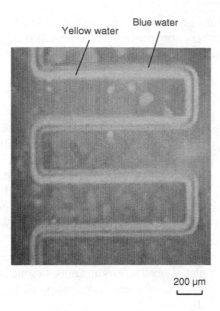

200 µm

Fig. 16. Non-mixing flow of two colored waters. Blue and yellow waters flew without being mixed. (Source: BIODEVICES 2014, p. 9).

giving flow-path differences between in-course and out-course tracks may be a measure to mix flows.

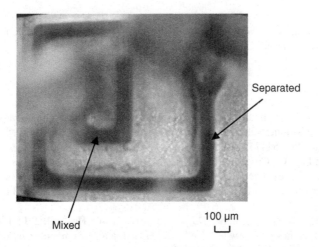

Separated

Mixed   100 µm

Fig. 17. Example of mixed liquids (Source: BIODEVICES 2014, p. 10).

On the other hand, as another mixing principle besides making turbulences in flows, diffusion is considered. If two statistic fluids are contacted, they are gradually mixed together by diffusion according to the diffusion coefficients and process times. For this reason, long flow paths and long-time mixing may be effective.

How to clarify the factors deciding the mixing characteristics and how to certainly mix the flowing fluids are very important research subjects to be investigated hereafter. Considering reported results [17, 18], vigorous efforts should be endeavored.

6   Conclusion

A new fabrication method of micro-fluidic devices was proposed. In the method, flow paths designed arbitrarily by customers or researchers are easily fabricated by single lithography processes. Flow-paths designed using versatile CAD tools are transferred to film reticles in quick turn-around times, and precisely replicated to thick resist films coated on silicon wafers. Because the wavelengths of exposure light rays are appropriately selected to reduce the light absorption in the resist film, flow paths with almost vertical sidewalls are formed. Thus, space patterns with flow-path shapes are directly used as flow-path grooves of micro-fluidic devices.

Wafer chips with flow-path patterns were sandwiched by vessel and lid plates and bound together using 4 couples of small screws and nuts. In spite of such easy and simple assembly, flow paths were successfully sealed depending on the elasticity of the thick resist film with the flow paths.

Using a device with simple flow paths including two inlet ports, a meandering mixing root, and an outlet port, it was investigated whether two fluids simultaneously injected from the two inlets were mixed or not. As a result, yellow and blue waters were not mixed. It is the next subjects to clarify the factors deciding the mixing of flows and to control the injected flows mixable.

Acknowledgements. This work was partially supported by Research Institute for Science and Technology of Tokyo Denki University, Grant Number Q13T-02.

References

1. Bouhadda, I., Sagazan, O., Bihan, F.L.: Suspended gate field effect transistor with an integrated micro-fluidic channel performed by surface micromachining for liquids sensing. Procedia Eng. **47**, 754–757 (2012)
2. Curto, V.F., Fay, C., Coyle, S., Byrne, R., O'Toole, C., Barry, C., Hughes, S., Moyna, N., Diamond, D., Lopez, F.B.: Real-time sweat pH monitoring based on a wearable chemical barcode micro-fluidic platform incorporating ionic liquids. Sens. Actuators B **171–172**, 1327–1334 (2012)
3. Kashkary, L., Kemp, C., Shaw, K.J., Greenway, G.M., Haswell, S.J.: Improved DNA extraction efficiency from low level cell numbers using a silica monolith based micro fluidic device. Anal. Chim. Acta **750**, 127–131 (2012)
4. Jacobs, T., Kutzner, C., Kropp, M., Brokmann, G., Lang, W., Steinke, A., Kienke, A., Hauptmann, P.: Combination of a novel perforated thermoelectric flow and impedimetric sensor for monitoring chemical conversion in micro fluidic channels. Procedia Chem. **1**, 1127–1130 (2009)

5. Avram, M., Iliescu, C., Volmer, M., Avram, A.: Microfluidic device for magnetic separation in lab-on-a-chip systems. In: 21st International Microprocesses and Nanotechnology Conference Digest of Papers, Microprocesses and Nanotechnology pp. 442–443 (2008)
6. Eun, D.S., Kong, D.Y., Chang, S.J., Yoo, J.H., Hong, Y.M., Shin, J.K., Lee, J.H.: Micro-PCR Chip with Nanofluidic Heat-Sink for Faster Temperature Changes. In: 21st International Microprocesses and Nanotechnology Conference Digest of Papers, Microprocesses and Nanotechnology 2008, pp. 448–449 (2008)
7. Lopez, F.B., Scarmagnani, S., Walsh, Z., Paull, B., Macka, M., Diamond, D.: Spiropyran modified micro-fluidic chip channel as photonically controlled self-indicating system for metal ion accumulation and release. Sens. Actuators B **140**, 295–303 (2009)
8. Campbel, K., McGrath, T., Sjölander, S., Hanson, T., Tidare, M., Jansson, Ö., Moberg, A., Mooney, M., Elliott, C., Buijs, J.: Use of a novel micro-fluidic device to create arrays for multiplex analysis of large and small molecular weight compounds by surface plasmon resonance. Biosens. Bioelectron. **26**, 3029–3036 (2011)
9. Chen, C.S., Chen, S.C., Liao, W.H., Chien, R.D., Lin, S.H.: Micro injection molding of a micro-fluidic platform. Int. Commun. Heart Mass Transf. **37**, 1290–1294 (2010)
10. Oakley, J.A., Robinson, S., Dyer, C.E., Greenman, J., Greenway, G.M., Haswell, S.J.: Development of a gel-to-gel electro-kinetic pinched injection method for an integrated micro-fluidic based DNA analyser. Anal. Chim. Acta **652**, 239–244 (2009)
11. Yang, R., Lu, B.R., Wang, J., Xie, S.Q., Chen, Y., Huq, E., Qu, X.P., Liu, R.: Fabrication of micro/nano fluidic channels by nanoimprint lithography and bonding using SU-8. Microelectron. Eng. **86**, 1379–1381 (2009)
12. Horiuchi, T., Watanabe, H., Hayashi, N., Kitamura, T.: Simply fabricated precise microfluidic mixer with resist flow paths sealed by an acrylic lid. In: Proceedings of The third International Conference on Biomedical Electronics and Devices, pp. 82-87 (2010)
13. Horiuchi, T., Yoshino, S.: Fabrication of precise micro-fluidic devices using a low-cost and simple contact-exposure tool for lithography, In: Proceedings of The 7th International Conference on Biomedical Electronics and Devices, pp. 5–11 (2014)
14. Martinez, A.W., Phillips, S.T., Wiley, B.J., Gupta, M., Whiteside, G.M.: FLASH: a rapid method for prototyping paper-based microfluidic devices. Lab Chip **8**(12), 2146–2150 (2008)
15. Martinez, A.W., Phillips, S.T., Whiteside, G.M., Carrilho, E.: Diagnostics for the developing world: microfluidic paper-based analytical devices. Anal. Chem. **82**, 3–10 (2010)
16. Horiuchi, T., Shinoda, D.: Feasibility of a micro-fluidic health care device measuring content of sodium chloride. In: Proceedings of The 5th International Conference on Biomedical Electronics and Devices, pp. 297–301 (2012)
17. Jain, M., Nandakumar, K.: Optimal patterning of heterogeneous surface charge for improved electrokinetic micromixing. Comput. Chem. Eng. **49**, 18–24 (2013)
18. Rahimi, M., Azimi, N., Parvian, F., Alsairafi, A.A.: Computational Fluid Dynamics modelling of micromixing performance in presence of microparticles in a tubular sonoreactor. Computer and Chemical Engineering **60**, 403–412 (2014)

Design of a Wearable Remote Neonatal Health Monitoring Device

Hiteshwar Rao[1]([✉]), Dhruv Saxena[1], Saurabh Kumar[3], G.V. Sagar[1],
Bharadwaj Amrutur[1,2], Prem Mony[4], Prashanth Thankachan[4],
Kiruba Shankar[4], Suman Rao[5], and Swarnarekha Bhat[5]

[1] Robert Bosch Center for Cyber Physical Systems, Bangalore, India
{rao.hiteshwar,dhruv.writeme}@gmail.com
http://www.cps.iisc.ernet.in
[2] Department of Electronics and Communication Engineering,
Indian Institute of Science, Bangalore, India
[3] Department of Instrumentation and Applied Physics,
Indian Institute of Science, Bangalore, India
[4] Division of Epidemiology and Population Health,
St. John's Research Institute, Bangalore, India
[5] Department of Neonatology, St. John's Medical College Hospital,
Bangalore, India

Abstract. In this text we present the design of a wearable health monitoring device capable of remotely monitoring health parameters of neonates for the first few weeks after birth. The device is primarily aimed at continuously tracking the skin temperature to indicate the onset of hypothermia in newborns. A medical grade thermistor is responsible for temperature measurement and is directly interfaced to a microcontroller with an integrated bluetooth low energy radio. An inertial sensor is also present in the device to facilitate breathing rate measurement which has been discussed briefly. Sensed data is transferred securely over bluetooth low energy radio to a nearby gateway, which relays the information to a central database for real time monitoring. Low power optimizations at both the circuit and software levels ensure a prolonged battery life. The device is packaged in a baby friendly, water proof housing and is easily sterilizable and reusable.

Keywords: Neonatal monitoring · Wearable sensors · Telemedicine · Body sensor network · Temperature measurement · Temperature sensors · Skin temperature · Noninvasive wearable wireless monitoring system · Biomedical measurements

1 Introduction

Neonatal Mortality Rate (NMR) [3] is defined as number of newborn deaths (that is within the first 28 days of life) per thousand births. The global neonatal deaths today account for more than 40 % of all child deaths before the age of five

© Springer International Publishing Switzerland 2015
G. Plantier et al. (Eds.): BIOSTEC 2014, CCIS 511, pp. 34–51, 2015.
DOI: 10.1007/978-3-319-26129-4_3

and is estimated to be more than 8 million [20]. More than 95 % of these deaths occur in developing nations like regions of Africa and South Asia. Reports [11] have shown that India has about 10 times higher NMR compared to the western world. NMR in India was 31 as of 2011, a 33 % decrease in NMR since 1990, yet taking into account its burgeoning population, approximately 1 million newborn died in 2010, nearly 30 % of the global neonatal deaths [2].

Recent research indicates that hypothermia is increasingly considered as a major cause of neonatal morbidity and mortality, especially in rural resource constrained settings [16,17]. Hypothermia for neonates is defined as an aberrant thermal state of diminution of their body's temperature below 36.5°C. Further decrease in body temperature causes respiratory depression, acidosis, decreases the cardiac output, decreases the platelet function, increases the risk of infection and may even lead to fatality without preemption [18]. WHO has classified hypothermia into following three categories depending on the body temperature [1].

- Mild hypothermia: 36.0 to 36.4°C
- Moderate hypothermia: 32.0 to 35.9°C
- Severe hypothermia: < 32°C

In newborns, hypothermia can be caused by loss of body heat to surroundings through conduction, convection, radiation or evaporation. Premature newborns are even more susceptible to these factors because of their low weight at the time of birth, they have a large 'surface area to weight ratio' with minimal subcutaneous fat. They have poorly developed shivering, sweating and vasoconstriction mechanisms and they are unable to retain their body's heat [18]. Hypothermia has a wider spread in the developing nations. In the rural context thermal care of newborn is often overlooked and hypothermia goes undetected. The most prevalent technique in rural settings on which caregivers rely is human touch, which is less sensitive and is not a reliable method. Commonly used mercury thermometers are often fragile and require some degree of training. Hence there is a need for an automated and robust way of measuring the neonate's temperature on a continuous basis, and be able to initiate intervention in a prompt manner as required. This has been the main motivation towards developing the temperature monitoring device and system.

The ever increasing cost for state-of-the-art medical facilities like NICU (Neonatal Intensive Care Unit) impedes the user from accessing it in rural areas. The problem is exacerbated in rural parts as India's population is still largely rural, with limited or no access to modern health care infrastructure. Mothers are often discharged earlier and infants are taken home right after delivery. In some cases delivery is even carried out at their residence without any medical professional or facilities at their disposal. We would like to capitalise the technical advancements of 21st century to perform remote neonatal healthcare monitoring in an economical manner.

In this paper, we propose a novel wearable monitoring device, designed and developed for the neonates in remote, rural and resource constraint settings. Our objective is to develop an ultra low power wireless skin temperature sensor, capable of monitoring newborns' body temperature unobtrusively and in real time over a span of the first few days to weeks. Sensed temperature data will be securely uploaded via a gateway device to a centralised database. Analytics on the temperature data will be run to determine the intervention needed in case of temperature excursions beyond normal levels. This system will be given to new mothers to take home after delivery.

However this will require solutions to a number of significant challenges like ultra low power sensing, device integration and packaging, ultra low power short haul communication and baby friendly design.

1.1   Requirements and Challenges

We conducted a user study in the NICU of St. John's Medical College Hospital in Bangalore, India. The study involved understanding the current techniques and equipment used in NICU to monitor the health of the neonates. This led to understandings in how the neonates are handled and how their vital parameters such as body temperature are measured. It also revealed concerns of the doctors regarding current equipment and the feasibility of use of such equipment in a remote rural setting. Issues like placement of the sensor on the body, ability of the device to be sterilised and other aspects which will be highlighted in the coming sections, were dealt with through brainstorming sessions and concept generation. Feedback from the doctors/neonatologists was taken time and again on the concepts generated which led to a more concrete list of requirements.

Safety. In an NICU setting neonates are kept under radiant warmers to regulate their body temperature. To achieve this, neonates' body temperature is constantly monitored by employing conventional temperature probes, attached to their skin with adhesive tape as shown in Fig. 1.

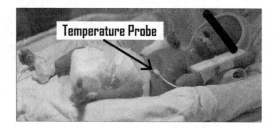

Fig. 1. Neonate kept under radiant warmer in NICU at St. John's Medical College Hospital.

The skin of neonates is extremely delicate and vulnerable to environmental stress. Research [24] has shown that increase in microbial growth under temperature probe can be harmful. Medical tape causes irritation and when it is removed it causes skin abrasion and damage [21]. Risk of the damage of internal organs involved with the tympanic and rectal temperature measurement limits their use for continuous monitoring. Thus the primary concern and requirement is their protection, safety and non-invasive monitoring.

For continuous measurement of body temperature the safest potential location for the sensor is over the right upper quadrant of the abdomen just below the rib-cage.

High Accuracy. For monitoring hypothermia or hyperthermia, temperature measurement accuracy should be same as that of medical grade NICU temperature probes. Thus an high accuracy of $0.1°C$ should be achieved in a remote rural setting.

Longer Battery Life. Once the device is installed and given to the user, neonates should be continuously monitored for the duration of first 2 to 3 weeks, with a sampling period of once every 15 min. Battery should last long enough without any need for charging or replacement. This calls for an ultra low power design to meet the requirement of longer battery life.

Robustness. The device needs to unobtrusively operate for several days. During the operation in remote regions, manual intervention for maintenance or repair is difficult to provide. Hence the device should be robust enough to cater to challenges like shock, vibration, and should not get reset accidentally. It should be hermetically sealed to protect the electronics from getting damaged in case liquids seep in during sterilisation or due to contact with body fluids (e.g. sweat, urine, faeces, etc.). Moreover, the device should be adjustable so that it can used on neonates of different abdominal girths, it should be aesthetically pleasant and should be baby friendly.

Detailed description of the solution to these requirements and challenges is provided in Sect. 2.3.

1.2 Related Works

In India, remote rural health monitoring is being enabled by many companies and government agencies. For instance, in [19], the company has developed a health kiosk and system called ReMeDi, which is deployed at the primary health center. The Kiosk allows a number of basic health tests to be conducted, the results of which are communicated over the cellular network to a central repository which keeps track of patient health data. This system is in use in a number of rural districts in Bihar and parts of Karnataka. Many other similar systems are being developed and deployed by various NGOs and startups across India.

Studies indicate that monitoring certain basic parameters, like temperature, could help indicate impending problems and hence with timely intervention, perhaps the mortality can be reduced. In this regard, an innovative product, for keeping babies warm, has been developed by a startup called Embrace [10].

In a related work, authors in [9] describe some sensors and packaging which has been developed for monitoring new borns. The sensors are embedded in a smart jacket, with careful attention paid to the baby friendliness of the design.

Another device [15] addresses a similar application where temperature of body can be measured and transmitted wirelessly to an android platform based device.The device has a battery life of only 48 hours and a relatively larger size as compared to our design.

The authors in [14], report an internet based health monitoring system for newborns. The parents fill up an online form with some data about their babies regularly. These include: Weight, body temperature, sleeping patterns, skin color, feeding etc. The remote nursing staff monitor these parameters and provide timely advice. The authors conducted a clinical research study for the efficacy of this system and found that it helped reduce the number of visits to the hospital by a factor of 3 for the babies being monitored via this system, as compared to a control group, which did not use it. This study encourages us to develop automated monitoring techniques like in [9] which will be more reliable and efficient than manually entering the data.

2   System Design

This section describes the design decisions involved in developing the hardware platform and the prototype device

2.1   Wireless Communication

There are many low-power wireless technologies like Bluetooth low-energy (BLE), Bluetooth classic, ANT, ZigBee, Wi-Fi, Nike+, IrDA, and the near-field communications (NFC) standards currently being employed in the field of healthcare.

For our application, the following critical key parameters drove the selection of the wireless interface: ultra-low-power, low cost, small physical size, application's network topology requirements and security of communication.

The authors in [5] do a power consumption analysis of BLE, ZigBee and ANT sensor nodes in a cyclic sleep scenario and find BLE to be the most energy efficient. We believe that in the next few years, millions of mobiles and computers will support BLE, thus enabling BLE based sensors to utilize these as gateways to the internet [4,12]. We already see commercial products with BLE like Fit-Bit [13], Pebble Watch and Hot Watch, and hence it encourages us to leverage the advantages of employing BLE as the short-range wireless communication technology for connecting the sensor to the gateway.

2.2   Hardware Platform

Sensor Platform. A monolithic design of sensor platform is required to minimize the form factor, facilitate ease of manufacturing and to reduce the overall cost of the product. Hence a custom made platform has been developed using a

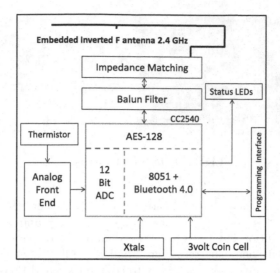

Fig. 2. Block diagram of the temperature sensor.

multilayer Printed Circuit Board. The assembled sensor platform has a dimension of 28 mm x 25 mm x 8 mm.

The sensor hardware platform as shown in Fig. 2a and consists of a Microcontroller(MCU) with integrated Bluetooth 4.0(BLE) and a 12-bit ADC (CC2540 from Texas Instruments), the NICU grade temperature sensor with its analog front end circuit, status LEDs, power supply and RF balun filter and antenna for wireless communication over 2.4 GHz ISM band. The microcontroller system has 256 KB programmable flash memory and 8-KB RAM and supports very low-power sleep modes, with sleep current as low as only 0.4 μA. There is built-in 128-bit hardware AES support for secure communication.

Sensor hardware can be programmed wirelessly and application programs can be downloaded on MCU's flash over the air via mobile or a gateway device. This feature enables us to dynamically control the sensor platform without having to disassemble the module and remove it from the package.

High precision MF51E NTC thermistors [8] are used for extremely accurate temperature measurements. These are especially designed and calibrated for medical equipment. The extremely small size of thermistor allows it to respond very quickly to small variations in temperature.

The temperature profile running on sensor meets the universal standard of body temperature profile defined by Bluetooth SIG [6]. Hence our sensor device can communicate to any authenticated host independent of the gateway platform being used.

Antenna Selection and Performance. The antenna for the sensor device has to balance between small size and efficiency. High efficiency antenna enables larger separation between the sensor device and the gateway - thus simplifying

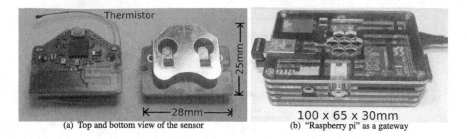

Fig. 3. Hardware platforms.

the usage scenario. On the other hand, larger size impacts the overall device size which is not desirable for a wearable device. A chip antenna offers a very small footprint solution but leads to a compromise of some critical parameters, such as, reduced efficiency, shorter range, smaller useful bandwidth, more critical and difficult tuning, increased sensitivity to components and PCB and increased sensitivity to external factors.

Detuning of chip antenna happens due to its proximity with ground plane, power source, plastic enclosure and the condition whether it is worn by the user or not. Our experiments with the chip antenna indicated an effective range of about 5 m. To overcome these issues, we have used an embedded Inverted F-antenna (IFA). Performance and characteristics of both the antennas are tabulated below Tables 1 and 3.

Table 1. Comparison of chip antenna and Inverted-F Antenna.

Parameters	Chip antenna	IFA
Range^a (m)	3-5	10-12
Bandwidth^b (MHz)	100	300
Efficiency^b	50 %	90 %
Reflection loss^b	> 50 %	< 10 %
Cost	Low	Nil
Size (mm)	8 x 6	25.7 x 7.5

a: Measured in closed indoor enviroment
b: Obtained from dataheet [25]

The Inverted-F Antenna has higher efficiency, longer range and a wider bandwidth than a chip antenna, though larger in size. For our device, the coin cell was another size limiting factor and hence we found this to be a good choice which provides high performance at a very low cost.

Low Power Analog Front End. Conventional thermistor interfacing techniques [7] use a linearization circuitry followed by a gain stage(G) and finally a

ADC(A) as shown in Fig. 4(b). We could eliminate most of these components in our approach shown in Fig. 4(a), by using the high precision thermistor, a high precision low tolerance low temperature coefficient resistor R_1 and the high resolution 12 bit ADC in the CC2540. Thanks to the well defined Temperature - Resistance characteristics of the thermistor the non-linearity can be taken care of by solving following log-polynomial Steinhart-Hart equation in software.

$$\frac{1}{T} = A + B\ln\left(R_{Th}\right) + C\left(\ln\left(R_{Th}\right)\right)^3 \tag{1}$$

A, B, and C are the Steinhart-Hart coefficients which are provided by the manufacturer [8]. This non linearity correction can be done either in the gateway or the sensor, thus incurring no power penalty on the sensor device itself.

The minimum voltage resolution for the 12 bit ADC operating at full range of $0-3$ V is 0.732 mV. For the use case in which the sensor module will be employed, the temperature range lies from $25°C-40°C$. The minimum accuracy requirement within this temperature range is 0.1°C which corresponds to a minimum change of 2.743 mV for the sensor module, well above the LSB of the ADC and hence eliminates the need for a separate gain stage.

As per the circuit shown in Fig. 4(a) analog voltage reference for 12 bit ADC ($M = 12$) is also supplied from digital I/O. Therefore the voltage at the input of ADC is given by

$$V_{ADC} = \frac{N_{ADC}}{2^M}V_{REF} = \frac{R_{Th}}{R_{Th} + R_1}V_{REF} \tag{2}$$

From above equation R_{Th} can be calculated independent of the voltage supplied by digital I/O.

$$R_{Th} = \frac{N_{ADC}}{2^M - N_{ADC}}R_1 \tag{3}$$

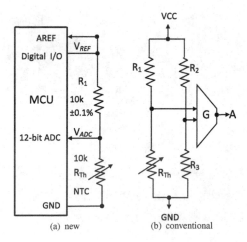

(a) new   (b) conventional

Fig. 4. Thermistor interfacing technique.

For the change in R_{Th} due to ±1 LSB variation of ADC and tolerance of R_1,

$$\Delta R_{Th} = \frac{R_{TH}\left(2^M\right)\Delta N_{ADC}}{\left(2^M - N_{ADC}\right)N_{ADC}} + \frac{\left(N_{ADC}\right)\Delta R_1}{\left(2^M - N_{ADC}\right)} \tag{4}$$

The proposed approach uses a single resistor R_1 with a low temperature coefficient of $\pm10\,ppm/°C$ and a tolerance of 0.1%. For $R_1 = 10K\Omega$, ΔR_1 is $\pm10\Omega$ due to tolerance and $\pm2\Omega$ due to temperature change of 20°C. Therefore, total $\Delta R_1 \approx \pm11\Omega$. ΔR_{Th} due the change in ±1 LSB of ADC is calculated to be $\pm10\Omega$ at 25°C and $\pm5\Omega$ at 42°C.

From Eq. 4 the total ΔR_{Th} due to ±1 LSB variation of ADC and tolerance of R_1 is found to be $\pm15\Omega$ at 25°C and $\pm12\Omega$ at 42°C. This corresponds to an error margin of $\pm0.03°C$ at 25°C and $\pm0.08°C$ at 42°C. Error margin is within our accuracy requirement. The thermal noise from the resistors are negligible.

The conventional approach suffers from increased errors due to the number of resistors used because each has its own temperature dependence and tolerance values. Another drawback of the earlier technique is that current is constantly consumed by the wheat-stone bridge and the gain stage, irrespective of the micro-controller being in sleep mode. In the proposed design, since the sensor interface

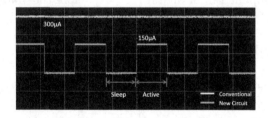

Fig. 5. Current consumption during active and sleep mode.

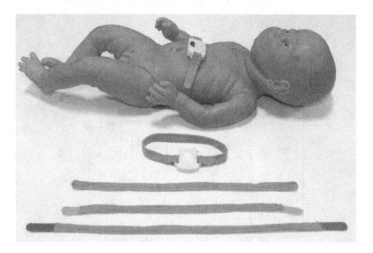

Fig. 6. Prototyped temperature sensor with belts

is powered from the micocontroller's GPIO, the current consumption can be significantly reduced by suitably programming the GPIO output during the sleep mode. This is illustrated in the measurements of both the conventional and the proposed approach in Fig. 5.

The use of well calibrated thermistor and a very low tolerance resistor eliminates the need for any other calibrations saving the cost and effort.

Gateway Platform. The sensor communicates with the gateway and the temperature data is stored on a centralized database over a secured internet backbone provided by GPRS/Wi-fi. We have developed the gateway using both a smartphone (Google Nexus 4) as well as a low cost platform "Raspberry pi" as shown in Fig. 3(b). Raspberry pi has a 700 MHz ARM core as an application processor and 512 MB RAM. Dual USB connectors are used for BLE and Wi-Fi dongles. A Linux based operating system is booted on it to run the application program. In the case of the smartphone an android application connects to the sensor and relays the information to the server.

2.3   Prototype and Implementation

Temperature Sensing Interface. The thermistor has a dimension of 1.6 mm x 4 mm and is required to be in thermal contact with the test surface to measure its temperature. However due to the requirements of the device to be robust and safe for the neonate (without any sharp/pointy objects sticking out of the device) the sensor was placed in contact with a thermal interface which would touch the body of the neonate on one side and house the sensor on the other side as depicted in Fig. 7.

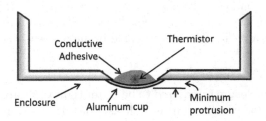

Fig. 7. Vertical cross-section of the enclosure.

The interface consists of an aluminium cup made by cold-working of an aluminium sheet. Aluminium was chosen for its easy formability as well as ability to retain its shape once it is cold-worked. Aluminium also has a very high thermal conductivity ranging from 200−250 W/m.K, which enables our thermal interface design to attain thermal equilibrium with the body of the patient in a short span of time. The aluminium cold-worked cup is only 0.1 mm thick to minimize the heat loss. To ensure complete thermal contact,the thermistor is glued on the inner side of the aluminium cup with a highly thermally conductive copper tape

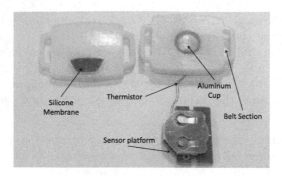

Fig. 8. Disassembled prototype.

which ensures efficient and quick conduction of heat from the aluminium surface to the sensing element. The cup is securely housed inside the casing while ensuring that it protrudes slightly from the bottom surface of the casing to enable reliable thermal contact with the body of the patient.

Power Switch. The device is designed to stay in deep sleep by default when not operational. To initiate connection establishment with a BLE gateway, we have provided a power switch, which needs to be pressed and held for 5 s or more. However since the same switch is also used to reset the device it is placed at a short depth inside the casing to minimise the chance of accidental reset. A silicone membrane is flushed with the top surface of the enclosure such that only if the membrane is pressed to a depth of 3 mm or more will the switch get activated. This requires concentrated force on the center of the membrane. Experiments were conducted to check if any sort of accidental pressing could lead to such a situation and it was concluded that only intentional pressing could achieve such a result. The silicone membrane also ensures that the casing is water tight.

Enclosure. The enclosure is required to be water tight to protect the electronics from liquids used for sterilization as well as from urine. Moreover the casing needs to be made with minimum number of parts for ease of production and low cost. The methods for making the casing water tight are either to make a simple lip interface in the design where the top and bottom parts can be pasted, or to go for a more complicated design which involves either snap fits or screw fittings and requires rubber gaskets to make the device water tight. The former method although cheaper to manufacture has the disadvantage that the casing can only be used one time. However since the device does not need battery replacement for several months, the former method seems more feasible given the fact that even an openable casing will be prone to damage since it will move from patient to patient and might eventually be required to be replaced. The custom designed enclosure as shown in the Fig. 6 is rapid-prototyped with the latest and higher resolution 3-D printing technology [23]. The 3-D printed prototype was drop tested from a height of 2 m without any damage to the prototype and enclosed circuitry. The actual production model however would be even more robust due to higher strength of commonly used injection molding materials like Polypropylene or Acrylonitrile butadiene styrene(ABS).

Belt Design. Respiration rate for newborns varies from 30−60 breaths per minute and during a respiration cycle the abdomen region expands and contracts. An increase in the height of the sternum of around 1/2 cm is considered as a normal expansion [22]. Hence the belt has to be designed with the requirements of being

- soft and elastic to accommodate the changes in abdominal circumference during breathing
- able to accommodate different sizes of neonates
- washable for sterilization and reuse

All the above requirements were met through a design where the belt is made out of a soft fabric. The belt has a thin elastic band inside the fabric which can be elongate upto 3 cms, hence it dynamically allows for expansion of the belt during breathing. The belt also has small loops in which fastening velcro can be attached. The fastening velcro is placed at the ends of the belt and when the belt folds on itself, the velcro hooks come in contact with the loops thereby fastening the belt. This allows for accommodation of different sizes of neonates. The fabric is completely washable and the belt can be washed and dried for reuse. Multiple belts can be assigned for a patient and these belts can be replaced daily to maintain hygiene.

Baby Friendly Design. The industrial design of the device reflects the face of a friendly abstract toy, where form follows function. All features of the face have functional relevance. To keep the number of parts to a minimum, light gates have been avoided by making the sections under which the LEDs are housed thinner than the overall casing (Figs. 8, 9 and 13).

The thin sections allow light from the LEDs to be seen well. These sections also give the appearance of eyes for the face. The membrane for the switch forms the mouth and the cavities through which the belt passes give an appearance of ears. This friendly face appearance of the design could help enforce confidence in parents that the device is friendly and is safe for their child. This design was arrived at after multiple iterations and discussions with the Neonatologists at SJRI. Initial prototypes were improved through feedback from the Neonatology team and features like exposed edges, sharp corners and crevices as well as exposed buttons were removed and the final design was presented which has been accepted by the Neonatology team.

Fig. 9. Baby friendly enclosure.

3 Evaluation

This section describes the evaluation method employed for testing our sensor platform. Experimental results indicate that the device has the required accuracy. Later in the section power calculation is performed to estimate the battery life.

3.1 Experimental Setup

As described in the Sect. 2.3 temperature sensing element is thermally interfaced with aluminium. To determine the response time and accuracy of the packaged module, experimental setup shown in the Fig. 10 is used.

A glass beaker filled with 300 ml of water is placed on a hot plate. An accurate RTD based temperature sensor is placed in the beaker. The temperature of water is controlled by giving a closed loop feedback to hot plate's heating element. The sensor device is immersed in water bath along with a bare thermistor. A precision digital thermometer is employed for setting reference temperature. Packaged devices were kept immersed in the water bath for 10 h of continuous operation

Fig. 10. Experimental setup.

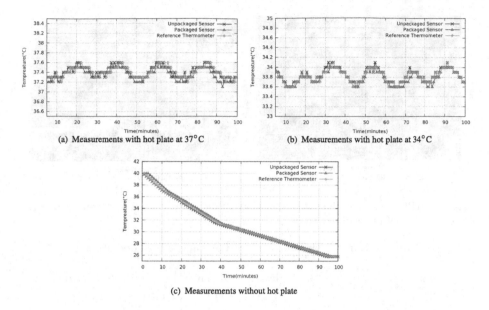

(a) Measurements with hot plate at 37°C

(b) Measurements with hot plate at 34°C

(c) Measurements without hot plate

Fig. 11. Responsiveness and accuracy measurement of the sensor.

and the temperature measurements were logged wirelessly. Temperature readings from all the sensors were periodically sampled once every two minutes.

3.2   Results

Responsiveness. Figures 11(a) and (b) show the temperature measurements during the experiment. A periodic oscillation of the temperature waveform was observed, indicating the action of the servo control loop of the hot plate. The period of the oscillations is approximately 40 min and is within $\pm 0.3°$C of the temperature configuration being used.

By definition, responsiveness is the time taken by the system to respond to an event or a change, hence these small changes in temperature provides an ideal environment to determine the response time.

Table 2. Response time and sensivity comparison.

Parameter	Bare	Reference	Packaged
Response	< 4 s	≈ 3 min	≈ 4 min

The bare thermistor is fastest in responsiveness followed by the reference thermometer and finally the packaged device. The observed response times are shown in Table 2, indicating that the packaged sensor would take approximately 3−4 min to attain thermal equilibrium with the surroundings. In our actual application scenario, we are required to take continuous temperature readings once every 15 min, hence the responsiveness is well within our requirement limits.

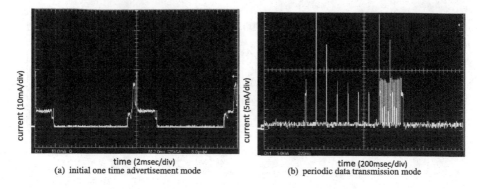

(a) initial one time advertisement mode   (b) periodic data transmission mode

Fig. 12. Oscilloscope plot showing current measured using precision current probe for different modes of operation.

Table 3. Current consumption for different mode of operation.

Mode	Imax	Iave	On Time	Energy(3 V@ 90 %η)
Sleep	1 μA	1 μA	899 s	1.1 mJ
Data transmission	30 mA	17 mA	1 s	56.6 mJ
Initial advertisement	30 mA	4.5 mA	180 s	2700 mJ

Accuracy. Using the same configuration for measuring the accuracy, water was heated upto 40°C and then was allowed to cool down to room temperature. Figure 11(c) shows the logged temperature reading. The variance (error bar) in the reading from the sensor device is $\pm0.04°C$. By comparing the data obtained from the sensor with a high precision digital thermometer it was calculated that the error was within 0.1°C for the temperature range of 25°C to 40°C.

3.3   Power Consumption

The sensor device establishes a connection with the near by gateway device through initial advertisement. After a successful connection event, the sensor is configured for periodic sleep and data transmission. Figure 12 and Table 3 shows the measured current and energy consumed by the device during these modes of operation.

The profile in Fig. 12(b) shows typical power consumption over a sensing cycle. Sleep current is measured using 6.5 digit 34411A Agilent multimeter and transient current is measured using hall effect current probe with TDS5104 Tektronix oscilloscope. The initial advertisement is the most energy expensive - however this action is expected to happen very rarely. The data transmission energy is also significant compared to sleep energy and hence needs to be minimised to ensure long battery life.

(a) disassembled view   (b) packaged sensor

Fig. 13. New prototype with inertial and temperature sensor.

Battery Life Estimation. The sensor device is configured to send temperature measurements once every 15 min. Thus there will be 96 data transmission cycles in a day. The device is powered by a 3 V CR2032 coin cell of 225 mAh given capacity. Considering the derated capacity to be 200 mAh, the total energy it can deliver is estimated to be 2160 Joules. Total energy consumed by the device during one full day of operation is estimated to be 5.54 Joules which results in a battery lifetime of 388 days (about a year).

4 Recent Developments

After continuous feedback from the doctors and parents of newborns, we realised that the size of the device was still a concern for the users and needed to be reduced. In an effort to make the device smaller, we have developed the new design shown in the figure. The designed pcb uses a chip antenna instead of the inverted-F antenna used in the previous design to reduce the antenna footprint. This helped in reducing the device size immensely with some loss in transmitting range which can be compensated by increasing the transmitting power if needed. An inertial sensor along with an op-amp has been added on board to facilitate the measurement of breathing rate. The new design has been developed bearing in mind that the baby friendly appeal of the product should not be lost. The casing design derives its inspiration from a soft pebble and the pcb has been designed to securely fit inside the casing. Other attributes such as water tightness and ability to be sterilized are maintained in the device.

5 Conclusions

Continuous, automated monitoring of vital parameters of neonates in remote rural areas has the potential of saving many lives each year. The current methods are not reliable and the technological intervention enabled by the proposed device can play a major role in getting these vital parameters to the doctors in real time. Critical requirements like reliability and robustness of the device are met through a methodical design approach using state-of-the-art technology like the integrated blue tooth low energy microcontroller, along with optimised sensor electronics and software, to ensure good performance and battery life. Human interface aspects have been incorporated into the device to allow for a friendly yet robust device which fulfils all the physical requirements for the device to be used comfortably on neonates in rural settings while addressing maintenance and sterilisation issues. The design also aims to connect emotionally with the stakeholders like parents of the neonates and health care providers to enforce confidence and a feeling of security. The current design is modular and can be extended beyond temperature measurement.

Acknowledgements. We acknowledge the Robert Bosch Center for Cyber Physical Systems at the Indian Institute of Science for funding support.

References

1. WHO: Thermal Control of the Newborn: a Practical Guide. Maternal and Safe Motherhood Programme (1993)
2. WHO: Newborn deaths decrease but account for higher share of global child deaths (2012)
3. WHO, UNICEF, and UN: Levels and Trends in Child Mortality Report (2012)
4. Omre, A.H.: Bluetooth low energy: wireless connectivity for medical monitoring. J. Diab. Sci. Technol. **4**(2), 457–463 (2010)
5. Artem, D., Steve, H., Stuart, T., Joshua, S.: Power consumption analysis of Bluetooth low energy, Zigbee, and ANT sensor nodes in cyclic sleep scenario. In: IEEE International Wireless Symposium, Beijing (2013)
6. Bluetooth: Health temperature profile (2012). https://developer.bluetooth.org/TechnologyOverview/Pages/HTP.aspx
7. Boano, C., Lasagni, M., Romer, K., Lange, T.: Accurate temperature measurements for medical research using body sensor networks. In: 14th IEEE International Symposium on Object/Component/Service-Oriented Real-Time Distributed Computing Workshops (ISORCW) 2011, pp. 189–198 (2011)
8. Cantherm: High precision ntc thermistors (2009). http://www.cantherm.com/products/thermistors/mf51e.html
9. Chen, W., Dols, S., Oetomo, S. B., Feijs, L.: Monitoring body temperature of newborn infants at neonatal intensive care units using wearable sensors. In: Proceedings of the Fifth International Conference on Body Area Networks, BodyNets 2010, pp. 188–194, New York. ACM (2010)
10. Embrace: A low cost infant warmer (2012). http://embraceglobal.org/
11. Health, G.: Causes of neonatal and child mortality in India: a nationally representative mortality survey. Lancet **376**(9755), 1853–1860 (2010)
12. Gomez, C., Oller, J., Paradells, J.: Overview and evaluation of bluetooth low energy: an emerging low-power wireless technology. Sensors **12**(9), 11734–11753 (2012)
13. Montgomery-Downs, H.E., Insana, S.P., Bond, J.A.: Movement toward a novel activity monitoring device. Sleep Breath **16**(3), 913–917 (2012)
14. Isetta, V., Agustina, L., Lopez Bernal, E., Amat, M., Vila, M., Valls, C., Navajas, D., Farre, R.: Cost-effectiveness of a new internet-based monitoring tool for neonatal post-discharge home care. J. Med. Internet Res **15**(2), e38 (2013)

15. iThermometer: Bluetooth digital thermometer for android devices (2012). http:// download.chinavasion.com/download/CVXX-A140.pdf
16. Kumar, V., Mohanty, S., Kumar, A.: Effect of community-based behaviour change management on neonatal mortality in shivgarh, Uttar Pradesh, India: a cluster-randomised controlled trial. Lancet **372**(9644), 1151–1162 (2008)
17. Kumar, V., Shearer, J.C., Kumar, A., Darmstadt, G.L.: Neonatal hypothermia in low resource settings: a review. J. Perinatol. **29**, 401–412 (2009)
18. Macfarlane, F.: Paediatric anatomy, physiology and the basics of paediatric anaesthesia (2006)
19. Neurosynaptics (2002). http://www.neurosynaptic.com/
20. Oestergaard, M.Z., Inoue, M.: Neonatal mortality levels for 193 countries in 2009 with trends since 1990: A systematic analysis of progress, projections, and priorities. PLoS Med **8**(8), e1001080 (2011)
21. Rutter, N.: Clinical consequences of an immature barrier. Semin. Neonatology **5**(4), 281–287 (2000)
22. Scavacini, A.S., Miyoshi, M.H., Kopelman, B.I., Peres, C.D.A.: Chest expansion for assessing tidal volume in premature newborn infants on ventilators. Jornal de Pediatria **83**, 329–334 (2007). Please check and confirm inserted author name and initials is correct in Ref. [22]
23. Stratasys High-quality, precise, multi-material prototypes in a compact system (2013). http://www.stratasys.com/3d-printers/design-series/precision/ objet260-connex
24. Susan, B., Debra, D., Lori, A.L.: Neonatal thermal care, part II: microbial growth under temperature probe covers. J. Neonatal Netw. J. Neonatal Nurs. **20**, 19–23 (2001)
25. TI: 2.4ghz inverted f- antenna (2008). http://www.ti.com/lit/an/swru120b/ swru120b.pdf

Novel Wireless Systems for Telemedicine and Body Area Networks Applications

Haider Khaleel[1(✉)], Chitranjan Singh[2], Casey White[1],
Hussain Al-Rizzo[2], and Seshadri Mohan[2]

[1] Department of Engineering Science, Sonoma State University,
Rohnert Park, CA, USA
hrkhaleel@ualr.edu
[2] Department of Systems Engineering, University of Arkansas at Little Rock,
Little Rock, AR, USA

Abstract. The demand for novel wireless system solutions is increasing exponentially in today's world. Wireless systems are extremely beneficial and applied in a wide spectrum of fields such as: personal communication, medicine, military, firefighting, entertainment, aeronautics, and Radio Frequency Identification (RFID). In this chapter, the design and analysis of two compact antennas aimed for telemedicine and Body Area Networks (BAN) are presented. The first design is based on a MIMO antenna array utilizing μ-negative metamaterial (MNG) structures that lead to low correlation between antennas when placed closely on a user's body. The antenna array resonates at 5.2 GHz. The second design is a mechanically flexible directional Yagi-Uda antenna resonating at 2.5 GHz for on-body communication. Both designs have the merits of light weight, low profile, mechanical robustness, compactness, and high efficiency. Such properties suggest that the proposed designs would be reasonable candidates for telemedicine and BAN applications that are constrained by limited physical space.

Keywords: Telemedicine · Antenna array · Metamaterial · Multiple input multiple output · Wireless body area networks

1 Introduction

Telemedicine has evolved rapidly over the past decade due to the increasing demand for remote health monitoring of post-surgery patients, recovery tracking, seniors, athletes, fire fighters and astronauts [1]. In telemedicine systems, important health parameters such as body temperature, blood pressure, and heart rate are transmitted wirelessly to remote monitoring stations (clinics, hospitals, etc....) [2].

Obviously, a reliable wireless scheme which is enabled by a use of antenna(s) is required for optimal performance of such systems. In this specific case, antennas are required to be small in size, lightweight, and robust with desired radiation characteristics. They also need to be comfortable and conformal to the shape of the body. Microstrip antennas offer a favorable advantage since they have a hemispherical radiation pattern, i.e. radiates away from the users body, which minimizes the exposure

© Springer International Publishing Switzerland 2015
G. Plantier et al. (Eds.): BIOSTEC 2014, CCIS 511, pp. 52–65, 2015.
DOI: 10.1007/978-3-319-26129-4_4

to harmful electromagnetic radiation. Furthermore, they offer a low profile solution, low cost, and ease of fabrication [3]. However, this class of antennas suffers from a very narrow bandwidth, hence, a low profile antenna with a directional radiation pattern and a wide bandwidth is essential in telemedicine applications. Several techniques have been proposed to achieve directional radiation pattern by adding a cavity or a shielding plane underneath the antenna, or using absorbers [4]. However these techniques lead to either an unacceptable increase in the antenna's height, or a more complicated manufacturing process. In [5], a PEC reflector was inserted between a human head and a folded loop antenna. This approach increases the return loss and decreases the antenna's efficiency. In [8], a Single Negative Metamaterial (SNG) is utilized to reduce the electromagnetic exposure, though efficient, it leads to a high profile system. Flexible Artificial Magnetic Conductor (AMC) based antenna is proposed in [9] for telemedicine application. A polyimide based printed antenna is integrated with an AMC ground plane which is utilized to minimize the Specific Absorption Rate (SAR) and the impedance mismatch caused by the human tissues proximity. Although the proposed design offers a low profile solution, it is relatively large in some applications.

On the other hand, there has been an increasing interest into on-body communication as part of WBAN research. On body mode is as important as off-body since it is essential for communication between sensors and devices located on or within the users body.

The gain from Multiple-Input Multiple-Output (MIMO) techniques in wireless sensor networks equipped with miniaturized sensor nodes cannot be fully exploited due to the difficulty encountered when placing traditional multiple antennas with sub-wavelength physical separation.

In this chapter, two antenna designs are proposed for telemedicine and body area networks applications. The first design is based on an antenna array utilizing μ-negative metamaterial (MNG) structures that lead to reduced mutual coupling between the antenna radiating elements. It is worth mentioning that mutual coupling reduction is essential to achieve optimal performance in Multiple-Input Multiple-Output (MIMO) communication schemes. The antenna array resonates at 5.2 GHz. The obtained correlation coefficient of 0.04 is low enough for realizing full diversity gain from using such an antenna array on sensor nodes in a telemedicine WLAN environment. Furthermore, Bit Error Rate (BER) simulation result for a 2×2 Alamouti diversity scheme in an IEEE 802.11a system is also presented. The second design presented is a highly directive, slot fed, vinyl based printed Yagi-Uda antenna operating at ISM band (2.4-2.483 GHz). Antennas with high directivity are often preferred for on-body communication in BAN systems.

Design and simulations are conducted using CST Microwave Studio which is based on the Finite Integration Technique. Results suggest that the proposed design would be a suitable candidate for telemedicine and BAN applications that are constrained by limited space.

2 Proposed Antenna Designs

2.1 Miniaturized MIMO Antenna Array with Reduced Correlation

2.1.1 Introduction

MIMO techniques employing multiple antennas at one or both ends of a wireless communication link have shown the potential to reach higher spectral efficiencies [11, 12]. By combining signals at transmit antennas and receive antennas, MIMO substantially improves either data rate using Spatial Multiplexing (SM) or reliability using diversity techniques. Space-Time Block Code (STBC), known as Alamouti scheme [13] for two transmit antennas, is the simplest technique with linear receiver complexity to improve the reliability of a wireless communication system. The capacity and reliability of MIMO systems depend on the Signal-to-Noise Ratio (SNR) and the correlation properties among the channel transfer functions of different pairs of transmit and receive antennas [14]. One of the basic requirements to realize the gains from MIMO systems is low correlation in the effective channel. The correlation comes from three sources, namely, the correlated fading channel, correlation among the transmit antennas, and correlation among the receive antennas. The correlation between two antennas depends on the coupling and isolation between them [17]. The coupling in turn is dependent on the physical separation between antennas and the matching network.

Several papers including [18] have previously presented analysis and simulation of the effect of correlation between antennas on the performance of Alamouti STBC scheme. The Alamouti scheme starts losing performance as the separation between the two antennas reaches somewhere between 0.1λ to 0.3λ. The size of traditional antenna is a major concern in placing multiple antennas on a miniaturized sensor node. The free space wavelength for the frequency band at 5.2 GHz, used in WLAN, is 5.77 cm. To obtain low correlation for WLAN applications with physical separation between antennas to be $\geq \lambda/2 = 2.88$ cm is not difficult. However, application of MIMO in wireless sensor networks, especially in medical applications, requires multiple antennas to be placed on tiny sensor nodes. To place multiple antennas with physical separation of $\geq \lambda/2$ on a sensor node is very challenging using traditional antenna technologies.

This section presents a unique antenna design and antenna array based on metamaterial that leads to correlation coefficient of 0.42 without an isolation structure and 0.04 with proper isolation between the antennas. We present the design, and radiation pattern of antennas, and BER performance of using them in a WLAN system.

2.1.2 System Model with Antenna Correlation

We considered a MIMO system with 2 transmit and 2 receive antennas having Alamouti STBC [11] encoder that maps the two consecutive symbols x1 and x2 to two antennas over two symbol periods at the transmitter. The correlation coefficient between the two transmit antennas Tx1 and Tx2 is α. The transmitted signals pass through a Rayleigh flat-fading wireless channel with a (2×2) channel matrix H with element $h_{ij} = \alpha_{ij} \exp(j\theta_{ij})$ representing the gain between transmit antenna j, receive antenna i, and additive white Gaussian noise (AWGN). The correlation coefficient between the two receive antennas Rx1 and Rx2 is β. The decoding is based on linear combining of signals received at the two antennas over two symbol periods. This scheme performs maximum likelihood detection of x1 and x2 using simple linear combining. The discrete channel model for a (2×2) MIMO system is given below in (1)-(3).

$$Y = C_R H C_T X + N; X = \begin{bmatrix} x_1 \\ x_2 \end{bmatrix}, Y = \begin{bmatrix} y_1 \\ y_2 \end{bmatrix} \tag{1}$$

$$H = \begin{bmatrix} h_{11} & h_{12} \\ h_{21} & h_{22} \end{bmatrix}, N = \begin{bmatrix} n_1 \\ n_2 \end{bmatrix} \tag{2}$$

$$C_T = \begin{bmatrix} 1 & \alpha \\ \alpha & 1 \end{bmatrix}, C_R = \begin{bmatrix} 1 & \beta \\ \beta & 1 \end{bmatrix} \tag{3}$$

The correlation matrices between antennas at the transmitter and receiver are given by CT and CR, respectively in (3). The simple implementation of linear combiner at the receiver would lead to a minimal increase in the complexity of receiver at a sensor node. The performance gain from using Alamouti scheme has a direct relationship with the coupling between the two antennas. The coupling depends on the physical separation between the antennas. When we consider a two transmit antenna scenario, the correlation coefficient for a uniform distribution of sources is given by (4). The assumption of uniform distribution holds quite well in an indoor environment. In (4), S11, S12, S21, S22, are scattering S-parameters for the antenna system. The S-parameters takes into account the relationship between voltage and current at the input and output terminals of the antenna, the loading, and matching network employed. The analytical derivation of S-parameters is quite involved and requires several approximations. The terms S12 and S21 contain the effect of mutual coupling between antennas. The correlation coefficient can be computed either using (4) after obtaining the S-parameters from full-wave 3-D electromagnetic simulation of antennas or through measurement of antenna prototypes or using (5) from the three-dimensional far-field radiation pattern.

$$\rho_{12} = \frac{\left| s_{11}^* s_{12} + s_{21}^* s_{22} \right|^2}{\left(1 - \left(|s_{11}|^2 + |s_{21}|^2 \right) \right) \left(1 - \left(|s_{12}|^2 + |s_{22}|^2 \right) \right)} \tag{4}$$

$$\rho_{12} = \frac{\int_{4\pi} \overline{G}_1 \overline{G}_2^* d\Omega}{\sqrt{\int_{4\pi} \overline{G}_1 \overline{G}_1^* d\Omega \int_{4\pi} \overline{G}_2 \overline{G}_2^* d\Omega}} \tag{5}$$

2.1.3   Antenna Design

As mentioned previously, the reduction of mutual coupling between closely-spaced antenna elements is essential to the performance of MIMO systems due to the fact that the mutual coupling affects the phase and distribution of the current, input impedance and radiation pattern in each antenna element which significantly reduces the capacity of the MIMO systems. Several techniques have been reported to reduce the mutual coupling between radiating elements in MIMO systems. Some of these techniques are based on the use of Electromagnetic Band Gap (EBG) structures [19], defected ground plane [20], and the use of μ-Negative (MNG) structures. In [21], MNG structures have been used to reduce the mutual coupling between two high profile monopoles, where the achieved reduction in mutual coupling was 20 dB. In this chapter, we utilize MNG to reduce the mutual coupling between two conformal micro strip radiating elements sharing a common substrate intended for applications that require compact space, i.e., wearable medical devices and miniaturized sensor nodes. When MNG structures are excited with a specific polarization, an electric current is induced through the loops of split ring resonators, as a consequence, the structures act as magnetic dipoles and an effective medium with a negative permeability over a certain frequency range is generated. As a result, the existence of real propagating modes is prevented within this medium [22]. This behavior is utilized to block the mutual coupling between the radiating elements of the proposed antenna. The proposed design consists of two elliptical shaped patch elements with a major axis of 14 mm and a minor axis of 4 mm placed on a 19 mm × 14.5 mm × 0.85 mm RO3006 substrate with a dielectric constant of 6.15 backed by a ground plane. Two identical elliptical radiating elements were designed to resonate at 5.2 GHz. The inter-element separation distance is 10.6 mm (0.18 λ), where λ is the free space wavelength. Parametric study was performed for the two coaxial feed locations to achieve optimal impedance matching. The optimized locations are 8 mm along the major axis and 3 mm along the minor axis for the first element, and 6 mm and 2.5 mm for the second element. The return loss is -20 dB with a -10 dB bandwidth of 50 MHz. A unit cell of a square ring resonator is designed to provide an effective negative permeability of -8 at 5.2 GHz, which provides a reasonable isolation between the two elements. Next, a set of 8 unit cells is arranged horizontally between the radiating elements, separated by 1.5 mm from each other. The unit cell consists of two square metal strip inclusions with opposite orientation printed on both sides of a 4.4 mm × 4.4 mm × 0.8 mm RO3006 sub strate. The gap in each ring is 0.3 mm. The height of the antenna along with the metamaterial structures does not exceed 6 mm which categorizes the design under low profile antennas. The front view and side view of the proposed antenna design along with corresponding dimensions are depicted in Fig. 1 and Table 1, respectively.

The antenna and MNG unit cell were designed and simulated using both time-domain and frequency-domain solvers of CST Microwave Studio which is based on the Finite Integration Technique (FIT). The simulated S-parameters for the proposed antenna are shown in Fig. 2.

From the transmission coefficient S12, we observe a large reduction in mutual coupling (-26 dB) at 5.2 GHz for the design with MNG structures compared to -4 dB for the design without MNG. We also notice a slight shift in the resonance frequency for the MNG case (around 20 MHz), which can be compensated for by slightly

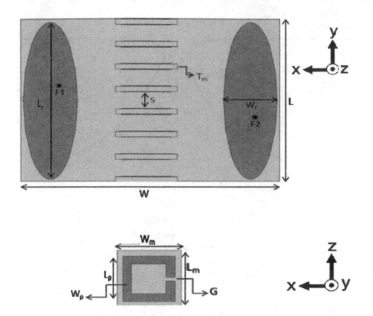

Fig. 1. Two-element antenna array based on elliptical patch radiating elements with MNG. Top view (top) and side view (bottom).

Table 1. Dimensions (mm) of the proposed antenna system.

L	14.5	W_m	4.4
W	19	L_m	4.4
L_r	14	T	0.85
W_r	4	L_p	3.5
S	1.5	W_p	0.6
T_m	0.45	G	0.3

adjusting the patch length in order to keep the patch resonance frequency identical in both cases. Moreover, we simulated the correlation coefficient for both cases (with and without MNG structures). The correlation coefficient for the MNG case is 0.04 versus 0.42 for the case without MNG. It is also worth mentioning that the simulated correlation coefficient has been extracted from the far-field analysis which is more accurate than the S-parameter method [24]. According to the S-parameters and correlation coefficient analysis, the design with MNG provides significant isolation between the radiating elements compared to the design without MNG with the same element spacing. The simulated E-plane and H-plane combined far-field radiation patterns at 5.2 GHz for the MNG case is presented in Fig. 3.

2.1.4   Error Performance Simulation

The setup for testing the performance of 2×2 Alamouti STBC scheme using the antenna array proposed in Sect. 3 is based on WLAN IEEE 802.11 a system with

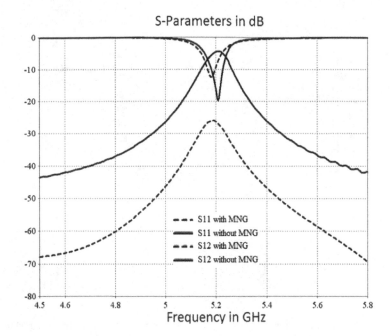

Fig. 2. Simulated S-parameter for both cases (with and without MNG).

5.2 GHz carrier frequency, QPSK modulation scheme and simulation is performed under flat-fading channel with Rayleigh Fading coefficients. We used two transmit antennas on one substrate at the transmitter and two receive antennas on another substrate at the receiver. The BER which is an essential parameter in telecommunication systems, is the percentage of transmitted bits that have errors relative to the total number of bits received in a specific transmission process.

The BER performance for (a) uncorrelated antennas at the transmitter (TX) as well as at the receiver (RX), (b) antenna array designed using MNG structure (correlation coefficient at both TX and RX is 0.04), and (c) antenna array designed without MNG structure (correlation coefficient at both TX and RX is 0.42), proposed in Sect. 3, is shown in Fig. 4.

We notice negligible performance loss for antenna array from MNG structure as compared to the performance for uncorrelated antennas.

2.2   Slot Fed Flexible Yagi-Uda Antenna

2.2.1   Introduction

Flexible and wearable devices are becoming increasingly popular for modern electronic devices. Body, personal, and local body area networks need antennas robust to interference caused by bending, flexing, and stand up to the abuse of everyday wear. One proposed way to overcome this is to use durable and flexible substrates such as vinyl or polyimide. Polyimide based antennas have been demonstrated [26] and simulations of

Farfield Gain Abs (Phi=90)

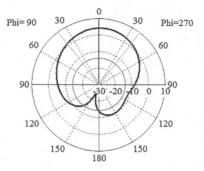

Theta / Degree vs. dB

(a)

Farfield Gain Abs (Phi=90)

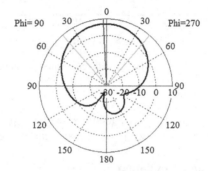

Theta / Degree vs. dB

(b)

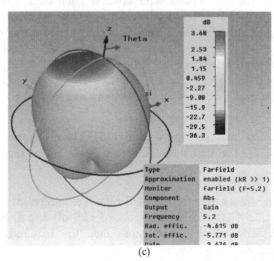

(c)

Fig. 3. simulated (a) E-plane (*YZ*), (b) H-plane (*XZ*) and (c) 3D radiation patterns for the proposed antenna design at 5.2 GHz.

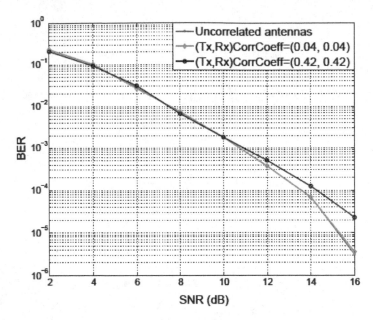

Fig. 4. BER performance of (2 × 2) MIMO with correlated transmit and correlated receive antennas.

two-dimensional flexible yagi-uda type antennas have been presented in [25]. In this section a flexible ISM band (2.4 GHz) Yagi-Uda antenna is presented.

2.2.2   Antenna Design and Simulation

Antenna Geometry. The antenna design presented is a uniplanar structure. The feed is a slot type coplanar waveguide, and the radiating element is structured as a dipole. There is a reflecting element and three parasitic elements that act as directors.

The dimensions, in millimeters, are as follows: $L_1 = 100$, $W_1 = 80$, $W_2 = 26.5$, $W_3 = 51.5$, $W_4 = W_5 = 46$, $L_2 = 6$, $L_3 = 13$, $L_4 = L_6 = L_8 = L_{10} = 2$, $L_5 = L_7 = L_9 = 19$, $G = 0.5$. An iterative design process was used to achieve desired $S_{1,1}$, and directivity results.

Simulation. Simulation was performed using the Time Domain solver of CST Microwave Studio [23]. The simulated E-cut, and H-cut are presented in Fig. 5 and Fig. 6 respectively.

As seen from the simulated reflection coefficient S11 and return loss in Fig. 7, the antenna resonates at 2.5 GHz with a return loss of 15.7 dB and a -10 dB impedance bandwidth of 145 MHz which covers more than the required ISM bandwidth (2.4-2.483 GHz) (Fig. 8).

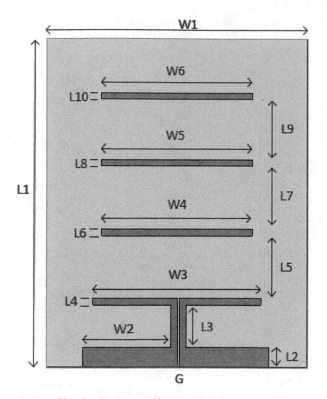

Fig. 5. Geometry of the proposed antenna.

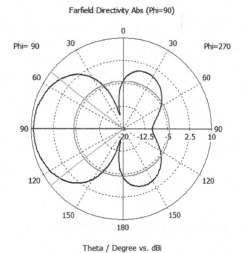

Fig. 6. Simulated E-plane radiation pattern.

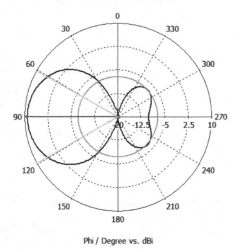

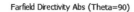

Fig. 7. Simulated H-plane radiation pattern

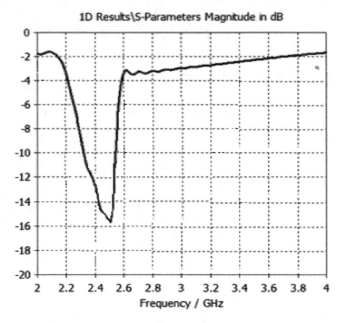

Fig. 8. Simulated reflection coefficient S11 for the proposed Yagi-Uda antenna.

3 Conclusion

In this chapter, the design and analysis of two novel antennas aimed for telemedicine and WBAN applications are presented. The first design is based on a MIMO antenna array utilizing μ-negative metamaterial structures to achieve low mutual coupling between the antenna radiating elements. The second design is a mechanically flexible directional Yagi-Uda antenna resonating at 2.5 GHz for on-body communication. Both designs have the merits of compactness, light weight, mechanical robustness, and high efficiency. Such properties suggest that the proposed designs would be reasonable candidates for telemedicine and WBAN applications that requires small size, low profile, cost effectiveness, and light weight.

References

1. Fong, B., Ansari, N., Fong, A.C.M.: Prognostics and health management for wireless telemedicine networks. Wireless Commun. IEEE **19**(5), 83–89 (2012)
2. Algaet, M.A., Bin Muhamad Noh, Z.A., Shibghatullah, A.S., Milad, A.A.: Provisioning quality of service of wireless telemedicine for e-health services. In: 2013 IEEE Conference on Information and Communication Technologies (ICT), pp. 199–202, 11-12 April 2013
3. Khaleel, H.R., Al-Rizzo, H.M., Rucker, D.G., Elwi, T.A.: Wearable Yagi microstrip antenna for telemedicine applications. Radio and Wireless Symposium (RWS) 2010 , pp. 280–283. IEEE, 10-14 January 2010
4. Haga, N., Saito, K., Takahashi, M., Ito, K.: Characteristics of cavity slot antenna for body-area networks. IEEE Trans. Antennas Propag. **57**(4), 837–843 (2009)
5. Rowe, W.S.T., Waterhouse, R.B.: Reduction of backward radiation for CPW fed aperture stacked patch antennas on small ground planes. IEEE Trans. Antennas Propag. **51**(6), 1411–1413 (2003)
6. Targonski, S.D., Pozar, D.M.: Aperture-coupled microstrip antennas using reflector elements for wireless communications. In: Proceedings of IEEE-APS Conference Antennas and Propagation for Wireless Communications, pp. 163–166, November 1998
7. Sanz-Izquierdo, B., Batchelor, J.C.: WLAN jacket mounted antenna. In: Antenna Technology: Small and Smart Antennas, Metamaterials, and Applications, pp. 57–60, June 2007
8. Islam, M.T., Faruque, M.R.I., Misran, N.: Reduction of Specific Absorption Rate (SAR) in the human head with ferrite material and metamaterial. Prog. Electromagnet. Res. C **9**, 47–58 (2009)
9. Raad, H., Abbosh, A., Al-Rizzo, H.H., Rucker, D.G.: Flexible and compact AMC based antenna for telemedicine applications. IEEE Trans. Antennas Propag. **61**(2), 524–531 (2013)
10. Khaleel, H. R., Al-Rizzo, H.M., Rucker, D.G., Al-Naiemy, Y.: Flexible printed monopole antennas for WLAN applications. In: 2011 IEEE International Symposium on Antennas and Propagation (APSURSI), pp. 1334–1337, 3-8 July 2011
11. Chou, H.-T., Cheng, H.-C., Hsu, H.-T., Kuo, L.-R.: Investigation of isolation improvement techniques for MIMO WLAN portable terminal applications. Prog. Electromagnet. Res. PIER **85**, 349–366 (2008)
12. Abouda, A.A., Haggman, S.G.: Effect of mutual coupling on capacity of MIMO wireless channels in high SNR scenario. Prog. Electromagnet. Res. PIER **65**, 27–40 (2006)

13. Alamouti, S.: A simple transmit diversity technique for wireless communications. IEEE J. Select. Areas Commun. **16**(8), 1451–1458 (1998)
14. Shiu, D.S., Foschini, G.J., Gans, M.J., Kahn, J.M.: Fading correlation and its effect on the capacity of multielement antenna systems. IEEE Trans. Commun. **48**, 502–513 (2000)
15. Shin, H., Lee, J.H.: Capacity of multiple-antenna fading channels: spatial fading correlation, double scattering, and keyhole. IEEE Trans. Inf. Theory **49**(10), 2636–2647 (2003)
16. Wallace, J.W., Jensen, M.A.: Mutual coupling in mimo wireless systems: a rigorous network theory analysis. IEEE Trans. Wireless Commun. **3**(4), 1317–1325 (2004)
17. Kyritsi, P., Cox, D., Valenzuela, R., Wolniansky, P.: Correlation analysis based on mimo channel measurements in an indoor environment. IEEE J. Sel. Areas Commun. **21**(5), 713–720 (2003)
18. Caban, S., Rupp, M.: Impact of transmit antenna spacing on 2x1 alamouti radio transmission. Electron. Lett. **43**(4), 198–199 (2007)
19. Ikeuchi, R., Hirata, A.: Dipole Antenna Above EBG Substrate for Local SAR Reduction. Antennas and Wireless Propagation Letters, IEEE **10**, 904–906 (2011)
20. Caloz, C., Okabe, H., Iwai, T., Itoh, T.: A simple and accurate model for microstrip structures with slotted ground plane. IEEE Microwave Wireless Comput Lett. **14**(4), 133–135 (2004)
21. Bait-Suwailam, M.M., Boybay, M.S., Ramahi, O.M.: Electromagnetic coupling reduction in high-profile monopole antennas using single-negative magnetic metamaterials for mimo applications. IEEE Trans. Antennas Propag. **58**(9), 2894–2902 (2010)
22. Pendry, J., Holden, A., Robbins, D., Stewart, W.: Magnetism from conductors and enhanced nonlinear phenomena. IEEE Trans. Microwave Theory Tech. **47**(11), 2075–2084 (1999)
23. CST microwave studio (2010, April). http://www.cst.com/Content/Products/MWS/Overview.aspx
24. Blanch, S., Romeu, J., Corbella, I.: Exact representation of antenna system diversity performance from input parameter description. Electron. Lett. **39**, 705–707 (2003)
25. Khaleel, H.R., Al-Rizzo, H.M., Rucker, D.G.: Compact polyimide-based antennas for flexible displays. IEEE J. Disp. Technol. **8**(2), 91–97 (2012)
26. Khaleel, H.R., Al-Rizzo, H.M., Rucker, D.G., Mohan, S.: A Compact Polyimide-Based UWB Antenna for Flexible Electronics. Antennas and Wireless Propagation Letters IEEE **11**, 564–567 (2012)
27. Khaleel, H.R., Al-Rizzo, H.M., Rucker, D.G.: Compact polyimide-based antennas for flexible displays. IEEE J. Disp. Technol. **8**(2), 91–97 (2012)
28. Abbosh, A.I., Babiceanu, R.F., Al-Rizzo, H., Abushamleh, S. Khaleel, H.R.: Flexible Yagi-Uda antenna for wearable electronic devices. In: IEEE International Symposium on Antennas and Propagation Society (2013)
29. Ding, Y., Jiao, Y., Fei, P., Li, B., Zhang, Q.: Design of a multiband quasi-Yagi-type antenna with CPW-to-CPS transition. IEEE Antennas Wireless Propag. Lett. **10**, 1120–1123 (2011)
30. Hsu, S.S., Wei, K.C., Hsu, C.Y., Ru-Chuang, H.: A 60-GHz millimeter-wave CPW-fed Yagi antenna fabricated by using 0.18-μm CMOS technology. IEEE Electron Device Lett. **29**(6), 625–627 (2008)
31. Kan, H.K., Abbosh, A.M., Waterhouse, R.B., Bialkowski, M.E.: Compact broadband coplanar waveguide-fed curved quasi-Yagi antenna. IET Microwave Antennas Propag. **1**(3), 572–574 (2007)
32. Ta, S.X., Choo, H., Park, I.: Wideband double-dipole Yagi-Uda antenna fed by a microstrip-slot coplanar stripline transition. Prog. in Electromagnet. Res. B **44**, 71–87 (2012)
33. Cai, R.N., Lin, S., Huang, G.L. et al.: Research on a novel Yagi-Uda antenna fed by balanced microstrip line. In: Proceedings China-Japan Joint Microwave Conference (CJMW) 2011, pp. 1–4, April 2011

34. Huang, H.C., Lu, J.C., Hsu, P.: A compact printed Yagi type antenna for GPS application. In: Proceedings Asia-Pacific Microwave Conference (APMC), pp. 1698–1701 Melbourne, December 2011
35. DeJean, G.R., Tentzeris, M.M.: A new high-gain microstrip yagi array antenna with a high front-to-back (F/B) Ratio for WLAN and millimeter-wave applications. IEEE Trans. Antennas Propag. **55**(2), 298–304 (2007)
36. Agarwal, K., Guo, Y.-X., Salam, B. Albert, L.C.W.: Latex based near-endfire wearable antenna backed by AMC surface. In: IEEE Microwave Workship Series on RF and Wireless Technologies for Biomedical and Healthcare Applications (IMWS-BIO), pp. 1–3, December 2013

Impact of Threshold Computation Methods in Hardware Wavelet Denoising Implementations for Neural Signal Processing

Nicola Carta[✉], Danilo Pani, and Luigi Raffo

DIEE - Deptartment of Electrical and Electronic Engineering, University of Cagliari,
Via Marengo 3, 09123 Cagliari, Italy
{nicola.carta,danilo.pani,luigi}@diee.unica.it

Abstract. Wavelet denoising effectiveness has been proven in neural signal processing applications characterized by a low SNR. This non-linear approach is implemented through the application of some thresholds on the detail signals coming from a sub-band decomposition. The computation of the thresholds could exhibit a high latency when involving some estimators such as the Median Absolute Deviation (MAD), which is critical for real-time applications. When a VLSI implementation is pursued for low-power purposes, such as in the neuroprosthetic field, these aspects cannot be overlooked. This paper presents an analysis of the main VLSI hardware implementation figures related to this specific aspect of the signal denoising by wavelet processing. Xilinx System Generator has been exploited as a design and co-simulation tool to ease the hardware development on off-the-shelf FPGA platforms. The MAD estimator has been both combinatorially and sequentially implemented, and compared against the sample standard deviation. The study reveals similar performance on the neural signals but dramatically worse implementation figures for the MAD. The combinatorial version of the MAD actually prevents an efficient implementation on medium-small devices. This result is important to perform a correct implementation choice for implantable real-time systems, where the device size is relevant for an usable realization.

Keywords: Wavelet denoising · Neural signal processing · FPGA · Design tools

1 Introduction

Wavelet denoising (WD) is a non-linear filtering technique usually adopted to remove the background noise added to the signal of interest, especially in presence of a Gaussian source whose spectrum overlaps the useful signal bandwidth. Its effectiveness has been proven in several biomedical signal processing applications [1,19], including neural signals denoising [5]. When a poor signal to noise ratio (SNR) is present, the adoption of WD can help in revealing even hidden

© Springer International Publishing Switzerland 2015
G. Plantier et al. (Eds.): BIOSTEC 2014, CCIS 511, pp. 66–81, 2015.
DOI: 10.1007/978-3-319-26129-4_5

events in the time domain [3]. Algorithmically, it consists of a sub-band decomposition of the signal, thresholding (introducing the non-linearity) and recomposition. Several methods for calculating the thresholds have been presented in the scientific literature, with different performance both in terms of quality (effectiveness) and efficiency. This issue is normally overlooked even though it is actually important when real-time performance is required.

Some specific applications such as neural signal processing for motor/sensory neuroprostheses could benefit from the adoption of WD [3]. In this case, in general, both real-time performance and low power dissipation are required, all the more so when the device is aimed to be implanted. From this perspective, Application Specific Integrated Circuits (ASICs) represent the most powerful implementation platform, even though their design requires highly specific skills and a longer development time [13]. Since the outcome of the design process is quite inflexible, taking into account the quickly mutable environment generated by the advancements of the research in biomedical signal processing, the adoption of tools for automatic creation of hardware description language (HDL) designs can speed up the prototyping phase on Field Programmable Gate Array (FPGA) or ASIC. Examples of such tools are ORCC[1] or Xilinx System Generator[2]. Without specific add-ons, these tools are generally rather ineffective for low-power design [15]. The integration with tools such as Simulink enables a faster development thanks to the possibility of considering the implementation at an higher level.

This paper presents an analysis of the main VLSI hardware implementation figures related to the specific aspect of the threshold estimation in wavelet denoising for neural signal processing. The threshold estimation stage, which must iteratively evaluate the average level of the noise affecting the signal of interest, is marginally considered in the largest part of the applications. Except when a fixed threshold is used [7], the estimation of the standard deviation of the noise is usually required. The Median Absolute Deviation (MAD), known to be a robust estimator of the dispersion in presence of outliers, is compared here to the sample standard deviation in terms of effectiveness and efficiency when the algorithm is implemented in hardware on an FPGA platform, in the light of a perspective development of an implantable neural signal processing ASIC. Both combinatorial and sequential versions of the MAD have been implemented, along with sample standard deviation and a pure software implementation on a MicroBlaze processor. The results in terms of area and latency reveal the poor scalability of the MAD implementations and the comparable effectiveness with simpler approaches, resulting from the evaluation onto an open neural signals database [17].

2   Wavelet Denoising of Neural Signals

WD has been used in neural signal processing since a long time to cancel the background noise that can be approximated to a Gaussian distributed

---

[1] orcc.sourceforge.net.

[2] www.xilinx.com/tools/sysgen.htm.

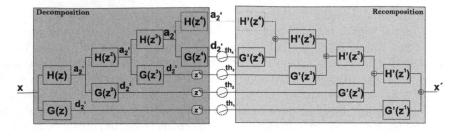

Fig. 1. Block Diagram of the Wavelet Denoising scheme using the à-trous approach.

random source [14]. For some wavelets, WD can be implemented using a system of quadrature mirror Finite Impulse Response (FIR) filters for each stage. The input signal is decomposed in its low-frequency and high-frequency bands, respectively called "approximation" and "detail". The approximation is split again in the same way repeatedly until the N−th level of decomposition has been reached and the detail signals are thresholded, before the recomposition is pursued, as can be seen in Fig. 1. The basic analysis elements are a low-pass $H(z)$ and an high pass $G(z)$ filters. Being the Nyquist frequency of the approximation one half of that of the incoming signal, the sample rate can be reduced (decimated approach) so that the same filters can be used in every level without any information loss [9]. Alternatively, the sample rate can be preserved upsampling in each level the filter coefficients of the previous one, in the so called *algorithme à trous* scheme [6]. In practice, the filters at the stage i are simply $H(z^i)$, $G(z^i)$ for the analysis and the mirrored versions for the synthesis. Such a redundant approach leads to the same time resolution in every level [11] and to shift invariance [4]. When the N-th level has been computed, all the details are thresholded either *hard* or *soft*, respectively whether the samples of the detail signals are simply cleared to zero if below the threshold or also the samples above threshold are modified by subtracting the value of the threshold itself. During recomposition, the sample-wise averaging between an approximation and the related detail can be implemented as a simple sum provided that the coefficients of the synthesis filters are multiplied by 0.5.

2.1 The Thresholding Aspects

The choice of the threshold influences the quality of the denoising so much that even data-specific approaches have been presented so far [12]. The threshold can be fixed [7] or adaptive [18], the same or different for all the details. In particular, adaptive thresholds are typically computed estimating the rms or the standard deviation σ of the signal at the different levels of the decomposition and then correcting them through a multiplicative factor. Different scaling factors have been derived and are preferred by different authors, as for the Minimax [3], Stein's Unbiased Risk [8] or Universal [2] methods. The last one, chosen in this work regardless the method used to estimate the standard deviation of the noise, is defined as:

$$\theta = \sigma\sqrt{2\log M} \tag{1}$$

where M is the length of the signal frame in terms of number of samples.

For the estimation of the standard deviation of the noise, due to the robustness to the presence of outliers[3], usually the preferred method is the MAD, defined as:

$$MAD = median_i\left(|X_i - median_j(X_j)|\right) \tag{2}$$

Furthermore, because of the high-pass nature of the detail signals, it is common to implement the MAD as simply the median of the absolute value of the details:

$$\overline{MAD} = median_j\left(|X_j|\right) \tag{3}$$

It has been proved that $MAD \approx 0.6745\sigma$. Block-on-line threshold adaptiveness can be guaranteed exploiting a sliding window approach with variable window length and overlap. For the sake of the comparison between the different threshold estimation techniques in terms of performance and hardware figures, the overlapping parameter is without effects. For the computation of the sample standard deviation exploiting this sliding window approach, starting from the technique used in [16] and thanks to the zero-mean nature of the high-pass detail signals, for each N new input samples in the window, the related sum of squares for the j-th decomposition level can be computed as:

$$s_j = \sum_{n=1}^{N} d_j^2[n] \tag{4}$$

and then used to determine σ for the 4-time larger windows as:

$$\sigma = \sqrt{\frac{1}{4N-1}\sum_{k=1}^{4} s_j} \tag{5}$$

Thanks to the sliding window approach, the threshold value is updated every N sampling periods ($M = 4 \times N$). The longer the observation window, the better the estimation accuracy, provided that instantaneous variations (neural spikes) do not influence the threshold computation.

3   Hardware System Design

From an hardware perspective, when the final goal is a low-power embedded architecture, the MAD estimation is effective for multiplier-free systems. Nevertheless, it pays the absence of actual mathematical computations with an algorithmic complexity associated to the required sorting of one half of the block of data. It is also onerous from the memory perspective when the block size

---

[3] In neural signal processing, the spikes corresponding to the action potentials of the active neurons can be considered as partly composed of outlier samples in recordings with an average SNR.

is huge. For such reasons, the MAD is more suited for off-line processing than for on-line systems [16,20] such as those required in brain-machine interfaces exploiting adaptive thresholds. Threshold adaptation is useful when the noise process in non-stationary, for instance in real-world scenarios when the subject moves in a real environment characterized by different noise sources spread over the space.

For the analysis of the hardware features associated to the different design choices, the *"algorithme à-trous"* has been selected using the simplest Haar wavelet to reduce the memory footprint. The approximation signal at the 4th level has been cleared, so that at 12 kHz the overall processing without the non-linearity introduced by denoising would restrict the bandwidth between 375 Hz and the Nyquist frequency [16]. The FIR filters have been implemented in the Transposed Direct-Form I, not implementing specific optimizations since the filters banks are the same regardless the chosen threshold estimation strategies. The Universal scaling has been selected, as already said. The overlap between adjacent windows has been fixed to three quarters of the window length.

Hereafter, we consider the two alternative solutions implemented for the estimation of the standard deviation of the noise:

– the $\overline{MAD}$ of the signal, using either a combinatorial or an iterative approach;
– the sample standard deviation σ of the signal.

In order to evaluate the different solutions from a hardware perspective, we adopted Xilinx System Generator to speed-up the hardware design and perform accurate co-simulations. The final design can be straightforwardly mapped onto an FPGA board for performance evaluation. In our tests, a Xilinx Virtex-5 LX330 has been chosen as target device for its considerable amount of available resources.

The estimation of the standard deviation of the noise has been performed as described in the previous section. For the $\overline{MAD}$ implementation, on System Generator the absolute value of each input sample is extracted, then the median is computed on the whole incoming window of M input rectified samples and the multiplication by a constant is performed to estimate the standard deviation.

As already said, the median value calculation requires the hardware implementation of a sorting algorithm which represents a costly operation from a hardware point of view. A first possible solution can be the unfolded sorter presented in [2], for which the Simulink model considering windows of $M = 8$ input samples is presented in Fig. 2. The basic sorting cell makes the comparison between two inputs A and B and swaps them if $A < B$. It is possible to demonstrate that, if the comparators work in parallel, $M - 1$ steps are sufficient to properly perform the sorting of M elements. The output is updated in a combinatorial way every time a sample arrives in input at the sampling frequency f_s, after the proper shift of the values saved into the registers needed to prepare the input samples for the processing. The $\overline{MAD}$ is computed as the arithmetic mean of the two central elements of the sorted array for an even number of samples.

This solution clearly presents scalability issues with the enlargement of the observation window. In this case, beyond the penalty associated to the huge

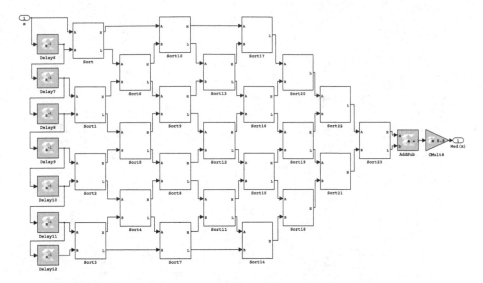

Fig. 2. The Simulink Model of the sorting algorithm for the median estimation using an unfolded combinatorial approach for windows with M=8 samples.

amount of hardware resources, the increasing internal critical path determined by the cascade of comparators largely limits the maximum operating frequency.

To overcome such problems, an iterative (folded) approach to the implementation of the sorting algorithm, able to reuse the same resources at each step, can be used. It is similar to the one proposed in [10] about the Burrows-Wheeler transform but in this work it has been adapted to the wavelet denoising case. In this case, the sorting strategy uses only two levels of comparators. At the beginning, the swaps are performed only for the registers related to odd adjacencies, activating only the first level of comparators. If the vector is not yet sorted, at the next iteration only the comparators of the second level are active, and so on until the sorting process is completed. It is possible to demonstrate that the number of necessary steps is $M/2$ if M is the number of samples to sort. Figure 3 shows the iterative scheme.

The *swp* signal coming out from the comparator block is used to specify that the two inputs have been swapped. The samples in input to the parallel sorter, belonging to each observation window, are temporarily saved into a single-port memory. Immediately after the last sample of the window has been saved in this memory, its content is copied into the registers and the sorting process can start. When all the *swp* signals are equal to 0 during the last necessary iteration, the input vector is correctly sorted. A finite state machine, one for each *Threshold Estimator* block (i.e. one for each decomposition level), is used to control the various phases of the process.

In order to compare the hardware characteristics of these models of threshold estimation based on the $\overline{MAD}$ against those of a traditional sample standard

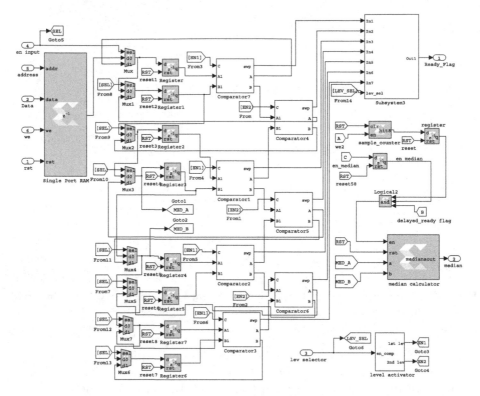

Fig. 3. The Simulink Model of the sorting algorithm for the median estimation using an iterative approach for windows with M=8 samples.

deviation as described above, an hardware model has been designed also for such an approach.

Figure 5 shows the Simulink model for this implementation. Every time an input sample arrives, it is squared and added to the current value of s_j. After N samples, the final value of s_j is saved in one of the 4 locations of the single-port RAM used as circular buffer in order to determine the correct value of σ over the sliding window. It involves the hardware implementation of the squared root as the final processing stage for the threshold estimation able to work with the proper internal fixed point representation. To this aim, we chose to exploit an iterative approach that is possible to demonstrate is able to converge in less than 64 iterations in case of input samples with a 16-bit fixed-point representation using 15 bits of fractional part, achieving a good level of accuracy. Its pseudocode is presented hereafter.

```
max = sqrt_in;
min = 0;
for(int i=0; i<ITERS; i++)
     avg = (max+min)/2;
     avg_2 = avg*avg;
```

```
    if(avg_2 > sqrt_in) max = avg;
    else min=avg;
sqrt_out = min;
```

It operates as a bisection research method, dividing by 2 at each iteration the range of possible values in which the result is searched until the algorithm converges to a specific output value. This algorithm determines the hardware implementation shown in Fig. 4.

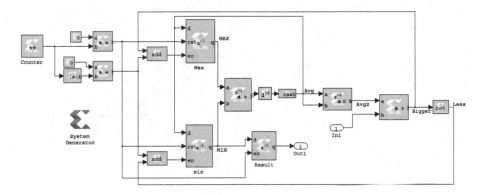

Fig. 4. The Simulink Model of the Squared Root block.

Regardless the chosen approach, the value of the threshold θ obtained through the Universal scaling approach is sent in input to the *Thresholder* block, able to apply the hard thresholding on the detail samples. It should be also considered that the value of θ is different for the various levels.

4   Results

Before analysing the results in terms of hardware resources which are necessary for the different solutions presented above, such solutions have been evaluated from a functional perspective. A publicly available dataset of simulated neural signals obtained from real physiological action potentials recorded from animals at the central nervous system level has been used [17]. The synthetic signals are obtained by linearly mixing an artificial sequence of real spikes from three neurons to other spikes at random times and amplitudes, representative of the background activity (of tunable intensity) of the neurons at a greater distance from the recording electrodes. The sampling frequency has been scaled to 12 kHz and the useful bandwidth is declared to be in the range 300 Hz - 3 kHz.

4.1   Performance Analysis

Figure 6 shows in the first row the neural signal with a low level of background noise used as input for the two hardware implementations based on the calculation of the $\overline{MAD}$ and of the sample standard deviation (the two versions of the

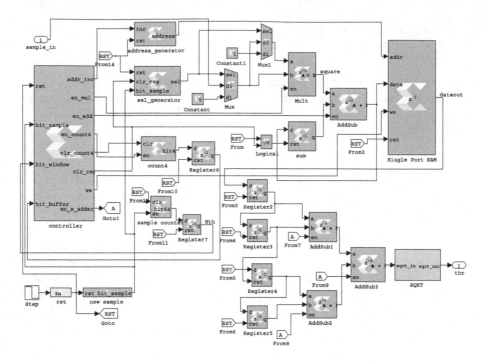

Fig. 5. The Simulink Model of the Threshold Estimator block based on the calculation of the sample standard deviation.

one implementing the $\overline{MAD}$ produce the same results). The next rows present the related outputs considering observation windows of $M = 4 \times 64$ samples.

It is possible to see that for both solutions, the wavelet denoising is able to remove, after an initial transient, the background noise added to the neural signal without cutting significant spikes. The same performance can be achieved using the same input signal but with a stronger background noise for which it is difficult to identify the various spikes on the raw signal, as can be shown in the first row of the Fig. 7. Even using neural signals with very low SNR, the two implementations behave similarly preserving the relevant spikes.

Then, we considered the possibility of enlarging the observation window in order to provide a more significant frame for computing the statistics on the signal. For example, Fig. 8 shows the outputs of the two solutions using $\overline{MAD}$ and sample standard deviation in the case of windows of $N = 128$ samples and a low level of background noise. The initial transient is obviously longer in comparison to the previous cases.

We also analysed the trend of the thresholds in output from the same decomposition level for different observation window lengths, for the two hardware solutions. The aim is to verify which is the minimum value of N, considering a sliding window length of $4 \times N$, that allows obtaining a good denoising.

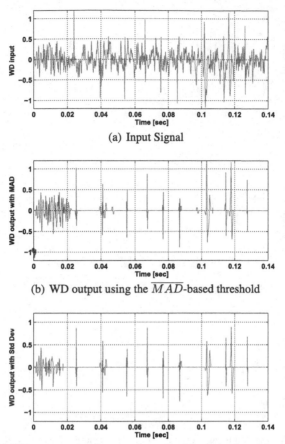

(a) Input Signal

(b) WD output using the $\overline{MAD}$-based threshold

(c) WD output using the sample standard deviation based threshold

Fig. 6. Wavelet Denoising input and outputs: low level noise, N=64 samples.

As can be noticed from Fig. 9, after a variable transient period according to the chosen value of N, the longer the observation window the better the stability of the threshold, not influenced by the presence of the neural spikes of interest. In fact, in the case of $N = 128$, the threshold estimation assumes an almost constant trend; the goal should be that of selecting the solution which provides the best compromise in terms of threshold estimation and required hardware resources.

4.2   Hardware Implementation Results

Synthesis results on a Xilinx FPGA Virtex-5 LX330 are presented in Table 1, which shows the percentage of available slices and Look-up Tables (LUTs) needed

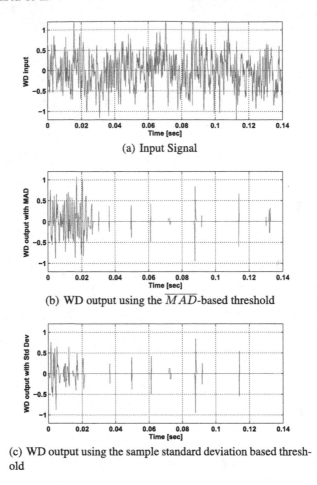

(a) Input Signal

(b) WD output using the $\overline{MAD}$-based threshold

(c) WD output using the sample standard deviation based threshold

Fig. 7. Wavelet Denoising input and outputs: high level noise, N=64 samples.

for the three considered threshold estimation blocks only, since the remainder of the wavelet denoising implementation is the same regardless of this stage.

The solution based on the combinatorial (unfolded) $\overline{MAD}$ implementation, as highlighted in Table 1, is absolutely inefficient, taking into account it has been presented only for $N = 8$, with the usual 4-times larger observation window. A rough estimation of the hardware resources required in case of $N = 32$ would lead to more than 330kLUT over the 207360 available ones, thus exceeding the considerable amount of physical resources on the target FPGA. The huge amount of LUTs, compared to the folded version, is incompatible with a real implementation in the context of this application, taking into account that the observation window length should be large enough to properly estimate the statistics of a signal sampled at 12 kHz.

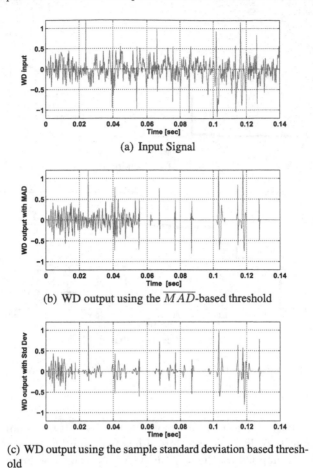

(a) Input Signal

(b) WD output using the $\overline{MAD}$-based threshold

(c) WD output using the sample standard deviation based threshold

Fig. 8. Wavelet Denoising input and outputs: low level noise, N=128 samples.

FPGA synthesis results demonstrate that the wavelet denoising solution based on the threshold estimation by the sample standard deviation allows minimizing the necessary hardware resources regardless the length of the observation window. Furthermore, the usage of slices and LUTs of the $\overline{MAD}$-based solution, even using a folded approach, is clearly incompatible with an efficient implementation of this processing stage, especially compared to the same data related to the sample standard deviation based implementations.

4.3 Latency Execution Evaluation

To evaluate the efficiency of the various approaches at the enlargement of the observation window, a cycle-accurate profiling has been performed. It has been carried out on the first-level detail obtained when processing the low-noise signal used for the performance evaluation. These results have been compared to the

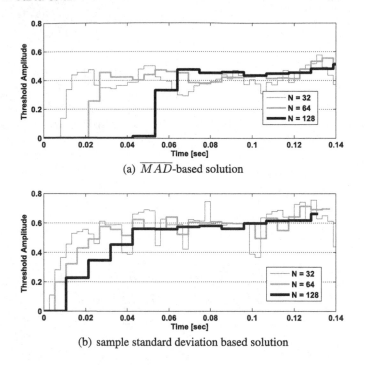

(a) $\overline{MAD}$-based solution

(b) sample standard deviation based solution

Fig. 9. Threshold variation over time using different lengths of the observation window.

corresponding values of execution latency of a totally-software solution. To this aim, an equivalent C code for each solution has been implemented on a MicroBlaze processor and mapped on the same FPGA, analysing the latency by means of a counter which is enabled when the last detail sample of the current data block is received and disabled when the threshold has been computed.

Obviously, when implementing the folded $\overline{MAD}$ approach, the latency is data-dependent and related to the number of swaps performed during the sorting of the absolute values passed in input to the processing stage, as can be seen in Table 2. This is not true for the sample standard deviation, which with the proposed approach always requires 68 cycles in hardware. For the $\overline{MAD}$, the number of swaps on a real signals follows a Gaussian distribution, as evaluated through the Lilliefors tests. For this reason, for different values of M we reported in Table 2 the minimum and maximum observed number of swaps, the mean and standard deviation. The range of variation, mean value and standard deviation of the execution latency is also provided for both the hardware and the software solutions, evaluating them on the whole detail signal and verifying even in this case a Gaussian distribution by the same statistic test. As can be seen, the number of execution cycles is always higher for the software implementation of the $\overline{MAD}$, and grows with the size of the observation window.

Table 1. FPGA synthesis results for the *Threshold Estimator* varying the length of the observation window.

	N	$f_{max}[MHz]$	Slice Registers	LUTs
Sample std	32	417.34	205 / 207360 (0.07 %)	142 / 207360 (0.04 %)
	64	416.61	206 / 207360 (0.07 %)	144 / 207360 (0.04 %)
	128	416.02	207 / 207360 (0.07 %)	147 / 207360 (0.04 %)
unfolded $\overline{MAD}$	8	417.08	558 / 207360 (0.27 %)	20270 / 207360 (9.77 %)
folded $\overline{MAD}$	32	242.78	2493 / 207360 (1.20 %)	7164 / 207360 (3.45 %)
	64	246.70	4932 / 207360 (2.38 %)	14377 / 207360 (6.93 %)
	128	221.42	9806 / 207360 (4.73 %)	28861 / 207360 (13.91 %)

Table 2. Comparison of execution latency for the various approaches with respect to a totally-software solution.

		$\overline{MAD}$			
		max	min	avg	std
M=32	Swaps	412	88	247	45
	lat. HW sol.[cycles]	139	59	114	19
	lat. SW sol.[cycles]	21070	16210	18595	679
M=64	Swaps	1498	531	997	134
	lat. HW sol.[cycles]	294	209	239	26
	lat. SW sol.[cycles]	82960	68455	75455	2004
M=128	Swaps	5426	2637	3987	381
	lat. HW sol.[cycles]	600	360	518	69
	lat. SW sol.[cycles]	325240	283405	303755	5712

5   Conclusions

Despite the ASIC implementation of digital signal processing algorithms is usually the preferred choice when low-power requirements and high processing speed are needed, the straightforward implementation of the best algorithmic solution exploited in the field could lead to unsatisfactory results. This paper analysed the particular case of the threshold computing for wavelet denoising algorithm. This part of the algorithm is usually marginally considered, but could be in the critical path when real-time update of the threshold is required.

In this paper, comparisons between a sample standard deviation and the widespread $\overline{MAD}$ reveals similar functional performance with dramatically better characteristic of the former in terms of hardware implementation, regardless the $\overline{MAD}$ is implemented as a combinatorial trellis as suggested by some authors or in a more efficient folded version. A latency analysis also reveals the superior performance of the sample standard deviation, jointly to its data-independence, which is an important aspect for real-time implementations. The paper also stresses the benefit of using hardware-software co-simulations tools such as Xilinx System Generator for rapid prototyping and verification on FPGA, which represents a value added for the research in rapidly evolving fields such as neural engineering.

Acknowledgements. The research leading to these results has received funding from the Region of Sardinia, Fundamental Research Programme, L.R. 7/2007 "Promotion of the scientific research and technological innovation in Sardinia", CRP-60544, ELoRA Project.

References

1. Anand, C.S., Sahambi, J.S.: Wavelet domain non-linear filtering for MRI denoising. Magn. Reson. Imaging **28**(6), 842–861 (2010)
2. Bahoura, M., Ezzaidi, H.: FPGA-implementation of discrete wavelet transform with application to signal denoising. Circuits Syst. Sig. Process. **31**(3), 987–1015 (2012)
3. Citi, L., Carpaneto, J., Yoshida, K., Hoffmann, K.P., Koch, K.P., Dario, P., Micera, S.: On the use of wavelet denoising and spike sorting techniques to process electroneurographic signals recorded using intraneural electrodes. J. Neurosci. Methods **172**(2), 294–302 (2008)
4. Cohen, A., Kovacevic, J.: Wavelets: the mathematical background. Proc. IEEE **84**(4), 514–522 (1996)
5. Diedrich, A., Charoensuk, W., Brychta, R., Ertl, A., Shiavi, R.: Analysis of raw microneurographic recordings based on wavelet de-noising technique and classification algorithm: wavelet analysis in microneurography. IEEE Trans. Biomed. Eng. **50**(1), 41–50 (2003)
6. Holschneider, M., Kronland-Martinet, R., Morlet, J., Tchamitchian, P.: A real-time algorithm for signal analysis with the help of the wavelet transform. In: Combes, P.J., Grossmann, P.A., Tchamitchian, P.P. (eds.) Wavelets, pp. 286–297. Springer, Heidelberg (1990)
7. Kuzume, K., Niijima, K., Takano, S.: FPGA-based lifting wavelet processor for real-time signal detection. Sig. Process. **84**(10), 1931–1940 (2004)
8. Mahmoud, M.I., Dessouky, M.I.M., Deyab, S., Elfouly, F.H.: Signal denoising by wavelet packet transform on FPGA technology. special issue of ubiquitous computing and communication. J. Bioinform. image **3**, 54–58 (2008)
9. Mallat, S.: Multifrequency channel decompositions of images and wavelet models. IEEE Trans. Acoust. Sign. Process. **37**(7), 2091–2110 (1989)
10. Martinez, J., Cumplido, R., Feregrino, C.: An FPGA-based parallel sorting architecture for the burrows wheeler transform. In: International Conference on Reconfigurable Computing and FPGAs, ReConFig 2005 (2005)
11. Martínez, J., Almeida, R., Olmos, S., Rocha, A., Laguna, P.: A wavelet-based ECG delineator: evaluation on standard databases. IEEE Trans. Biomed. Eng. **51**(4), 570–581 (2004)
12. Medina, C., Alcaim, A., Apolinario Jr., J.A.: Wavelet denoising of speech using neural networks for threshold selection. Electron. Lett. **39**, 1869–1871 (2003)
13. Montani, M., Marchi, L.D., Marcianesi, A., Speciale, N.: Comparison of a programmable DSP and FPGA implementation for a wavelet-based denoising algorithm. In: Proceeding of IEEE 46th Midwest Symposium on Circuits and Systems. vol. 2, pp. 602–605 (2003)

14. Oweiss, K.G., Anderson, D.J.: Noise reduction in multichannel neural recordings using a new array wavelet denoising algorithm. Neurocomputing **38–40**, 1687–1693 (2001)
15. Palumbo, F., Carta, N., Pani, D., Meloni, P., Raffo, L.: The multi-dataflow composer tool: generation of on-the-fly reconfigurable platforms. Journal of Real-Time Image Processing pp. 1–17 (2012)
16. Pani, D., Usai, F., Citi, L., Raffo, L.: Impact of the approximated on-line centering and whitening in OL-JADE on the quality of the estimated fetal ECG. In: Proceedings of the 5th International IEEE/EMBS Conference on Neural Engineering (NER), pp. 44–47 (2011)
17. Quiroga, R.Q., Nadasdy, Z., Ben-Shaul, Y.: Unsupervised spike detection and sorting with wavelets and superparamagnetic clustering. Neural Comput. **16**(8), 1661–1687 (2004)
18. Radovan, S., Saša, K., Dejan, K., Goran, D.: Optimization and implementation of the wavelet based algorithms for embedded biomedical signal processing. Comput. Sci. Inf. Syst. **10**, 502–523 (2013)
19. Singh, B.N., Tiwari, A.K.: Optimal selection of wavelet basis function applied to ECG signal denoising. Digit. Sign. Proc. **16**(3), 275–287 (2006)
20. Zhang, M., Deng, R., Ma, Z., Zhang, M.: A FPGA-based low-cost real-time wavelet packet denoising system. In: Proceedings of 2011 International Conference on Electronics and Optoelectronics (ICEOE). vol. 2, pp. V2–350-V2-353 (2011)

Development of Concurrent Object-Oriented Logic Programming Platform for the Intelligent Monitoring of Anomalous Human Activities

Alexei A. Morozov[1,4]([✉]), Abhishek Vaish[2], Alexander F. Polupanov[1,4],
Vyacheslav E. Antciperov[1], Igor I. Lychkov[3],
Aleksandr N. Alfimtsev[3], and Vladimir V. Deviatkov[3]

[1] Kotel'nikov Institute of Radio Engineering and Electronics of RAS, Moscow, Russia
[2] Indian Institute of Information Technology, Allahabad, India
[3] Bauman Moscow State Technical University, Moscow, Russia
[4] Moscow State University of Psychology & Education, Moscow, Russia
morozov@cplire.ru

Abstract. The logic programming approach to the intelligent monitoring of anomalous human activity is considered. The main idea of this approach is to use first order logic for describing abstract concepts of anomalous human activity, i.e. brawl, sudden attack, armed attack, leaving object, loitering, pickpocketing, personal theft, immobile person, etc. We have created a research led software platform based on the Actor Prolog concurrent object-oriented logic language and a state-of-the-art Prolog-to-Java translator for examining the intelligent visual surveillance. A method of logical rules creation is considered in relation to the analysis of anomalous human behavior. The problem of creation of special built-in classes of Actor Prolog for the low-level video processing is discussed.

Keywords: Anomalous human activity · Intelligent visual surveillance · Object-oriented concurrent logic programming · Actor Prolog

1 Introduction

Human activity recognition is a rapid growing research area with important application domains including security and anti-terrorist issues [1,6,7]. Recently logic programming was recognized as a promising approach for dynamic visual scenes analysis [4,8,17–19]. The idea of the logic programming approach is in usage of logical rules for description and analysis of people activities. To approach the problem, knowledge about object co-ordinates and properties, scene geometry, and human body constraints is encoded in the form of certain rules in a logic programming language and is applied to the output of low-level object/feature detectors. There are several studies based on this idea. In [4] a system was designed for recognition of so-called long-term activities (such as fighting and meeting) as temporal combinations of short-term activities (walking, running, inactive, etc.) using a logic programming implementation of the Event Calculus.

© Springer International Publishing Switzerland 2015
G. Plantier et al. (Eds.): BIOSTEC 2014, CCIS 511, pp. 82–97, 2015.
DOI: 10.1007/978-3-319-26129-4_6

The ProbLog state-of-the-art probabilistic logic programming language was used to handle the uncertainty that occurs in human activity recognition. In [19] an extension of predicate logic with the bilattice formalism that permits processing of uncertainty in the reasoning was proposed. The VidMAP visual surveillance system that combines real time computer vision algorithms with the Prolog based logic programming had been proposed by the same team. S. O'Hara [18] communicated the VERSA general-purpose framework for defining and recognizing events in live or recorded surveillance video streams. According to [18], VERSA ensures more advanced spatial and temporal reasoning than VidMAP and is based on SWI-Prolog. F.A. Machot et al. [8] have proposed real time complex audio-video event detection based on the Answer Set Programming approach. The results indicate that this solution is robust and can easily be run on a chip.

Research indicates that conventional approaches to human behavior recognition include low-level and high-level stages of video processing. In this paper, we addressed the problem of the high-level semantic analysis of people activity. We have created a research led software platform based on the Actor Prolog concurrent object-oriented logic language [9–14] and a state-of-the-art Prolog-to-Java translator [15] for implementation of the logical inference on video scenes. The Prolog-to-Java translator provides means for a high-level concurrent programming and a direct access to the low-level processing procedures written in Java.

In the case of simple human behavior, a set of logic program rules can be created manually on the basis of a priori knowledge of the particular behavior features, for example, speed of moving, but in the case of complex spatio-temporal behavior, special methods of automatic logical rules creation are to be developed.

We have described our first experiments in the area of human activity recognition in Sect. 2. The problem of creation of special built-in classes of the Actor Prolog logic language for the low-level video processing is discussed in Sect. 3. A method of logical rules creation based on a hierarchy of fuzzy finite state automata is briefly considered in Sect. 4.

2   Logical Analysis of Manually Marked Videos

On the first stage of the research, we have performed several experiments on analysis of manually marked videos that is traditional approach in the area. The CAVIAR data sets [5] were used. The CAVIAR data sets are annotated using the XML-based Computer Vision Markup Language (CVML). The structure of CVML is simple enough, so we read it using the 'WebReceptor' built-in class of the Actor Prolog for XML/HTML parsing. The CVML annotations contain information about co-ordinates of separate persons and groups of persons in videos. So, our experiments have pursued the following goals:

1. To check if the Actor Prolog system is fast enough to process videos in real time even without performing low-level analysis.
2. To check if there is enough information about the positions of persons for accurate estimation of the velocity and the acceleration of separate personages in the video scene.

Fig. 1. An example of CAVIAR video with a case of abrupt motions.

Fig. 2. The logic program has recognized that two persons were fighting.

The latter issue is important because the accurate estimation of the velocity and/or acceleration opens a way for the recognition of so-called abrupt motions of objects [4]. This kind of motions is necessary for recognition of several long-term activities (such as fighting or sudden attack), though recognition of abrupt motions is not usually provided by standard low-level analyzing procedures. The abrupt motions are not marked in the CAVIAR annotations as well.

An example of abrupt motion recognition is shown in Figs. 1 and 2. A program written in Actor Prolog uses given co-ordinates of two persons to estimate the distance between them and the 2-nd derivative of the co-ordinates to detect abrupt motions.

A logical rule describes an abnormal behavior (fighting) as a conjunction of two conditions:

1. Several persons have met sometime and somewhere.
2. After that they implement abrupt motions.

The text of the logic program is not given here for brevity. After recognition of these two conditions, the logic program has decided that there was a case of a scuffle and has indicated the fighting persons by a red rectangle (see Fig. 2).

This example demonstrates a possibility of recognition of video scenes semantics using the logical inference on results of the low-level recognition of separate objects; however one can see the following bottle-neck of the approach.

Manually defined co-ordinates of the objects were used for estimation of their acceleration and nobody can guarantee that automatic low-level procedures will provide exact values of co-ordinates that are good enough for numerical differentiation. So, the discussion on the high-level recognition procedures is impossible without consideration of underlying low-level recognition methods.

The second issue of this example is whether it is useful to separate the recognition process into concurrent sub-processes implementing different stages of the high-level logical inference. Working intensities of different sub-processes are different. For example, the differentiation of co-ordinates requires more computational resources and another sub-process that implements recognition of people behavior could wait for the results of differentiation.

3 Advanced Logic Analysis of Video Scenes

On the next stage of the research, we have implemented experiments on video analysis based on the automatically extracted information about co-ordinates and velocity of blobs in video scenes.

3.1 Implementation of Base Low-Level Video Processing Procedures

A promising approach for implementation of the low-level recognition procedures in a logic language is usage of the OpenCV computer vision library and we are planning to link Actor Prolog with the JavaCV library that is a Java interface to OpenCV. Nevertheless, Java has enough standard tools to solve simple image processing/recognition problems and we have started our experiments with pure Java.

We have created low-level Java procedures [16] that implement several basic recognition tasks:

1. Background subtraction;
2. Discrimination of foreground blobs;

Fig. 3. A low-level procedure discriminates trajectories (violet lines) of objects and moments of their interactions (green circle marks and blue links).

3. Tracking of the foreground blobs over time;
4. Detection of interactions between the blobs.

The first experiments have demonstrated clearly that the exact estimation of an object velocity was impossible without taking into account the interactions of objects (see Fig. 3), because of edge effects of differentiation in the interaction points.

After implementation of the object interactions check, we have got tracks that were accurate enough to determine whether a person is walking or running. In the next section, we will describe an approach to lower boundary estimation of blob velocity and discuss its possible application to the detection of anomalous behavior of people.

3.2 A Fast Algorithm for Estimation of Object Velocity

At this stage of research, we use standard method of recovering physical co-ordinates of objects in a scene, based on computing inverse matrix of projective transformation by co-ordinates of four defining points. A well-known disadvantage of this method is so-called ground plane assumption, that is, one cannot compute co-ordinates of body parts that are situated outside from a pre-defined plane. Usually, this pre-defined plane is a ground one and we can estimate properly the co-ordinates of person's shoes only. Generally speaking, this problem cannot be avoided in the framework of single camera approach, nevertheless, our idea is in usage of object velocity (but not co-ordinates) for the anomalous behavior detection and this point is exploited in the following algorithm.

We consider simplified rectangle blobs describing moving objects in the scene (see example in Fig. 5). Co-ordinates of every corner of the blob are recovered using the inverse matrix of the projective transformation. Then, one compares the co-ordinates of corresponding corners of the blob in consecutive frames and calculates the first derivative of their co-ordinates. The idea is that only the corners situated in the ground plane give realistic estimations of velocity and other corners give greater values because upper parts of body visually correspond to more distant points in the ground plane. So, we exploit this property of projective transformation and accept the lower boundary estimation of object velocity as a minimal value of velocities (V_{11}, V_{12}, V_{21}, and V_{22}) of four blob corners:

$$V \approx min(abs(V_{11}), abs(V_{12}), abs(V_{21}), abs(V_{22}))$$

Note, that the algorithm does not recover the direction of blob movement. The precision of the estimation of the blob velocity is not very high too, because of the approximate nature of the algorithm. Moreover, the automatic detection of blob shapes often produce illegal co-ordinates of blob corners because of common problems with shades, obstacles, digital noise etc., and this issue is an additional source of errors in the velocity estimation.

We have applied a median filtering to eliminate outliers in the velocity function. For instance, in the example in Fig. 4, the seven point median filter ensures an estimation of blob velocity that is good enough for discrimination of running and walking persons in the scene.

We have implemented this algorithm of velocity estimation in the library [16] of low-level methods of image analysis of the Actor Prolog system and use it in our experiments with the intelligent visual surveillance.

3.3 Creation of a Built-In Class of Actor Prolog

We have developed a special built-in class of the Actor Prolog language that uses formerly described low-level recognition procedures. The '*ImageSubtractor*' class of Actor Prolog implements the following tasks:

1. Video frames pre-processing including 2D-gaussian filtering, 2D-rank filtering, and background subtraction.
2. Recognition of moving blobs and creation of Prolog data structures describing the co-ordinates of the blobs in each moment.
3. Recognition of tracks of blob motions and creation of Prolog data structures describing the co-ordinates and the velocity of the blobs. The tracks are divided into separate segments; there are points of interaction between the blobs at the ends of a segment.
4. Recognition and ejection of immovable and slowly moving objects. This feature is based on a simple fuzzy inference on the attributes of the tracks (the co-ordinates of the tracks and the average velocities of the blobs are considered).
5. Recognition of connected graphs of linked tracks of blob motions and creation of Prolog data structures describing the co-ordinates and the velocity of the blobs.

We consider two tracks as linked if there are interactions between the blobs of these tracks. In some applications, it is useful to eject tracks of immovable and slowly moving objects from the graphs before further processing of the video scenes.

3.4 An Example of Anomalous Behavior Detection

Let us consider an example of logical inference on video. The input of the logic program written in Actor Prolog is the *Fight_RunAway*1 CAVIAR [5] dataset sample (the sequence of JPEG files is used). The program will use no additional information about the content of the video scene, but only co-ordinates of four defining points in the ground plane (the points are provided by CAVIAR). The total text of the logic program is not given here for brevity; we will discuss only the program structure and main stages of data analysis.

The logic program creates two concurrent processes with different priorities (see [12] for details about Actor Prolog model of asynchronous concurrent computations). The first process has higher priority and implements video data gathering. This process reads JPEG files and sends them to the instance of the

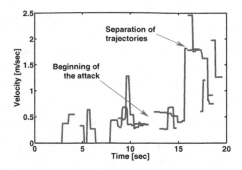

Fig. 4. An example of estimation of velocities of blobs in a visual scene (see Fig. 3). The X-axis denotes time in seconds and the Y-axis denotes lower boundary estimation of blob velocities (m/sec). One can recognize walking persons (before the beginning of the attack) and running persons (after the separation of the trajectories of persons) in the diagram.

'*ImageSubtractor*' predefined class that implements all low-level processing of video frames. The sampling rate of the video is 25 frames per second, so the process loads a new JPEG file every 40 ms.

The second concurrent process implements logical analysis of collected information and outputs results of the analysis. The analysis of video frames requires more computational resources, but it does not suspend the low-level analysis, because the second process has less priority. The analysis includes extraction of blobs, tracking of the blobs over time, detection of interactions between the blobs, creation of connected graphs of linked tracks of blobs, and estimation of average velocity of blobs in separate segments of tracks (see Fig. 4). This information is received by the logic program in a form of Prolog terms describing the list of connected graphs.

The '*ImageSubtractor*' class uses the following data structures for describing connected graphs of tracks (note, that the DOMAINS, the PREDICATES, and the CLAUSES program sections in Actor Prolog have traditional meaning developed in the Turbo/PDC Prolog systems):

```
DOMAINS:
ConnectedGraph  = GraphEdge*.
GraphEdge    = {
        frame1: INTEGER,
        x1: INTEGER,
        y1: INTEGER,
        frame2: INTEGER,
        x2: INTEGER,
        y2: INTEGER,
        inputs: EdgeNumbers,
        outputs: EdgeNumbers,
        identifier: INTEGER,
```

```
        coordinates: TrackOfBlob,
        mean_velocity: REAL
        }.
EdgeNumbers = EdgeNumber*.
EdgeNumber  = INTEGER.
TrackOfBlob = BlobCoordinates*.
BlobCoordinates = {
        frame: FrameNumber,
        x: INTEGER,
        y: INTEGER,
        width: INTEGER,
        height: INTEGER,
        velocity: REAL
        }.
```

That is, connected graph is a list of underdetermined sets [9] denoting separate edges of the graph. The nodes of the graph correspond to points where tracks cross, and the edges are pieces of tracks between such points. Every edge is directed and has the following attributes: numbers of first and last frames ($frame1$, $frame2$), co-ordinates of first and last points ($x1$, $y1$, $x2$, and $y2$), a list of edge numbers that are predecessors of the edge ($inputs$), a list of edge numbers that are followers of the edge ($outputs$), the identifier of corresponding blob (an integer $identifier$), a list of sets describing the co-ordinates and the velocity of the blob in different moments of time ($coordinates$), and an average velocity of the blob in this edge of the graph ($mean\_velocity$).

Fig. 5. A logical inference has found a possible case of a sudden attack in the graph of blob trajectories. Rectangle blobs are depicted by yellow lines, blob trajectories are depicted by red lined, moments of interactions between blobs are depicted by green circles and blue links (Color figure online).

The logic program checks the graph and looks for the following pattern of interaction among several persons: if two or more persons met somewhere in the scene, and one of them has walked (not run) before this meeting, and one of them has run (not walked) after this meeting, the program considers this scenario as a kind of a running away and a probable case of a sudden attack or a theft. So, the program alarms if this kind of sub-graph is detected in the total connected graph of tracks. In this case, the program draws all tracks of the inspected graph in red and outputs the "Attention!" warning in the middle of the screen (see Fig. 5).

One can describe formally the concept of a running away using defined connected graph data type.

```
PREDICATES:
is_a_running_away(
  ConnectedGraph,
  ConnectedGraph,
  ConnectedGraphEdge,
  ConnectedGraphEdge,
  ConnectedGraphEdge) - (i,i,o,o,o);
```

We will define the $is\_a\_running\_away(G, G, P1, E, P2)$ predicate with the following arguments: G is a graph to be analyzed (the same data structure is used in the first and the second arguments), E is an edge of the graph corresponding to a probable incident, $P1$ is an edge of the graph that is a predecessor of E, $P2$ is an edge that is a follower of E. Note that G is an input argument of the predicate and $P1$, E, and $P2$ are output ones. Here is an Actor Prolog program code with brief explanations:

```
CLAUSES:
is_a_running_away([E|_],G,P1,E,P2):-
  E == {inputs:I,outputs:O|_},
  O == [_,_|_],
  walking_person(I,G,P1),
  running_person(O,G,P2),!.
is_a_running_away([_|Rest],G,P1,E,P2):-
  is_a_running_away(Rest,G,P1,E,P2).
walking_person([N|_],G,P):-
  get_edge(N,G,E),
  is_a_walking_person(E,G,P),!.
walking_person([_|Rest],G,P):-
  walking_person(Rest,G,P).
running_person([N|_],G,P):-
  get_edge(N,G,E),
  is_a_running_person(E,G,P),!.
running_person([_|Rest],G,P):-
  running_person(Rest,G,P).
get_edge(1,[Edge|_],Edge):-!.
get_edge(N,[_|Rest],Edge):-
  N > 0,
  get_edge(N-1,Rest,Edge).
```

In other words, the graph contains a case of a running away if there is an edge E in the graph that has a predecessor $P1$ corresponding to a walking person and a follower $P2$ that corresponds to a running person. It is expected also that E has more than one follower (it is a case of a branching in the graph)[1].

```
is_a_walking_person(E,_,E):-
  E == {mean_velocity:V|_},
  V <= 0.5,!.
is_a_walking_person(E,G,P):-
  E == {inputs:I|_},
  walking_person(I,G,P).
```

That is, the graph edge corresponds to a walking person if the average blob velocity in this edge is less or equal to 0.5 m/s, or the edge has a predecessor that corresponds to a walking person.

```
is_a_running_person(E,_,E):-
  E == {mean_velocity:V|_},
  V >= 1.0,!.
is_a_running_person(E,G,P):-
  E == {outputs:O|_},
  running_person(O,G,P).
```

The graph edge corresponds to a running person if the average velocity in this edge is more or equal to 1 m/s, or the edge has a follower corresponding to a running person.

Note that aforementioned rules use plain numerical thresholds to discriminate walking and running persons for brevity. Better discrimination could be ensured by a kind of a fuzzy check, which can be easily implemented using arithmetical means of standard Prolog.

This example illustrates the possible scheme of a logic program implementing all necessary stages of video processing including video information gathering, low-level image analysis, high-level logical inference on the video scene, and reporting the results of the intelligent visual surveillance.

4   A Method of Logical Rules Creation

The logical rules considered in the previous section were created manually on the basis of a priori knowledge of the particular behavior, but we would like to create logical rules automatically in cases of complex spatio-temporal behavior. In this section, we describe the method of logical rules creation [3] based on a hierarchy of fuzzy finite state automata.

Let $T = \{t_i | t_i \in N\}$ be a discrete set of time instances with constant intervals $\Delta t = t_{i+1} - t_i$ between consecutive time instances, where $[t_s, t_e] = \{t | t_s \leq t \leq t_e\}$ is a time interval T. Suppose that each 0th level feature (a speed or a position)

---

[1] Note, that in the Actor Prolog language, the operator == corresponds to the ordinary equality = of the standard Prolog.

of each moving object θ from a set $\{\theta^1, \theta^2, \ldots, \theta^l\}$ at a time instance t equals $y_{i_0}(\theta_t), i_0 \in \{1, \ldots, m_0\}$, that we call a feature sample. Samples $Y_{i_0}[\theta_{t_s}, \theta_{t_e}] = \langle y_{i_0}(\theta_{t_s}), \ldots, y_{i_0}(\theta_{t_e}) \rangle, i_0 \in \{1, \ldots, m_0\}$ of a single 0th level feature at several consecutive time instances $t_s, \ldots, t_e$ during a $[t_s, t_e]$ time interval are called a trend.

Let us consider the following situation. Two persons walk alongside two roads that are perpendicularly directed towards their meeting point (intersection). While person A is far from the intersection, person B slows down waiting for person A. When person A enters the intersection, person B accelerates and runs into person A.

Let us formalize the persons' behavior. Let persons A and B walk along perpendicular lines with the intersection point O. Let xOy be a rectangular co-ordinate system such that the Ox axis corresponds to the A person and the Oy axis corresponds to the B person (Fig. 6). Let us consider each person as a rectangle and the co-ordinates of the centroid of the rectangle as the co-ordinates of the person. Suppose that persons move strictly along the co-ordinate axes and current positions of persons A and B can be determined by single co-ordinates $y_s(\theta_t^A)$ and $y_s(\theta_t^B)$ respectively. The $y_s(\theta_t^A)$, $y_s(\theta_t^B)$ co-ordinates of the persons A and B are considered as first features. The $y_v(\theta_t^A)$, $y_v(\theta_t^B)$ speed values of the persons are considered as second features.

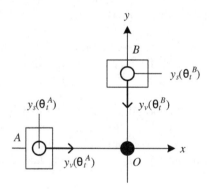

Fig. 6. Two persons in the rectangular co-ordinate system.

In order to estimate velocities one should specify an observation time interval $[t_s, t_e] = \{t | t_s \leq t \leq t_e\}$ and time instances $t_i \in [t_s, t_e]$. Linguistic variables are specified and behavior template models are defined as in [3]. Let $position(\theta^A)$, $speed(\theta^A)$ and $position(\theta^B)$, $speed(\theta^B)$ be linguistic variables that describe positions and speed values of persons A and B. The $position(\theta^A)$ and $position(\theta^B)$ linguistic variables assume linguistic values $far(\theta^X)$, $near(\theta^X)$, and $inside(\theta^X)$. The $speed(\theta^A)$ and $speed(\theta^B)$ linguistic variables assume linguistic values $high(\theta^X)$ and $low(\theta^X)$.

Fuzzy sets corresponding to linguistic values $far(\theta^X)$, $near(\theta^X)$, $inside(\theta^X)$ and $high(\theta^X)$, $low(\theta^X)$ are shown in Fig. 7. Fuzzy sets shown in Fig. 7 are used

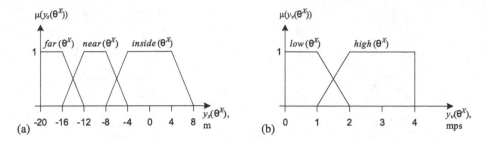

Fig. 7. (a) Fuzzy sets corresponding to linguistic values $far(\theta^X)$, $near(\theta^X)$, and $inside(\theta^X)$; (b) Fuzzy sets corresponding to linguistic values $low(\theta^X)$ and $high(\theta^X)$.

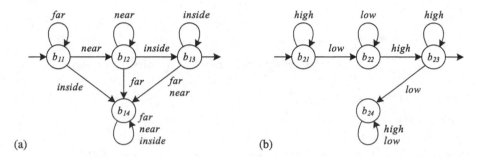

Fig. 8. (a) The $M_{pos(\theta^X)}$ first level automaton; (b) The $M_{speed(\theta^A)}$ first level automaton.

to define first level template automata $M_{pos(\theta^X)}$, $M_{speed(\theta^A)}$, and $M_{speed(\theta^B)}$ that describe position and speed of persons A and B.

An automaton $M_{pos(\theta^X)}$ shown in Fig. 8(a) determines a sequence of the linguistic values $[far(\theta^X), near(\theta^X), inside(\theta^X)]$ of the variable $position(\theta^X)$. The automaton graph is based on a chain of allowed states $b_{11} - b_{12} - b_{13}$ corresponding to the values of the determined sequence. b_{11} is the initial state (marked by the input arrow in Fig. 8(a)) and b_{13} is the final state of the automaton (the output arrow in Fig. 8(a)). State transitions are specified below. If the input linguistic value corresponds to the current state, then automaton retains its current state. If the input linguistic value corresponds to the next allowed state, then the automaton moves to that state. Automaton moves to the b_{14} denied state if an input linguistic value violates the allowed linguistic values sequence of the automaton. Note that the automaton cannot leave the denied state.

Automata $M_{speed(\theta^A)}$ and $M_{speed(\theta^B)}$ are presented in Figs. 8(b) and 9(a).

Let us define the second level template automaton that describes the joint persons' behavior. Let $condition(\theta^A, \theta^B)$ be a linguistic variable that assumes linguistic values $safe(\theta^A, \theta^B)$, $warning(\theta^A, \theta^B)$, and $unsafe(\theta^A, \theta^B)$, where

$$safe(\theta_t^A, \theta_t^B) = [pos(\theta_t^A) = far(\theta^A)] \wedge [pos(\theta_t^B) = far(\theta^B)] \qquad (1)$$

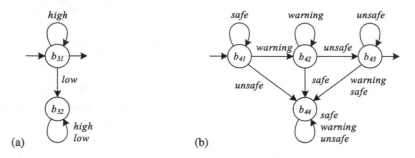

Fig. 9. (a) The $M_{speed(\theta^B)}$ first level automaton; (b) The $M_{condition(\theta^A, \theta^B)}$ second level automaton.

$$warning(\theta_t^A, \theta_t^B) = ([pos(\theta_t^A) = near(\theta^A)] \wedge [speed(\theta_t^B) = high(\theta^B)]) \qquad (2)$$
$$\vee ([speed(\theta_t^A) = high(\theta^A)] \wedge [pos(\theta_t^B) = near(\theta^B)])$$

$$unsafe(\theta_t^A, \theta_t^B) = [pos(\theta_t^A) = inside(\theta^A)] \wedge [pos(\theta_t^B) = inside(\theta^B)] \qquad (3)$$

Each linguistic value of the $condition(\theta^A, \theta^B)$ linguistic variable corresponds to a composite fuzzy set. The multidimensional domain of the composite fuzzy set is a Cartesian product of the domains of the corresponding fuzzy sets [3]. According to Eq. (1), the domain of the $safe(\theta_t^A, \theta_t^B)$ linguistic value is defined as follows:

$$\mathrm{dom}[safe(\theta_t^A, \theta_t^B)] = \mathrm{dom}[far(\theta^A)] \times \mathrm{dom}[far(\theta^B)],$$

where $\mathrm{dom}[E]$ is the domain of a fuzzy set E and $\times$ is the Cartesian product.

A membership function is expressed as follows:

$$a = b \wedge c \Longrightarrow R_a(y_b, y_c) = \min\{R_b(y_b), R_c(y_c)\} \qquad (4)$$
$$a = b \vee c \Longrightarrow R_a(y_b, y_c) = \max\{R_b(y_b), R_c(y_c)\} \qquad (5)$$
$$a = \neg b \Longrightarrow R_a(y_b) = 1 - R_b(y_b) \qquad (6)$$

where a, b, and c are fuzzy sets; R_E is a membership function of a fuzzy set E, determined on the $\mathrm{dom}[E]$ domain; $y_b \in \mathrm{dom}[b]$ and $y_c \in \mathrm{dom}[c]$ are feature values from corresponding domains.

According to Eq. (4), the membership function of a composite fuzzy set specified on the conjunction of two fuzzy sets is equal to minimum value of the membership functions of these fuzzy sets. The second level template automaton $M_{condition(\theta^A, \theta^B)}$ that describes the joint persons' behavior is shown in Fig. 9(b).

The computation scheme of the recognition process is presented in Fig. 10. It includes five units for evaluation and processing of linguistic variables arranged in two levels. Each unit computes value of corresponding linguistic variable and inputs it to the corresponding automaton.

The recognition of the situation is implemented in the following way. Initially, all first and second level automata have initial states. Then feature samples for

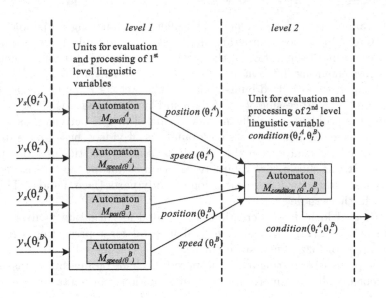

Fig. 10. The computation scheme of recognition.

consecutive time instances $t_i \in [t_s, t_e]$ are passed by turn into the first level units for evaluation and processing of the first level linguistic variables. The first level units compute values of linguistic variables and pass them into the second level unit for evaluation and processing of the second level linguistic variable $condition(\theta^A, \theta^B)$. The first and the second level automata may change their states during the operation. Situation is recognized if all first and second level automata have moved to their final states after the end of the processing of all feature samples. Situation is not recognized if one automaton has not moved to the final state at least.

The computation scheme based on the fuzzy finite automata can easily be converted to a logic program; one can use standard techniques of transforming finite state machines into the logic programs [2].

5   Conclusions

We have created a research led software platform based on the Actor Prolog concurrent object-oriented logic language and a state-of-the-art Prolog-to-Java translator for examining the intelligent visual surveillance. The platform includes the Actor Prolog logic programming system and an open source Java library of Actor Prolog built-in classes [16]. It is supposed to be complete for facilitation of research in the field of intelligent monitoring of anomalous people activity and studying logical description and analysis of people behavior.

Our study has demonstrated that translation from a concurrent object-oriented logic language to Java is a promising approach for application of the logic programming to the problem of intelligent monitoring of people activity; the Actor Prolog logic programming system is suitable for this purpose and ensures essential separation of the recognition process into concurrent subprocesses implementing different stages of high-level analysis.

In the paper, a specialized built-in class of the Actor Prolog language implementing simple pre-processing of video data and low-level analysis of video scenes concerning the problem of intelligent monitoring of people activity was demonstrated. We have implemented a simple analysis of videos based on automatically extracted information on the co-ordinates and velocities of blobs in the video scene. It was shown that robust recognition of abrupt motions is impossible without accurate low-level recognition of body parts (face, hands). This is a subject of further studies.

A method of logical rules creation is proposed for situation analysis in the environment of moving objects. A formal method for representing situations using hierarchy of fuzzy finite state automata was considered. Future work will include comprehensive testing of the proposed methods on massive datasets and development of fully automatic method for situation representation using real feature trends.

Acknowledgements. We acknowledge a partial financial support from the Russian Foundation for Basic Research, grant No 13-07-92694, and Department of Science and Technology, Govt. of India, grant No DST-RFBR P-159.

References

1. Aggarwal, J., Ryoo, M.: Human activity analysis: a review. ACM Comput. Surv. (CSUR) **43**(3), 16:1–16:43 (2011)
2. Bratko, I.: Prolog Programming for Artificial Intelligence. Addison-Wesley, Boston (1986)
3. Devyatkov, V.: Multiagent hierarchical recognition on the basis of fuzzy situation calculus. Vestnik, Journal of the Bauman Moscow State Technical University, Natural Science & Engineering, pp. 129–152. Vestnik MGTU, Moscow (2005)
4. Filippou, J., Artikis, A., Skarlatidis, A., Paliouras, G.: A probabilistic logic programming event calculus (2012). http://arxiv.org/abs/1204.1851
5. Fisher, R.: CAVIAR Test Case Scenarios. The EC funded project IST 2001 37540 (2007). http://homepages.inf.ed.ac.uk/rbf/CAVIAR/
6. Junior, J., Musse, S., Jung, C.: Crowd analysis using computer vision techniques. A survey. IEEE Signal Process. Mag. **27**(5), 66–77 (2010)
7. Kim, I., Choi, H., Yi, K., Choi, J., Kong, S.: Intelligent visual surveillance–a survey. Int. J. Control Autom. Syst. **8**(5), 926–939 (2010)
8. Machot, F., Kyamakya, K., Dieber, B., Rinner, B.: Real time complex event detection for resource-limited multimedia sensor networks. In: Workshop on Activity Monitoring by Multi-camera Surveillance Systems (AMMCSS), pp. 468–473 (2011)
9. Morozov, A.A.: Actor Prolog: an object-oriented language with the classical declarative semantics. In: Sagonas, K., Tarau, P. (eds.) IDL 1999, pp. 39–53. Paris, France (1999). http://www.cplire.ru/Lab144/paris.pdf

10. Morozov, A.A.: On semantic link between logic, object-oriented, functional, and constraint programming. In: MultiCPL 2002. Ithaca (2002). http://www.cplire.ru/Lab144/multicpl.pdf
11. Morozov, A.A.: Development and application of logical actors mathematical apparatus for logic programming of web agents. In: Palamidessi, C. (ed.) ICLP 2003. LNCS, vol. 2916, pp. 494–495. Springer, Heidelberg (2003)
12. Morozov, A.A.: Logic object-oriented model of asynchronous concurrent computations. Pattern Recognit. Image Anal. **13**(4), 640–649 (2003). http://www.cplire.ru/Lab144/pria640.pdf
13. Morozov, A.A.: Operational approach to the modified reasoning, based on the concept of repeated proving and logical actors. In: Salvador Abreu, V.S.C. (ed.) CICLOPS 2007, pp. 1–15. Porto, (2007). http://www.cplire.ru/Lab144/ciclops07.pdf
14. Morozov, A.A.: Visual logic programming method based on structural analysis and design technique. In: Dahl, V., Niemelä, I. (eds.) ICLP 2007. LNCS, vol. 4670, pp. 436–437. Springer, Heidelberg (2007)
15. Morozov, A.A.: Actor Prolog to Java translation (in Russian). IIP-9, pp. 696–698. Torus Press Moscow, Budva (2012)
16. Morozov, A.A.: A GitHub repository containing source codes of Actor Prolog built-in classes (including the Vision package) (2014). https://github.com/Morozov2012/actor-prolog-java-library
17. Morozov, A.A., Vaish, A., Polupanov, A.F, Antciperov, V.E., Lychkov, I.I., Alfimtsev, A.N., Deviatkov, V.V.: Development of concurrent object-oriented logic programming system to intelligent monitoring of anomalous human activities. In: Jr., A.C., Plantier, G., Schultz, T., Fred, A., Gamboa, H. (eds.) BIODEVICES 2014, pp. 53–62. SCITEPRESS (2014). http://www.cplire.ru/Lab144/biodevices2014.pdf
18. O'Hara, S.: VERSA–video event recognition for surveillance applications. M.S. thesis, University of Nebraska at Omaha (2008)
19. Shet, V., Singh, M., Bahlmann, C., Ramesh, V., Neumann, J., Davis, L.: Predicate logic based image grammars for complex pattern recognition. Int. J. Comput. Vis. **93**(2), 141–161 (2011)

Bioimaging

Diagnosing Alzheimer's Disease: Automatic Extraction and Selection of Coherent Regions in FDG-PET Images

Helena Aidos[(⊠)], João Duarte, and Ana Fred

Instituto de Telecomunicações, Instituto Superior Técnico,
Universidade de Lisboa, Lisbon, Portugal
{haidos,jduarte,afred}@lx.it.pt

Abstract. Alzheimer's Disease is a progressive neurodegenerative disease leading to gradual deterioration in cognition, function and behavior, with unknown causes and no effective treatment up to date. Techniques for computer-aided diagnosis of Alzheimer's Disease typically focus on the combined analysis of multiple expensive neuroimages, such as FDG-PET images and MRI, to obtain high classification accuracies. However, achieving similar results using only 3-D FDG-PET scans would lead to significant reduction in medical expenditure. This paper proposes a novel methodology for the diagnosis Alzheimer's Disease using only 3-D FDG-PET scans. For this we propose an algorithm for automatic extraction and selection of a small set of coherent regions that are able to discriminate patients with Alzheimer's Disease. Experimental results show that the proposed methodology outperforms the traditional approach where voxel intensities are directly used as classification features.

Keywords: Support vector machines · ROI · Feature extraction · Image segmentation · Mutual information

1 Introduction

One of the most common forms of dementia is Alzheimer's disease (AD), a progressive brain disorder that has no known cause or cure. It is a disease that slowly leads to memory loss, confusion, impaired judgment, personality changes, disorientation and the inability to communicate. An early detection is very important for an effective treatment, especially in the Mild Cognitive Impairment (MCI) stage, to slow down the progress of the symptoms and to improve patients' life quality. MCI is a condition where a person has mild changes in thinking abilities, but it does not affect daily life activities. People with MCI are more likely to develop AD, even though recent studies suggest that a person with MCI may revert back to normal cognition on its own [1].

Neuroimages allow the identification of brain changes and have been used for automated diagnosis of AD and MCI [16,20]. Due to the high variability of the pattern of brain degeneration in AD and MCI, the analysis of brain images

© Springer International Publishing Switzerland 2015
G. Plantier et al. (Eds.): BIOSTEC 2014, CCIS 511, pp. 101–112, 2015.
DOI: 10.1007/978-3-319-26129-4_7

is a very difficult task. Moreover, attempts are being made to develop tools to automatically analyze the images and, consequently, diagnosis AD and MCI conditions [13,14].

Most of the techniques developed have focused on analyzing small parts of the brain like hippocampus [10] or the gray matter volume [9]. However, these techniques have some limitations by the fact that the brain atrophy affects many and different regions in different stages of the disease. Therefore, researchers are focusing their techniques in analyzing the pattern of the entire brain. However, this leads to the "curse of dimensionality" because a brain image, like the fluorodeoxyglucose positron emission tomography (FDG-PET), contains thousands of voxels (or features). Dimensionality reduction and feature selection techniques are therefore fundamental for achieving high accuracy predictors for the diagnosis of Alzheimer's disease.

Some techniques are based in the segmentation of the brain into Regions of Interest (ROIs), which are associated with atrophy caused by the disease. Then, voxel intensities from each ROI are used as features [12,21]. Some other dimensionality reduction techniques from Machine Learning field [11,15], and feature selection techniques [4,6] have been applied to the diagnosis of AD.

In this paper, we propose a methodology to automatically extract features that represent interesting regions of the brain and, consequently, reducing the dimensionality of the space. One of the advantages of this methodology is that brain images, like FDG-PET, do not need to be pre-processed in order to remove the background and the scalp. This is due to the choice of the clustering algorithm, which is a variant of the DBSCAN (density-based spatial clustering of applications with noise) called XMT-DBSCAN [17]. Another advantage is that the space we obtain is approximately 100× smaller when compared to the original one, consisting of voxel intensities. This happens because each region (cluster) obtained by the clustering algorithm is represented by a feature, which is a weighted mean of the voxel intensities of that region.

This paper is organized as follows: Sect. 2 explains each step of the proposed methodology and Sect. 3 presents the dataset used in this paper as well the results obtained for the proposed methodology and for the classification task using the voxel intensity. Conclusions are drawn in Sect. 4.

2   The Proposed Methodology

In order to analyze the FDG-PET scans for each task: AD versus CN (Cognitive Normal), MCI versus CN and AD versus MCI, we propose the methodology shown in Fig. 1. We start by segmenting each 3-D image (a FDG-PET scan from a subject), followed by a construction of a probability matrix indicating the degree of belonging of each voxel to a region found by the clustering/segmentation algorithm. Then, we perform a feature extraction step using the voxel intensities and the probability matrix, obtaining a feature space representation for each problem. Finally, feature selection is applied and the subjects are classified, using support vector machines.

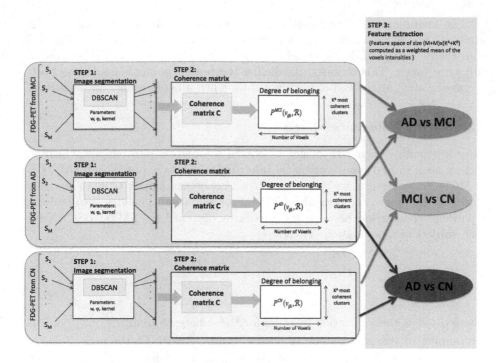

Fig. 1. The proposed methodology.

2.1   Step 1: Image Segmentation

Over the years, several 3-D segmentation methods have been developed such as region growing, watershed, among others [2,18]; watershed algorithm [3] is the most widely used. However, watershed tends to over-segment the 3-D images when the data is dense and non-homogeneous, or generate under-segmentation results in the case of dense regions with irregular shapes of objects. Since our FDG-PET scans are noisy images that have regions with different sizes, densities and irregular shapes, we propose to use a version of the DBSCAN algorithm, namely the XMT-DBSCAN [17], to segment the 3-D images.

XMT-DBSCAN is an extension of the original DBSCAN but has a few differences. Firstly, the local density of a voxel (a pixel in DBSCAN) is computed in the sub-window with size $ws = (2w+1) \times (2w+1) \times (2w+1)$ centered in the voxel, instead of the ball with radius eps. In our methodology, the local density is computed as

$$density(v_{ijk}) = \frac{\sum_{all-elements} I^w_{v_{ijk}} \odot K^w}{a_k}, \tag{1}$$

where $\odot$ is the element-wise product of two equally sized data cubes, K^w is a cubic Gaussian kernel with standard deviation equal to $ws/(4\sqrt{2\log(2)})$, $I^w_{v_{ijk}}$ is the sub-window from the intensity image, and a_k is the number of non-zero values in K^w.

The identification of the voxels as core points, border points and noise is similar to the original DBSCAN. Another modification to the original DBSCAN is in the definition of density-reachable chain [8], which is modified to contain only core voxels. This means that labeling the border points is made in a post-processing step, at the end of the algorithm, when all core points are identified.

2.2 Step 2: Coherence Matrix

After segmenting each 3-D image, we obtain a partition into regions (clusters) and we need to find some consensual information for each population (AD, CN or MCI). In that sense, we construct a block coherence matrix $\mathbf{C}$, with as many blocks as the squared number of subjects of a population. The idea is to perform a pairwise comparison between the partitions obtained by XMT-DBSCAN for each subject of a population. Therefore,

$$\mathbf{C}(\mu(l,i),\mu(p,j)) = \frac{|C_i^l \cap C_j^p|}{\sqrt{|C_i^l| \cdot |C_j^p|}}, \tag{2}$$

where $\mu(l,i)$ is the indexation function for the coherence matrix $\mathbf{C}$, $|C_i^l \cap C_j^p|$ is the number of voxels belonging to both C_i^l and C_j^p, with C_i^l the region/cluster i from subject l and C_j^p the region/cluster j from subject p. The indexation function is given by

$$\mu(l,i) = i + \sum_{j=1}^{l-1} m_j, \tag{3}$$

with m_j the number of clusters in the partition of subject j, i.e., $\mu(l,i)$ gives the index corresponding to cluster i of subject l, where each partition of a subject has m_j clusters. Figure 2 shows an example of a coherence matrix.

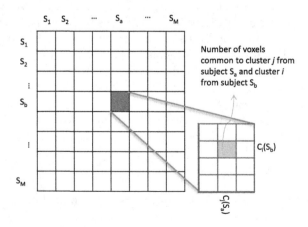

Fig. 2. Coherence matrix.

The matrix $\mathbf{C}$ shows the degree of overlapping of each pair of clusters. Since we want a region that is common in most of the subjects, we consider that values under 50 % of overlapping are discarded.

We start by searching the most coherent cluster in matrix $\mathbf{C}$ and obtain a region $\mathcal{R}$ corresponding to the union of all clusters with an overlapping over 50 % to the most coherent cluster found. Inside the region $\mathcal{R}$, we compute the probability (for a certain population) of each voxel belong to $\mathcal{R}$ as

$$P^A(v_{ijk}, \mathcal{R}) = \frac{\sum_{C_k \in \mathcal{R}} 1_{\{v_{ijk} \in C_k\}}}{\sum_{C_k \in \mathcal{R}} 1_{\{C_k \in \mathcal{R}\}}}, \tag{4}$$

where C_k is the k-th cluster of region $\mathcal{R}$, $1_{\{v_{ijk} \in C_k\}}$ is 1 if $v_{ijk} \in C_k$, and 0 otherwise; v_{ijk} is a voxel in the 3-D image and $A \in \{AD, CN, MCI\}$. The numerator of the previous equation is a count of the number of clusters in $\mathcal{R}$ where the voxel belongs, and the denominator is just the number of clusters in $\mathcal{R}$. This process is repeated until no coherent clusters are left in matrix $\mathbf{C}$. Therefore, P^A is $K \times N$ matrix, with K the number of regions and N the number of voxels in the 3-D image.

2.3   Step 3: Feature Extraction

So far we have found regions containing relevant information for each population. Now we want to discriminate AD vs CN, CN vs MCI and AD vs MCI. This means that we will construct a feature space for each of these problems using the voxels intensities from two populations and the regions found in step 2 corresponding to the same two populations.

Consider that M^A is the number of subjects from population A and M^B the number of subjects from population B. Also, K^A and K^B are the number of regions found in step 2 for population A and B, respectively. We want to construct a feature space $\mathbf{F}$ with $M^A + M^B$ samples and dimension $K^A + K^B$, in the following way

$$\mathbf{F}(\alpha(p,r), \beta(q,s)) = \frac{\sum_{v_{ijk}} I(v_{ijk} \in S_p^r) \cdot P^s(v_{ijk}, \mathcal{R}_q)}{\sum_{v_{ijk}} P^s(v_{ijk}, \mathcal{R}_q)}, \tag{5}$$

with $r, s \in \{\text{population } A, \text{population } B\}$. $I(v_{ijk} \in S_p^r)$ is the intensity of voxel v_{ijk} from subject p in population r and $P^s(v_{ijk}, \mathcal{R}_q)$ is the probability that voxel v_{ijk} belongs to region $\mathcal{R}_q$ in population s. $\alpha(p,r)$ and $\beta(q,s)$ are indexation functions given by

$$\alpha(p,r) = \begin{cases} p & \text{if } r = \text{ population } A \\ M^A + p & \text{if } r = \text{ population } B \end{cases}$$

and

$$\beta(q,s) = \begin{cases} q & \text{if } s = \text{ population } A \\ K^A + q & \text{if } s = \text{ population } B \end{cases}$$

respectively. $\alpha(p, r)$ is the indexation for subjects and $\beta(q, s)$ the indexation for regions.

Equation (5) is equivalent to compute a weighted mean of the intensity of a subject, where some voxels contribute more than others, obtaining a feature space for each classification task.

2.4   Step 4: Feature Selection

Typically, the number of voxels in a FDG-PET image is very high and some of those voxels are unimportant for the task in hand. So, it is very important to reduce the dimensionality of the space through feature selection. We use mutual information (MI) to rank the features and choose the ones with higher value.

Consider that x_i is the i-th element of a vector representing a feature **x**, and y a target value or label. The MI between the random variable x_i and y is given by

$$MI(i) = \sum_{x_i} \sum_{y} P(x_i, y) \log \frac{P(x_i, y)}{P(x_i)P(y)}. \tag{6}$$

The probability density functions for MI were estimated through the use of histograms.

2.5   Step 5: Classification

After selecting the most relevant features for each of the three diagnostic problems, we classify subjects using three different classifiers: support vector machine (SVM) algorithm with a linear kernel [7], k-nearest neighbor (k-NN) and Naive Bayes.

3   Experiments

3.1   Dataset

In this study, we used FDG-PET images for AD, MCI and CN subjects, retrieved from the ADNI database. The subjects were chosen to obey a certain criteria: the Clinical Dementia Rating (CDR) should be 0.5 or higher for AD patients, 0.5 for MCI patients and 0 for normal controls. This selection results in a dataset composed by 59, 142 and 84 subjects for AD, MCI and CN, respectively. Since our task is classification using the SVM algorithm, we decided to balanced the classes by using a random sub-sampling technique. Thus, 59 subjects from each MCI and CN groups were selected randomly. Table 1 summarizes some clinical and demographic information in each group.

The FDG-PET images have been pre-processed to minimize differences between images: each image was co-registered, averaged, reoriented (the anterior-posterior axis of each subject was parallel to the AC-PC line), normalized in its

Table 1. Clinical and demographic characteristics of each group. Age and MMSE (Mini Mental State Exam) values are means ± standard deviations.

Attributes	AD	MCI	CN
Number of subjects	59	59	59
Age	78.26 ± 6.62	77.71 ± 6.88	77.38 ± 4.87
Sex (% of males)	57.63	67.80	64.41
MMSE	19.60 ± 5.06	25.68 ± 2.97	29.20 ± 0.92

intensity, and smoothed to uniform standardized resolution. A more detailed description of the pre-processing is available in the ADNI project webpage[1].

The complete $128 \times 128 \times 60$ FDG-PET images were reduced using the Gaussian pyramid technique [5] to obtain a $64 \times 64 \times 30$ 3-D image. These complete images were used, which means that no background or extra-cranial voxels were excluded. We left those voxels because the image segmentation step will automatically discard them and only the relevant voxels will be labeled.

3.2 Experimental Setup

The FDG-PET image of the brain of each individual needs to be segmented with XMT-DBSCAN, the segmentation algorithm proposed in the methodology. In Sect. 2, we state that XMT-DBSCAN has two parameters: window size w and φ which is a threshold to identify core and border voxels (see [8] for more details). Those parameters were set to 2 and 3 for w, and φ takes values from $\{0.3, 0.5, 0.7\}$.

In the feature selection step we discretized the probability density functions through histograms with 8 bins and, after ranking the features according to the MI, we choose the ones with higher value. We consider several number of features selected by the MI, according to Table 2.

The final step of the proposed methodology consists in classifying subjects using a classifier. We set the cost of misclassification in the linear SVM as $\{2^{-16}, 2^{-14}, 2^{-12}, 2^{-10}, 2^{-8}, 2^{-6}, 2^{-4}, 2^{-2}, 2^0, 2^2, 2^4\}$ and the k parameter for k-NN to $\{1, 3, 5, 7\}$. We used a standard internal 10-fold cross-validation strategy to choose all the parameters (parameters for the segmentation part and the classification part), according to the best accuracy. This process was repeated 20 times and an average accuracy over the test sets was computed for each classifier [19].

We compare the proposed methodology with the one consisting of the voxel intensities, called VI. In that strategy, we first need to pre-process the FDG-PET images to remove the background and the scalp. Afterwards, steps 4 and 5 of the proposed methodology are applied. The number of selected features used to classify the subjects are shown in Table 2.

---

[1] http://adni.loni.usc.edu/methods/pet-analysis/pre-processing/.

Table 2. Number of features used to tested the feature selection step. The maximum number of features used corresponds to the complete feature space, as stated by columns 2–4, depending on the problem.

Parameter space	Max. features			Number of selected features
	AD vs CN	MCI vs CN	AD vs MCI	
$w = 2, \varphi = 0.3$	1436	1413	1433	50, 100, 250, 500, 1000, Max. features
$w = 2, \varphi = 0.5$	1037	1057	1050	50, 100, 250, 500, 1000, Max. features
$w = 3, \varphi = 0.3$	476	476	502	50, 100, 250, Max. features
$w = 3, \varphi = 0.5$	332	328	342	50, 100, 250, Max. features
$w = 3, \varphi = 0.7$	260	284	268	50, 100, 250, Max. features
Voxel intensity	36209			50, 100, 250, 500, 1000, 2500,
				5000, 10000, 25000, Max. features

3.3 Results

Figure 3 shows the accuracy of the three classifiers for the proposed methodology for each problem and the voxel intensity methodology.

For AD vs CN, Naive Bayes has a decrease of performance when the number of features increased. While SVM and k-NN remain constant and near 85 % of accuracy. For MCI vs CN, k-NN is the worst classifier with an accuracy below 70 % for the proposed methodology and below 65 % for VI methodology. SVM and Naive Bayes have a constant accuracy around 75 % for the proposed methodology. Moreover, VI methodology only achieves higher accuracies for more than 5000 features.

However, the most promising results of the proposed methodology are when discriminating AD from MCI subjects. In this problem, SVM has an accuracy around 72 %, higher than the VI methodology, even if we increased the number of features, the accuracy for VI is always below the proposed methodology. k-NN has an accuracy below 70 %, but it is also higher than the VI, except if we use more than 250 features. In this case, k-NN in the proposed methodology has lower accuracy than VI. The best classifier for AD vs MCI is Naive Bayes, it has an accuracy near 75 %, and it slightly decreased for more than 250 features. However, the accuracy for this classifier in the proposed methodology is always higher than for the VI.

In all the problems considered (AD vs CN, MCI vs CN and AD vs MCI), our methodology outperforms VI with only a few features, e.g., 50 features. The three classifiers are quite similar, but k-NN is the worst classifier among the three. Figure 4 presents the sensitivity and specificity for both methodologies and for each problem and classifier, considering 50 features.

Overall, the proposed methodology has a sensitivity similar to the VI methodology, and in most cases the inter-quartile range is smaller than VI. Moreover, in AD vs CN using k-NN and Naive Bayes, the median sensitivity for the proposed methodology is slightly higher than VI. However, the specificity is overall better for the proposed methodology compared to the VI methodology. The specificity is only similar in AD vs CN for k-NN and Naive Bayes.

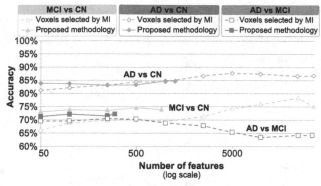

(a) Support vector machines

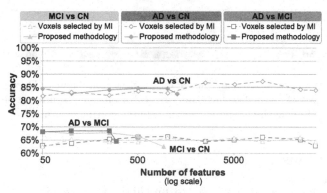

(b) k-nearest neighbor

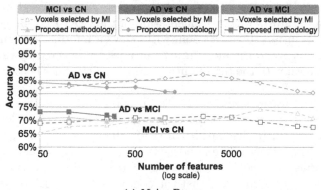

(c) Naive Bayes

Fig. 3. Average accuracy of 20×10 nested cross-validation of the proposed methodology from each problem compared to voxels intensities chosen through mutual information and classified using SVM, k-NN and Naive Bayes.

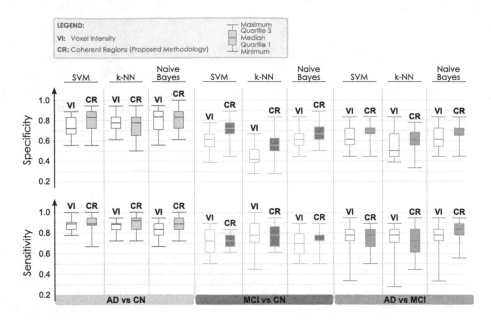

Fig. 4. Boxplots for sensitivity and specificity for each problem (AD vs CN, MCI vs CN, and AD vs MCI), each classifier (support vector machine, k-nearest neighbor, and Naive Bayes) and using the 50 features with highest mutual information.

4   Conclusions

This paper proposes a methodology to find interesting regions in the brain to efficiently discriminate subjects with Alzheimer's disease from the ones with mild cognitive impairment and from normal ones. The proposed methodology has several stages: starts with a segmentation of the FDG-PET image, followed by a grouping of clusters to form regions with relevant information. Those regions form a feature space and the most important ones are selected by ranking their mutual information with the target output. Finally a classifier is used to identify the subjects.

For number of features under 100, the proposed methodology outperforms another strategy consisting in ranking the mutual information of features with the target output, where the features are only the voxels intensities. Moreover, by comparing using all the space in both strategies for AD vs MCI, our methodology outperforms the other strategy, using a small number of features. Another advantage of this methodology is that the complete FDG-PET image was used, since the segmentation algorithm can identify background and extracranial voxels, which means we do not need to pre-process the images to remove those voxels.

Acknowledgements. This work was supported by the Portuguese Foundation for Science and Technology grants PTDC/SAU-ENB/114606/2009 and PTDC/ EEI-SII/2312/ 2012. Data used in the preparation of this article were obtained from the Alzheimer's Disease Neuroimaging Initiative (ADNI) database.

References

1. Alzheimer's Association: 2013 Alzheimer's disease facts and figures. Alzheimer's & Dementia: J. Alzheimer's Assoc. **9**(2), 208–245, 2013
2. Arbeláez, P., Maire, M., Fowlkes, C., Malik, J.: Contour detection and hierarchical image segmentation. IEEE Trans. Pattern Anal. Mach. Intell. **33**(5), 898–916 (2011)
3. Beucher, S., Lantuejoul, C.: Use of watersheds in contour detection. In: Proceedings of the International Workshop on Image Processing: Real-time Edge and Motion Detection/Estimation. Rennes, France, September 1979
4. Bicacro, E., Silveira, M., Marques, J.S., Costa, D.C.: 3D image-based diagnosis of Alzheimer's disease: Bringing medical vision into feature selection. In: 9th IEEE International Symposium on Biomedical Imaging: from Nano to Macro (ISBI 2012), Barcelona, Spain, pp.134–137. IEEE, May 2012
5. Burt, P.J.: Fast filter transform for image processing. Comput. Graph. Image Process. **16**(1), 20–51 (1981)
6. Chaves, R., Ramirez, J., Gorriz, J.M., Lopez, M., Alvarez, I., Salas-Gonzalez, D., Segovia, F., Padilla, P.: SPECT image classification based on NMSE feature correlation weighting and SVM. In: Yu, B. (ed.) Proceedings of the IEEE Nuclear Science Symposium Conference Record (NSS/MIC 2009), Orlando, FL, USA, pp. 2715–2719. IEEE, October 2009
7. Cortes, C., Vapnik, V.: Support-vector networks. Mach. Learn. **20**, 273–297 (1995)
8. Ester, M., Kriegel, H.P., Sander, J., Xu, X.: A density-based algorithm for discovering clusters in large spatial databases with noise. In: Simoudis, E., Han, J., Fayyad, U.M. (eds.) Proceedings of the 2nd International Conference on Knowledge Discovery and Data Mining (KDD 1996), Portland, Oregon, USA, pp. 226–231. AAAI Press, August 1996
9. Fan, Y., Batmanghelich, N., Clark, C., Davatzikos, C.: Spatial patterns of brain atrophy in MCI patients, identified via high-dimensional pattern classification, predict subsequent cognitive decline. NeuroImage **39**, 1731–1743 (2008)
10. Gerardin, E., Chételat, G., Chupin, M., Cuingnet, R., Desgranges, B., Kim, H.S., Niethammer, M., Dubois, B., Lehéricy, S., Garnero, L., Eustache, F., Colliot, O.: Multidimensional classification of hippocampal shape features discriminates Alzheimer's disease and mild cognitive impairment from normal aging. NeuroImage **47**(4), 1476–1486 (2009)
11. Lopez, M., Ramirez, J., Gorriz, J.M., Salas-Gonzalez, D., Alvarez, I., Segovia, F., Chaves, R.: Multivariate approaches for Alzheimer's disease diagnosis using bayesian classifiers. In: Yu, B. (ed.) Proceedings of the IEEE Nuclear Science Symposium Conference Record (NSS/MIC 2009), Orlando, FL, USA, pp. 3190–3193. IEEE, October 2009
12. Mikhno, A., Nuevo, P.M., Devanand, D.P., Parsey, R.V., Laine, A.F.: Multimodal classification of dementia using functional data, anatomical features and 3D invariant shape descriptors. In: 9th IEEE International Symposium on Biomedical Imaging - Proceedings (ISBI 2012), Barcelona, Spain, pp. 606–609. IEEE, May 2012

13. Morgado, P., Silveira, M., Marques, J.S.: Efficient selection of non-redundant features for the diagnosis of Alzheimer's disease. In: 10th IEEE International Symposium on BiomedicalImaging: from Nano to Macro - Proceedings (ISBI 2013), San Francisco, CA, USA, pp. 640–643. IEEE, April 2013
14. Ramírez, J., Górriz, J.M., Salas-Gonzalez, D., Romero, A., López, M., Álvarez, I., Gómez-Río, M.: Computer-aided diagnosis of Alzheimer's type dementia combining support vector machines and discriminant set of features. Inf. Sci. **237**, 59–72 (2013)
15. Segovia, F., Górriz, J.M., Ramírez, J., Salas-Gonzalez, D., Álvarez, I., López, M., Chaves, R.: A comparative study of feature extraction methods for the diagnosis of Alzheimer's disease using the ADNI database. Neurocomputing **75**, 64–71 (2012)
16. Silveira, M., Marques, J.S.: Boosting Alzheimer disease diagnosis using PET images. In: Proceedings of the 20th International Conference on Pattern Recognition (ICPR 2010), Instambul, Turkey, pp. 2556–2559, August 2010
17. Tran, T.N., Nguyen, T.T., Willemsz, T.A., van Kessel, G., Frijlink, H.W., van der Voort Maarschalk, K.: A density-based segmentation for 3D images, an application for X-ray micro-tomography. Anal. Chim. Acta **725**, 14–21 (2012)
18. Tripathi, S., Kumar, K., Singh, B.K., Singh, R.P.: Image segmentation: a review. Int. J. Comput. Sci. Manage. Res. **1**(4), 838–843 (2012)
19. Varma, S., Simon, R.: Bias in error estimation when using cross-validation for model selection. BMC Bioinf. **7**, 91 (2006)
20. Ye, J., Farnum, M., Yang, E., Verbeeck, R., Lobanov, V., Raghavan, N., Novak, G., DiBernardo, A., Narayan, V.: Sparse learning and stability selection for predicting MCI to AD conversion using baseline ADNI data. BMC Neurol. **12**, 46 (2012)
21. Zhang, D., Wang, Y., Zhou, L., Yuan, H., Shen, D.: Multimodal classification of Alzheimer's disease and mild cognitive impairment. NeuroImage **55**(3), 856–867 (2011)

Model Based Quantification of Tissue Structural Properties Using Optical Coherence Tomography

Cecília Lantos[1,2,3]($\boxtimes$), Rafik Borji[1], Stéphane Douady[2],
Karolos Grigoriadis[1], Kirill Larin[4], and Matthew A. Franchek[1]

[1] Department of Mechanical Engineering, University of Houston, Houston, USA
{clantos,rborji2,mfranchek}@uh.edu
[2] Laboratory of Matter and Complex Systems,
Paris Diderot University, Paris, France
cecilia.lantos@univ-paris-diderot.fr
[3] Department of Hydrodynamic Systems,
Budapest University of Technology and Economics, Budapest, Hungary
[4] Department of Biomedical Engineering, University of Houston, Houston, USA

Abstract. Optical Coherence Tomography is an optical imaging technique providing subsurface structural images with a resolution at histological level. It has been widely studied and applied in both research and clinical practice with special interest in cancer diagnosis. One of the major queries today is to represent the images in a standardized way. To this aim the qualitative image recorded on the tissue will be transformed into a quantitative model. The solution provided here is able to diagnose healthy versus cancerous tissue independently from the measurement settings.

Keywords: Optical Coherence Tomography · Tissue characterization Medical diagnostics · Cancer · Liposarcoma · Model-based · Imaging

1 Introduction

In clinical practice, diagnosis of cancer from structural features is validated by histological analysis. Imaging modalities are useful by the pathologist to complete the diagnosis. These modalities reveal the tissue topology and cellular components without excising the tissue, and it gives a good direction for micro-biopsy [1]. Although functional methods can reveal cancer, structural visualization remains the gold standard diagnostic technique in oncology [2].

Characteristics of tissue structural properties are studied non-invasively with different imaging modalities, such as Magnetic Resonance Imaging, Computed Tomography, Ultrasound, Optical Coherence Tomography [2, 3]. Each imaging modality works at different scales and uses a specific frequency range that is a function of the sensing modality, desired resolution and depth of penetration.

These techniques can be coupled for multidisciplinary analysis of the tissue providing different information detected from backscattered waves from the internal

© Springer International Publishing Switzerland 2015
G. Plantier et al. (Eds.): BIOSTEC 2014, CCIS 511, pp. 113–134, 2015.
DOI: 10.1007/978-3-319-26129-4_8

structure. The outputs of these backscattered signals are grayscale images with different resolution and imaging depth revealing the subsurface structure [2, 3]. In this study Optical Coherence Tomography (OCT) was used to analyze tissue structural properties. OCT records images based on near infrared (NIR) laser light scattered back from the tissue [4, 5].

Instead of qualitative interpretation of OCT images, the diagnosis process employs mathematical models whose coefficients are adapted to match the OCT dataset. Based on the resulting model coefficients, the topological changes in the tissue are rigorously quantified. The presented adaptive model based approach is a general solution to tissue morphology quantification enabling accurate cancer detection and diagnosis. Moreover, the imaging based adaptive model is applicable to clinical tools for physiologists and pathologists. To illustrate the details of the method, a statistical model is used to capture the scattering properties of OCT datasets thereby distinguishing various tissue types, namely normal adipose tissue, well-differentiated and de-differentiated liposarcoma tissue. From this application the proposed adaptive model based approach is a general solution, broadened toward the analysis of other type of cancer since the diagnosis is based on morphology. This imaging-based model will serve as a quantitative clinical tool for physiologists.

2   Tissue Structure Recorded with Optical Coherence Tomography

2.1   Liposarcoma Recorded with Optical Coherence Tomography

Optical Coherence Tomography (OCT) is a well-known structural imaging method applied on biological material, in particular for cancer diagnosis and boundary detection between healthy and cancerous lesion [4, 5]. OCT has a better resolution (3–10 μm) compared to other diagnostic methods, revealing the subsurface structure at histological level in a 1–3 mm deep region under the surface, and has proved to be the most suitable imaging method applied on liposarcoma [6–8].

Liposarcomas (LS) are the most common soft tissue sarcomas (STS) with primary retroperitoneal malignant tumors [9, 10]. According to the WHO (World Health Organization) classification LS is part of the Adipocytic Tumors.

In this study, an adaptive model based solution to tissue characterization is demonstrated on normal adipose tissue detection, intermediate (locally aggressive) tumor detection, the so-called well-differentiated liposarcoma (WDLS) detection and de-differentiated liposarcoma (DDLS) one type of malignant tumor having the risk to metastasize, detection [11]. Due to the distinct prognosis and treatment the correct diagnosis of these liposarcoma types is crucial [7].

Analysis of tissue subsurface structure is recognized by medical professionals as a trustworthy means to discriminate normal fat tissue from cancerous lesion called sarcoma. Analysis of genetic content is required to prove the diagnosis in some specific cases [10, 12].

Tissue samples were excised from human patients' abdomen/retroperitoneum at the University of Texas M. D. Anderson Cancer Center (UTMDACC). Protocols for tissue processing were approved by the UTMDACC and University of Houston Biosafety Committees. Normal fat tissue, well-differentiated liposarcoma and de-differentiated liposarcoma were acquired. Histological diagnosis and classification of samples was performed by a UTMDACC sarcoma pathologist (Fig. 1).

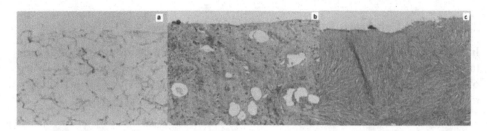

Fig. 1. Histological images (Magnification 10x) of (a) Normal Fat Tissue (b) Well-Differentiated Liposarcoma with extensive mitotic change (c) Highly Fibrotic De-Differentiated Liposarcoma (Color figure online).

The tissue was put in sterile phosphate buffered saline then stored in refrigerator until imaged with the OCT system. We record the tissue on a Spectral-Domain (SD) OCT measuring rig in the BioOptics Laboratory at the University of Houston. A supraluminescent laser diode (Superlum, S840-B-I-20) generates a broadband laser signal with center wavelength at λ_0 = 840 nm, spectral bandwidth at FWHM $\Delta\lambda$ = 50 nm and output power at 20 mW (Fig. 2) [6].

Fig. 2. Cross-Section OCT Images (B-scan composed of 500 A-lines); Logarithmic Response. (a) Normal Fat Tissue (b) WDLS with Extensive Myxoid Change (c) Highly Fibrotic DDLS. White bar = 500 μm.

The images show the cross-section of Human Normal Fat tissue (Fig. 2a), WDLS (Fig. 2b) and DDLS (Fig. 2c). This 2D cross-section called B-scan is composed of 500 adjacent A-lines. One A-line (1D) shows the backscattered intensity variation in function of depth from a laser footprint of $\sim$8 μm in focal plane in air. The region is 3 mm wide scanned with a galvanometer mirror, with backscattered light collected from a region of up to $\sim$1 mm in depth. The exact input power to the sample arm, and focus position is not noted.

A gray-scale image, obtained from OCT, shows the tissue internal structure. These images are different for each tested tissue (healthy vs. cancerous). The laser clearly reveals the large adipose cells seen in Normal Fat Tissue. WDLS has extensive myxoid change including vasculature not seen on this image, but still has some adipose cells with varying size, which is a means of WDLS identification. The part of DDLS imaged here resembles fibrotic tissue.

2.2   Optical Coherence Tomography Working Principle

In this work, qualitative information from OCT images is transformed into a quantitative parametric description of the tested tissue. The model is based on the variability of the A-lines in the cross-section at a given region. A single A-line of the different tissue types is presented in arbitrary unit on Fig. 3, and in dB scale on Fig. 4.

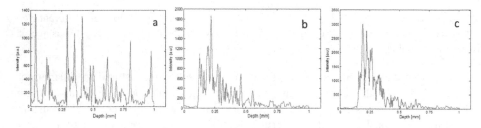

Fig. 3. OCT A-line in arbitrary unit (a) Normal Fat Tissue (b) WDLS with Extensive Myxoid Change (c) Highly Fibrotic DDLS.

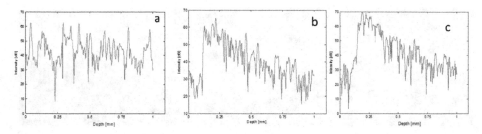

Fig. 4. OCT A-line in dB scale (a) Normal Fat Tissue (b) WDLS with Extensive Myxoid Change (c) Highly Fibrotic DDLS.

1D imaging (Intensity as a function of depth at a single point) is obtained by applying Digital Signal Processing methods on the data recorded on a line scan camera (Basler Sprint L104 K-2 k, 2 k pixel resolution, 29.2 kHz line rate), detecting 2048 wavelength intensity values between 800–890 nm. The signal is digitized using an analog-to-digital converter (NI-IMAQ PCI-1428).

The intensity detected on the line-scan camera is the cross-correlation of the broadband laser light electric field split in a Michelson interferometer and scattered back from a reference mirror and from the sample layers at each frequency component (*j*) [4, 5].

$$I(\lambda_j) = \left(E_R + \sum_k E_{Sk}\right)\left(E_R + \sum_k E_{Sk}\right)^* \tag{1}$$

The broadband laser light is decomposed into its spectral components when passing through a diffraction grating (Wasatch Photonics, 1200 grooves/mm). A lens focuses the decomposed NIR light to the camera. Each wavelength experiences constructive interference at different angles and path-length differences producing interference fringes which contain the scattering information from each tissue layer [4, 5].

Digital Signal Processing (DSP) methods are applied on the detected intensity as function of wavelength to correct for the laser variability, to compensate for the dispersion of light in the OCT elements (e.g. waveguides) and tissue, and to resample the data from wavelength (λ) to wavenumber space (k). A Fourier transform is applied to get the intensity function $I(z)$ as function of depth [13].

$$I(z) = FT\{I(k)\} \tag{2}$$

The measurement setting and DSP are carried out in Labview, whereas postprocessing of intensity functions $I(z)$ is performed in Matlab.

In Figures portraying the A-lines, the flat regions in Figs. 3–4 represent the light scattering in air, due to the path-length difference between the mirror of the reference arm and the tissue surface. Therefore, the abscissa should be truncated, and only the actual tissue region from which light scatters back should be considered.

The A-line examples are featured by a distinguishable periodicity and light attenuation in function of depth. Both factors represent structural distinction. The periodicity is due to light reflections from the cell boundary, and the light attenuation is due to stronger scattering in denser tissue. The structural resolution is reduced by speckle noise.

Furthermore the light attenuation, and degraded resolution is enhanced because of the Sensitivity roll-off due to the physical limitation of the spectrometer and camera array width, DSP techniques and light source spectral shape, that the camera records the same sample point at lower intensity from farther path-length differences. This sensitivity roll-off curve affects the intensity in function of path-length difference independently from the studied material [14].

The novelty of our study is to develop a mathematical model-based approach instead of visual grading of the structure to be able to differentiate tissue types independently from the measurement settings, exposure time, path-length difference between mirror and tissue surface, focus position, and sensitivity roll-off.

3 State of the Art of Tissue Structural Properties Quantification Using Optical Coherence Tomography

Over the past decades, several approaches based on experiments and/or theoretical assumptions have been elaborated to differentiate quantitatively between the various tissue types, specifically between healthy and cancerous tissue images recorded with OCT.

The attenuation of backscattered NIR laser light in function of depth (z) in the biological material theoretically follows an exponential function [4, 5]:

$$I(z) = I_0 e^{-u_t z} \tag{3}$$

defined by the scattering coefficient u_t characterizing different tissue types. Taking the logarithm of the intensity variation will provide the slope of the intensity attenuation, which can be measured now from the figure after averaging or filtering the A-lines. This implies the abstraction of the tissue structure, and has been applied on liposarcoma [8] and healthy nodes vs. nodes containing malignant cells [15].

This equation is *theoretical* and is based on single-scattering assumption that is valid at the superficial layers and in the focal volume [16–18]. For fixed focusing systems the position of the focal plane in the tissue and the depth of focus affect the detected light and the intensity slope [16, 17, 19, 20]. Although multiple-scattering can also contribute to the OCT signal, the single-scattering model with correction of the confocal Point Spread Function *(h(z))* can provide an accurate slope characteristics [16–18, 21]:

$$I(z) = h(z)I_0 e^{-u_t z} \tag{4}$$

$$h(z) = \left(\left(\frac{z - z_f}{z_R} \right)^2 + 1 \right)^{-1} \tag{5}$$

where z_f is the position of the focal plane, and z_R is the "apparent" Rayleigh length of the laser beam [16, 17, 20].

The corrected signal slope was calculated in normal and malignant ovarian tissue [19] focusing at the tissue surface, and in lymph nodes [22] focusing within the tissue. In the latter case besides the confocal optics the depth-dependent intensity loss scanning was also corrected based on calculation on a benchmark material.

Beside the slope calculation, a second parameter, namely the standard deviation around the slope was analyzed first in the breast in [23]. A third parameter, the frequency content after Fourier transforming the A-line was also extracted in [24]. Similar analysis was applied on liposarcoma and leiomyosarcoma in [25] with two parameters. All three parameters (slope, standard deviation and frequency content) were used on the same dataset in [26].

The input power does not affect the slope characteristics in contrary to the other quantification approaches, where the measurements are recorded with a fixed input power to the sample arm [23, 24, 27–29]. The problem of focus is treated only in [23, 24] by avoiding the use of focusing lens.

For the breast tissue in [27] the A-line pattern is quantified by periodicity analysis in addition to the frequency content. The mean distance between peaks distinguishes tissue types based on distinct cell size and density, periodic response and frequency oscillation. In [30] this analysis is developed further by extracting more parameters.

These quantification procedures were performed first on a single A-line. However the diagnosis is more accurate when these parameters are extracted from a set of adjacent A-lines.

Cancer has been also diagnosed based on fractals due to its disordered feature and irregularity (e.g. in the breast [28]). Fractal analysis in OCT was applied first time on distinct artery layers without visible structure [29]. The results reveal that the texture characteristics in the medium due to random speckle phenomena characterize tissue types [31, 32].

The techniques proposed in literature have not been applied in clinical practice yet. In this paper the main interest is to study tissue structural properties of OCT using Digital Signal Processing Techniques. Image processing methods and complex modeling techniques are not addressed in this study.

Cancerous fat tissue is much denser than healthy fat tissue. Since light scattering occurs mainly at the interfaces, scattering is much stronger in cancerous tissue. The attenuation of light is higher in the dense tissue, which is detectable with the attenuation coefficient u_t, although the slope analysis discards all structural information.

The inhomogeneous normal adipose tissue is distinguished with periodic scattering at the cell boundaries and the back reflection loses the periodicity as the adipose cells dedifferentiate in the cancerous tissue. An analysis of the structure's periodicities and frequency content is sensitive to this, as well as fractal analysis. Speckle analysis can provide additional information regarding invisible structure, but it is strongly dependent on the measurement settings.

4 Data Analysis to Quantify Tissue Structural Properties

4.1 OCT A-lines and B-scans Post-processing

The aim of our study is to develop a simple analysis technique based on a parametric method that captures the structural features from the strength of scattering. In this section the post-processing steps to determine the model in the Fourier-domain signatures that are derived from OCT data is explained.

A 3×3 mm^2 area from each of the three tissue samples (Normal Fat, WDLS, DDLS) is imaged where 500 B-scans are recorded with 500 A-lines per B-scan. We will focus on the statistical properties of the backscattered intensity signals.

For the computation, the tissue surface is numerically straightened to obtain correct intensity values at each depth position. For each B-scan, A-line sections are translated to different scanning positions, so that the depth variable has a common origin in all cases.

The canny edge detector from Matlab Image Processing Toolbox is applied on the B-scans after median filtering of the images. This can be used to align the scans, but does not yield the absolute position of the surface with respect to common origin. Before proceeding to the next step, all B-scans are screened to verify if straightened properly.

In order to find the tissue surface on the straightened images the first derivatives of the mean of the A-lines in one B-scan are calculated from the uncorrected images.

Then the tissue surface is defined at the highest derivative point, where the A-lines are truncated, and only the region backscattered from the tissue section will be analyzed.

After straightening and truncating the OCT images the statistical analysis has to be able to compare the data measured at different experimental setup. The normalization process will be defined by reducing the effects of the measurement setup, the sensitivity roll off from farther path-length differences, and the light attenuation effect.

At each depth position the mean and standard deviation of the intensity signals is calculated. Then, attenuation effects are removed from the data by dividing the Intensity Values or the standard deviation of the A-lines by the mean from each backscattered intensity response, so as to normalize every scan line.

However the normalization process compensates the exposure time setting, and light attenuation effect, there is still the problem of the focus position, and the sensitivity roll-off dependent on the path-length difference between the tissue surface position and the reference mirror which are varying at each measurement.

This sensitivity curve can be calculated from experiments with a second mirror in the sample arm using dynamic focusing. We shift the mirror toward depth and we place the mirror to the perfect focus position for recording, so as the camera sensitivity curve is obtained, and normalized by the maximum intensity value (black and blue dots on Fig. 5).

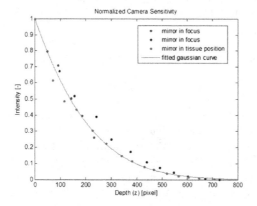

Fig. 5. Normalized Sensitivity roll-off; Black and Blue dots mark measurements from a mirror in the sample arm shifted in depth at focus position, red dots mark measurement from the mirror shifted in depth in the sample arm at fixed focus (Color figure online).

The normalized curve in Fig. 5 is independent from the measurement settings, but it is affected by the focus position which varies at each measurement, and remains fix during recording.

In the experiment setup, new measurements are made with the same mirror shifting in depth in the sample arm, but the backscattering signal was recorded with fixed focusing, which was set up from a former measurement with tissue in the sample arm.

The normalized values of these measurements are shown with red dots in Fig. 5.

However the experiments are based on specular reflection, in the diffuse tissue sample the "apparent" Rayleigh length doubles, so the focal region broadens, but according to certain analyses it can be neglected [16, 17, 20].

Since in our experiment the sensitivity roll-off curve is hardly affected by the intensity variation of the focal region (the focal plane being farther then the imaged region), the images are corrected according to the normalized sensitivity roll-off curve computed from fixed focusing, to eliminate the errors coming from the intensity variations because of the oblique tissue surface [14].

Matlab Curve Fitting Toolbox computes the parameters by Least Squared fitting procedure following an assumed Gaussian decay. Two methods will be compared below, based on the standard deviation per mean ratio and the mean normalized intensity values.

4.2 Standard Deviation Over Mean

In the first case the tissue characterization will be defined from the Probability Density Functions (PDF) of the STD/MEAN curves. The three-parameter Generalized Extreme Value (GEV) Distribution fits the histograms well due to its high flexibility:

$$y = f(x|k, \mu, \sigma) = \left(\frac{1}{\sigma}\right) exp\left(-\left(1 + k\frac{(x - \mu)}{\sigma}\right)^{-\frac{1}{k}}\right)\left(1 + k\frac{(x - \mu)}{\sigma}\right)^{-1-\frac{1}{k}} \quad (6)$$

where x is the std/mean of the intensity values, y is the distribution, k is the shape, σ is the scale, μ is the location parameter.

A 150 pixel corresponding to a depth of 0.659 mm is considered for analysis. This depth represents the threshold after which DDLS does not reflect light at the wavelength working range and camera settings (Fig. 6a).

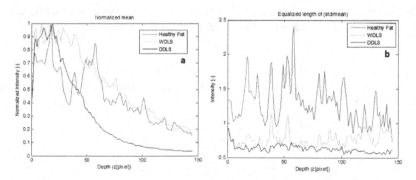

Fig. 6. (a) Averaged B-scan, mean intensity value at each depth position on Normal Fat, WDLS, DDLS; (150 pixels from the tissue surface) The curves here are normalized according to maximum value only for representation. (b) Standard Deviation over Mean at each depth position in the same region (Color figure online).

The Region of Interest (ROI) is then defined on the curves for analysis. This analysis relies on the use of a windowing scheme, in which sections of the intensities as function of depth are evaluated separately.

After evaluation of the data in each window region at each B-scan via the parameters of the GEV distribution, a window size of 40 pixel = 0.1758 mm is chosen, beginning from the tissue surface. This new method turned out to be independent on the surface scattering effect.

To depict the accuracy of the results, 160 WDLS, DDLS and 200 Normal Fat B-scans were analyzed. Figure 7 shows the mean and STD of the GEV parameters on the Gaussian corrected curves. This method proved to differentiate well between the three tissue types.

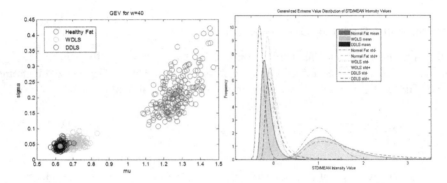

Fig. 7. Left: GEV Distribution parameters calculated from the std/mean ratio on the σ-μ plane. 200 B-scans for Normal Fat Tissue (red) 160 B-scans for WDLS (cyan) and DDLS (black). Green point marks the Center of Gravity. Right: Histogram of the GEV Distribution, mean and standard deviation are presented (Color figure online).

The curve coefficients discriminate between the healthy and cancerous tissues, but there is less distinction between WDLS and DDLS (Table 1).

Table 1. GEV parameters calculated from the STD/MEAN ratio of the intensity values at each depth position, mean and standard deviation on 200 B-scans of Normal Fat, and 160 B-scans of WDLS and DDLS.

STD/MEAN	k	σ	μ
Baseline (Normal Fat)	0.0007	0.2151	1.2796
	+0.2347	+0.0579	+0.0659
Deviation 1 (WDLS)	−0.0128	0.0529	0.7093
	+0.2443	+0.0120	+0.0359
Deviation 2 (DDLS)	0.0857	0.0493	0.6502
	+0.1673	+0.0128	+0.0584

The deviation from the baseline tissue is estimated based on the baseline tissue parameter (b) and the deviated tissue parameter (d). Details are shown in Table 2.

Table 2. Comparison of the GEV parameters calculated from the STD/mean ratio of the intensity values at each depth position, mean and standard deviation on 200 B-scans of Baseline Tissue and 160 B-scans of Deviation 1&2.

STD/MEAN	$\Delta k = \frac{k_d - k_b}{k_b}$	$\Delta\sigma = \frac{\sigma_d - \sigma_b}{\sigma_b}$	$\Delta\mu = \frac{\mu_d - \mu_b}{\mu_b}$
Baseline (Normal Fat)	0	0	0
	±335.2857	±0.2692	±0.0515
Deviation 1 (WDLS)	−19.2857	−0.7541	−0.4457
	[−368.2857; 329.7143]	[−0.8099; −0.6983]	[−0.4737; −0.4176]
Deviation 2 (DDLS)	121.4286	−0.7708	−0.4919
	[−117.5714; 360.4286]	[−0.8303; −0.7113]	[−0.5375; −0.4462]

The coefficients' occupied space in 3D (σ, μ, k) are represented in Fig. 8 for different tissues.

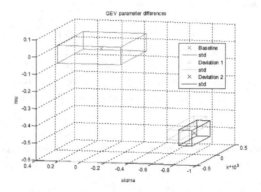

Fig. 8. Comparison of the GEV parameters represented at each axe of the 3D coordinate system calculated from the STD/mean ratio of the intensity values at each depth position, mean and standard deviation on 200 B-scans of Baseline Tissue and 160 B-scans of Deviation 1&2 (Color figure online).

4.3   Normalized Intensity Variation

A second method is developed to analyze the same data set. Instead of calculating the STD/MEAN, all the measured intensity values are now considered, and normalized by the mean intensity at each depth position. The same windowing process was applied on the A-lines and B-scans, and the optimal window size of 40 pixels beginning from the surface has been verified to provide the largest separation between healthy vs. cancerous tissue parameters on the σ-μ plane. The mean and STD of the GEV parameters characterizing the different tissue types are shown in Fig. 9 and Table 3.

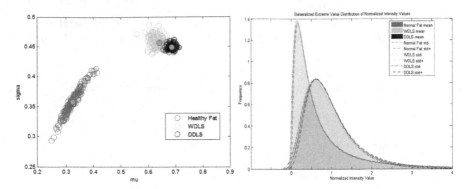

Fig. 9. Left: GEV Distribution parameters on the σ-μ plane calculated from the normalized intensity values. 200 B-scans for Normal Fat Tissue (red) 160 B-scans for WDLS (cyan) and DDLS (black). Green point marks the Center of Gravity. Right: Histogram of the GEV Distribution, mean and standard deviation are presented (Color figure online).

Table 3. GEV parameters calculated from the mean-normalized intensity values in ROI, mean and standard deviation on 200 B-scans of Normal Fat, and 160 B-scans of WDLS and DDLS.

I/MEAN	k	σ	μ
Baseline (Normal Fat)	0.8209	0.3532	0.3191
	+0.0647	+0.0181	+0.0254
Deviation 1 (WDLS)	0.1905	0.4561	0.6381
	+0.0358	+0.0100	+0.0229
Deviation 2 (DDLS)	0.1447	0.4462	0.6700
	+0.0684	+0.0044	+0.0364

Table 4. Comparison of the GEV parameters calculated from the mean-normalized intensity values in ROI, mean and standard deviation on 200 B-scans of Baseline Tissue and 160 B-scans of Deviation 1&2.

ΣI/MEAN	$\Delta k = \frac{k_d - k_b}{k_b}$	$\Delta\sigma = \frac{\sigma_d - \sigma_b}{\sigma_b}$	$\Delta\mu = \frac{\mu_d - \mu_b}{\mu_b}$
Baseline (Normal Fat)	0	0	0
	±0.0788	±0.0512	±0.0796
Deviation 1 (WDLS)	−0.7679	0.2913	0.9997
	[−0.8115; −0.7243]	[0.2630; 0.3196]	[0.9279; 1.0715]
Deviation 2 (DDLS)	−0.8237	0.2633	1.0997
	[−0.9071; −0.7404]	[0.2508; 0.2758]	[0.9856; 1.2137]

Similarly to the first method the curve coefficients characterize well the healthy and cancerous tissue but WDLS and DDLS coefficients are not sufficiently distinguished. The deviation from the baseline tissue is detailed in Table 4.

The coefficients' occupied space in 3D (σ, μ, k) are represented in Fig. 10 for different tissues, representing results similar to the first method.

The two proposed statistical analyses proved to be efficient methods in segregating tissue types (healthy vs. cancerous) accurately. The method is independent on the measurement settings as the results are normalized by the mean of the intensity values at each depth position, and errors due to path-length differences are corrected.

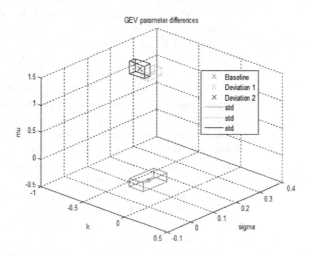

Fig. 10. Comparison of the GEV parameters represented at each axe of the 3D coordinate system calculated from the mean-normalized intensity values in ROI, mean and standard deviation on 200 B-scans of Baseline Tissue and 160 B-scans of Deviation 1&2 (Color figure online).

These statistical tools were applied also on the images without correction. As a result, only the shape factor parameter was affected considerably in the case where std/mean ratio is calculated.

The std/mean tool is also more sensitive to the way the tissue surface is found. In fact the histogram points are obtained by calculating std/mean from each depth position (40 pixels) whereas the second proposed method makes use of all data points in the ROI (40 × 200 or 40 × 160 pixel points). Thus the correction step is necessary to get absolute parameters that describe the tissue types.

The proposed approach can be enhanced to better distinguish WDLS and DDLS by incorporating additional factors such as the mean intensity values at each depth position. The only drawback in the measurements is the manual control of the tissue position under the laser light to get a visible subsurface structure.

Table 5. Normal and Cancerous Fat Tissue samples analyzed statistically.

Class	Label	Properties
NORMAL FAT TISSUE	K37	Normal Fat Tissue, composed of clusters of large adipocytes and fibrous connective tissue
	K28	
	K19	
	K22	
WDLS	K36	WDLS with significant myxoid components analyzed above. Myxoid components are frequent in WDLS especially in A/RP, and so it resembles myxofibrosarcoma, though the scattered atypical adipocytes represent the diagnosis of Well-Differentiated LS. Today the classification between these two Sarcoma types is clear, myxoid fibrosarcoma usually does not appear in the A/RP, and contains fine capillaries and small oval cells which distinguish from WDLS
	K25	Fibrous atypical lipomatous tumor, with moderate cellularity. In Lipomatous Tumor the adipose cells lose their original shape and size (atypical lipoma) and the septa is thickened
	K47	Atypical lipomatous tumor (WDLS) with small mitotic changes. In contrast to the other histological subtypes, the B-scans of this measurement in the 3 mm x 3 mm x1 mm Volume show various morphology. Depending on the B-scan some large adipose cells are still present or disappear, which will affect the computation
	K58	Lipomatous tumor with mild cytological atypia. The tissue components are mature adipocytes with little size variation, and some slightly atypical cells scattered in the tissue. The diagnosis was not clear from the histological images. Fluorescence imaging was required to differentiate from Lipoma. The corresponding histology is made on another tissue sample (K49) with similar diagnosis to K58
DDLS	K35	DDLS with highly fibrotic changes already represented in the data analysis
	K12	Spindle cell sarcoma with osteoid and bone formation, consistent with DDLS [no bone formation is seen in this sample on the image]
	K33	DDLS with extensive myxoid change, arising in well-differentiated liposarcoma (atypical lipomatous tumor). Tumor involves mesentery adjacent to bowel wall (intestinal margins is negative)
	K26b	section reveals cellular pleomorphic fibromyxoid areas which contain mitotic figures focally. These findings are compatible with the diagnosis of DDLS

5   Data Analysis Tools' Sensitivity to Different Tissues

In the development of the data analysis tools only one histological subtype of WDLS and DDLS is described, however they can represent several patterns [33, 34]. The comparison of the different histological subtypes is studied below based on the statistical analysis of the mean normalized intensity values.

Four samples per each tissue type are recorded different time and enumerated in chronological order. The dataset of the 4 Normal Fat tissue samples were cut from

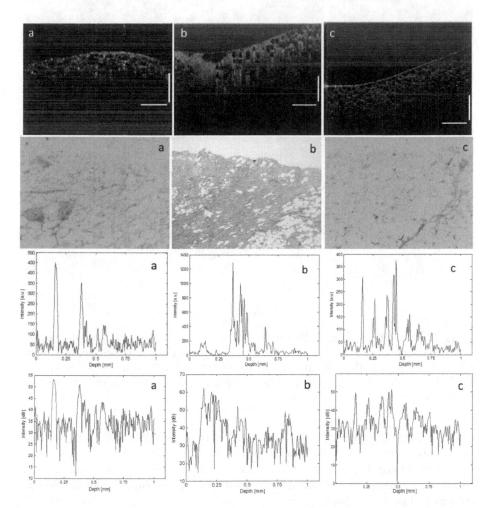

Fig. 11. First raw: OCT images with a white scale bar of 500 µm, second raw: Histology at 4 × Magnification (exact scale is not known), third raw: A-line example in arbitrary unit, fourth raw: A-line example in dB scale. (a) K25 Fibrous Atypical Lipotamous Tumor (b) K47 Atypical Lipomatous Tumor/WDLS with small mitotic change (c) K58 Lipomatous Tumor with mild cytological atypia (Color figure online).

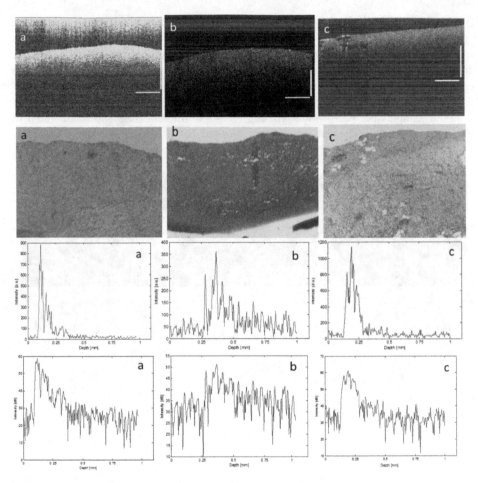

Fig. 12. First raw: OCT images with a white scale bar of 500 μm, second raw: Histology at 4 × Magnification (exact scale is not known), third raw: A-line example in arbitrary unit, fourth raw: A-line example in dB scale. (a) K12 Spindle cell sarcoma (DDLS) with osteoid formation (b) K33 DDLS with extensive myxoid change (c) K26b cellular pleomorphic fibromyxoid areas with mitotic figures (Color figure online).

Abdomen/Retro-Peritoneum (A/RP), and named K19, K22, K28 and K37, the last one was studied in the data analysis developed above. Basically there is no difference between the morphology, only at the surface obliqueness and regularity during OCT recording. In contrary, the WDLS, and DDLS samples represent different histological subtypes and topology (Table 5, Figs. 11, 12).

The mean normalized intensity values were calculated on 60 B-scans per tissue sample on the 40-pixel deep ROI. The GEV Distribution parameters (k, σ, μ) were calculated for each B-scan, and plotted in a 3D scatterplot in the k, σ, μ coordinate system (Fig. 13).

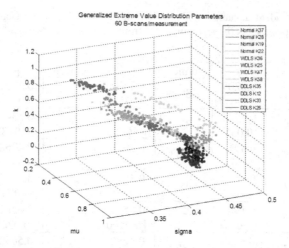

Fig. 13. Scatterplot representing 12 tissue samples, 4 tissue samples per Normal Fat Tissue, WDLS and DDLS, 60 data points per sample are represented (Color figure online).

The quantification method will consider the mean and standard deviation of the parameters to distinguish tissue types (Table 6).

The Generalized Extreme Value Probability Density Function of the mean curves of each sample is shown in Fig. 14.

The mean and standard deviation of the parameters in a 3D Coordinate System is captured to classify Normal Fat Tissue, WDLS and DDLS. The classification of the three tissue types is furthermore separated to the 4 samples per tissue type, plotting the data points representing the 12 tissue samples on the k, σ, μ parameter axes in a 3D coordinate system (Fig. 15).

A two-sample t-test was applied to the measurements to prove a statistically significant separation of the parameters between the tissue types at a 5 % of significance level. The results show the clear distinction between Normal Fat Tissue and DDLS considering all the three parameters. Probabilities of failing diagnosing are: P (k) = 0.29 %, P(σ) = 0.36 %, P(μ) = 0.2 %. The classification between WDLS and DDLS is proved according to k and μ parameters, and sigma parameter is approved to show differences between Normal Fat Tissue and WDLS or DDLS. Probabilities of failing diagnosis between WDLS and DDLS are: P(k) = 1.19 %, P(σ) = 28.41 %, P (μ) = 1.08 %, and between WDLS and Normal Fat Tissue are: P(k) = 18.12 %, P (σ) = 4.73 %, P(μ) = 19.59 %.

The quantification method based on the statistical properties of the tissue structure characteristics has shown its ability to differentiate healthy vs. cancerous tissue type. DDLS subtypes represent closely the same parameters regarding the absolute mean value and small standard deviation parameters. The missing adipose cells is a common feature of these DDLS subtypes, the tissue is mainly composed of smaller cells making it denser. The scattering properties detected with OCT reveal these characteristics, and our quantification method is able to confine this malignant cancer.

Table 6. Mean, standard deviation, and standard deviation per mean of Generalized Extreme Value Distribution parameters (k, σ, μ) computed from 4 samples per Normal Fat Tissue, WDLS and DDLS.

class	I/MEAN	k	std/mean	σ	std/mean	μ	std/mean
N O R M A L	K37	0.8209 +0.0647	7.88 %	0.3532 +0.0181	5.12 %	0.3191 +0.0254	7.96 %
	K28	0.3995 +0.0313	7.83 %	0.4092 +0.0086	2.10 %	0.5346 +0.0183	3.42 %
	K19	0.4915 +0.1334	27.14 %	0.3998 +0.0245	6.13 %	0.4768 +0.0694	14.56 %
	K22	0.5131 +0.0432	8.42 %	0.3817 +0.0108	2.83 %	0.4625 +0.0250	5.41 %
W D L S	K36	0.1905 +0.0358	18.79 %	0.4561 +0.0100	2.19 %	0.6381 +0.0229	3.59 %
	K25	0.3296 +0.0818	24.82 %	0.4295 +0.0090	2.10 %	0.5692 +0.0450	7.91 %
	K47	0.5256 +0.1792	34.09 %	0.4333 +0.0405	9.35 %	0.4550 +0.0776	17.05 %
	K58	0.4646 +0.0330	7.10 %	0.3971 +0.0098	2.47 %	0.4889 +0.0182	3.72 %
D D L S	K35	0.1447 +0.0684	47.27 %	0.4462 +0.0044	0.99 %	0.6700 +0.0364	5.43 %
	K12	0.1169 +0.0441	37.74 %	0.4452 +0.0041	0.92 %	0.6857 +0.0220	3.21 %
	K33	0.0792 +0.0496	62.66 %	0.4377 +0.0050	1.15 %	0.7064 +0.0282	4.00 %
	K26b	0.0857 +0.0444	51.58 %	0.4448 +0.0079	1.78 %	0.7006 +0.0262	3.74 %

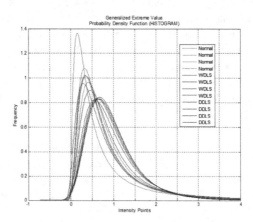

Fig. 14. GEV Distribution separating Normal Fat Tissue (red), WDLS (cyan) and DDLS (black). 4 samples per tissue type are presented (Color figure online).

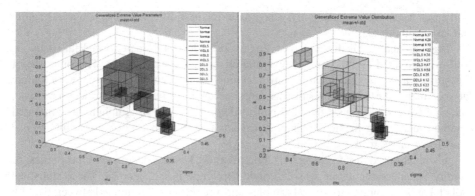

Fig. 15. Classification of Normal Fat Tissue, WDLS and DDLS. 4 tissue samples per tissue type are analyzed. Middle of the boxes represent the mean, the edges represent the standard deviation of the parameters k, σ, μ (Color figure online).

WDLS subtypes represent big variation in terms of all the 3 parameter range, and standard deviation per subtype. These results were expected from the histological findings and the OCT images, showing completely different morphology. Tissue types containing mainly adipose cells are overlapped with the Normal Fat Tissue region, and the denser subtypes are shifted toward the DDLS region. The big standard deviation of the parameters is due to the big variation of the morphology in the adjacent B-scans.

The morphology of the Normal Fat Tissue samples does not support the large scattering of the parameters. The error analysis reveals a big sensitivity of the tissue types in case of the tissue surface was not perpendicular to the laser beam during recording. Normal Fat Tissue is especially sensitive to the data analysis. For each tissue type the k parameter is sensitive to errors coming from surface obliqueness, saturation points or improper tissue surface straightening, and even more sensitive in the denser tissue types. Since the data analysis is based on the variations of parameter μ-σ, the quantification method provides accurate results.

6   Conclusion

Our objective was to study the response of tissue to a near infrared laser excitation, and, specifically, to characterize differences between healthy and cancerous tissue. The morphology of the subsurface is depicted based on the backscattered near infrared light. Parametric models of these backscattered signal characteristics are derived and linked statistically to the optical properties of normal adipose tissue, well-differentiated liposarcoma and de-differentiated liposarcoma.

The accurate diagnosis at early stage of cancer, as well as the recognition of the tumor boundary in tissue is highly important. However OCT has been well-recognized as a powerful method for cancer detection from tissue morphology, the diagnosis from these images is subjective and not obvious. This ongoing research intends to fill the need for an objective and quantitative means of data analysis. The goal of the current study was to develop a quantitative diagnostic method differentiating between healthy and cancerous tissue.

The data analysis is developed on images recorded on human normal fat tissue vs. well-differentiated (WD) and de-differentiated liposarcoma (DDLS). The outcome of this study was the development of statistical analysis to evaluate OCT images of human fat specimens. An accurate result was found to quantify healthy vs. cancerous tissue. The analysis can be applied in real-time for diagnosis due to its simplicity as compared to other quantifying method. This practical advantage gives a good possibility to use these tools in surgical evaluation.

A model-based tissue-characterization method based on structural properties of healthy vs. cancerous tissue was described on single tissue samples. These statistical tools have been applied to several samples and proved to differentiate between healthy and cancerous tissues.

Further refinement will allow to detect tumor boundary, diagnose other type of cancer (e.g. breast cancer) where structural analysis is required for diagnosis, or to monitor quantitatively tumor progression during cancer therapy.

Acknowledgements. The authors wish to acknowledge the very important contribution made by Shang Wang and Narendran Sudheendran. This work was supported by the Hungarian-American Fulbright Commission, the French Ministry of Research and the University of Houston. This work is the continuation of a poster presentation held on November 16, 2012 at the MEGA Research day of the University of Houston, and the Conference Proceedings of BIOSTEC – BIOIMAGING held in Angers, France March 3-6, 2014.

References

1. Kind, M., Stock, N., Coindre, J.M.: Histology and imaging of soft tissue sarcomas. Eur. J. Radiol. **72**, 6–15 (2009)
2. Rembielak, A., Green, M., Saleem, A.: Pat Price, Diagnostic and therapeutic imaging in oncology. Cancer Biol. Imaging Med. **39**(12), 693–697 (2011)
3. Morris, P., Perkins, A.: Diagnostic imaging, Physics and Medicine 2. Lancet **379**, 1525–1533 (2012)
4. Drexler, W., Fujimoto, J.G.: Optical Coherence Tomography: Technology and Applications. Springer, Berlin (2008)
5. Brezinski, M.E.: Optical Coherence Tomography, Principles and Applications. Academic Press Elsevier Inc., New York (2006)
6. Carbajal, E.F., Baranov, S.A., Manne, V.G.R., Young, E.D., Lazar, A.J., Lev, D.C., Pollock, R.E., Larin, K.V.: Revealing retroperitoneal liposarcoma morphology using optical coherence tomography. J. Biomed. Opt. **16**(2), 020502 (2011)
7. Lahat, G., Madewell, J.E., Anaya, D.A., Qiao, W., Tuvin, D., Benjamin, R.S., Lev, D.C., Pollock, R.E.: Computed tomography scan-driven selection of treatment for retroperitoneal liposarcoma histologic subtypes. Cancer **115**(5), 1081–1090 (2009)
8. Lev, D., Baranov, S.A., Carbajal, E.F. Young, E.D., Pollock, R.E., Larin, K.V.: Differentiating retroperitoneal liposarcoma tumors with optical coherence tomography. In: Proceedings SPIE 7890, Advanced Biomedical and Clinical Diagnostic Systems IX, 78900U (2011)

9. Unal, B., Bilgili, Y.K., Batislam, E., Erdogan, S.: Giant dedifferentiated liposarcoma: CT and MRI findings. Eur. J. Radiol. Extra **5**, 47–50 (2004)
10. Singer, S., Antonescu, C.R., Riedel, E., Brennan, M.F.: Histologic subtype and margin of resection predict pattern of recurrence and survival for retroperitoneal liposarcoma. Ann. Surg. **238**(3), 358–370 (2003)
11. Fletcher, C.D.M., Rydholm, A., Singer, S., Sundaram, M., Coindre, J.M.: Soft tissue tumours: epidemiology, clinical features, histopathological typing and grading. WHO Classification of Soft Tissue Tumors (2006)
12. Dei Tos, A.P.: Liposarcoma: new entities and evolving concepts. Ann. Diagn. Pathol. **4**(4), 252–266 (2000)
13. Gora, M., Karnowski, K., Szkulmowski, M., Kaluzny, B.J., Huber, R., Kowalczyk, A., Wojtkowski, M.: Ultra-high-speed swept-source OCT imaging of the anterior segment of human eye at 200 kHz with adjustable imaging range. Opt. Express **17**(17), 14880–14894 (2009)
14. Bajraszewski, T., Wojtkowski, M., Szkulmowski, M., Szkulmowska, A., Huber, R., Kowalczyk, A.: Improved spectral optical coherence tomography using optical frequency comb. Opt. Express **16**(6), 4163–4176 (2008)
15. McLaughlin, R.A., Scolaro, L.: Parametric imaging of cancer with optical coherence tomography. J. Biomed. Opt. **15**(4), 046029 (2010)
16. Lee, P., Gao, W., Zhang, X.: Performance of single-scattering model versus multiple-scattering model in the determination of optical properties of biological tissue with optical coherence tomography. Appl. Opt. **49**(18), 3538–3544 (2010)
17. Faber, D.J., van der Meer, F.J., Aalders, M.C.G.: Quantitative measurement of attenuation coefficients of weakly scattering media using optical coherence tomography. Opt. Express **12**(19), 4353–4365 (2004)
18. Thrane, L., Yura, H.T., Andersen, P.E.: Analysis of optical coherence tomography systems based on the extended Huygens-Fresnel principle. J. Opt. Soc. Am. A **17**(3), 484–490 (2000)
19. Yang, Y., Wang, T., Biswal, N.C., Wang, X., Sanders, M., Brewer, M., Zhu, Q.: Optical scattering coefficient estimated by optical coherence tomography correlates with collagen content in ovarian tissue. J. Biomed. Opt. **16**(9), 090504 (2011)
20. van der Meer, F.J., Faber, D.J., Sassoon, D.M.B., Aalders, M.C., Pasterkamp, G., van Leeuwen, T.G.: Localized measurement of optical attenuation coefficients of atherosclerotic plaque constituents by quantitative optical coherence tomography. IEEE Trans. Med. Imaging **24**(10), 1369–1376 (2005)
21. van Leeuwen, T.G., Faber, D.J., Aalders, M.C.: Measurement of the axial point spread function in scattering media using single-mode fiber-based optical coherence tomography. IEEE J. Sel. Top. Quantum Electron. **9**(2), 227–233 (2003)
22. Scolaro, L., McLaughlin, R.A., Klyen, B.R., Wood, B.A., Robbins, P.D., Saunders, C.M., Jacques, S.L., Sampson, D.D.: Parametric imaging of the local attenuation coefficient in human axillary lymph nodes assessed using optical coherence tomography. Biomed. Opt. Express **3**(2), 366–378 (2012)
23. Iftimia, N.V., Bouma, B.E., Pitman, M.B., Goldberg, B.D., Bressner, J., Tearney, G.J.: A portable, low coherence interferometry based instrument for fine needle aspiration biopsy guidance. Rev. Sci. Instrum. **76**, 064301 (2005)
24. Goldberg, B.D., Iftimia, N.V., Bressner, J.E., Pitman, M.B., Halpern, E., Bouma, B.E., Tearney, G.J.: Automated algorithm for differentiation of human breast tissue using low coherence interferometry for fine needle aspiration biopsy guidance. J. Biomed. Opt. **13**(1), 014014 (2008)

25. Wang, S., Sudheendran, N., Liu, C.-H., Manapuram, R.K., Zakharov, V.P., Ingram, D.R., Lazar, A.J., Lev, D.C., Pollock, R.E., Larin, K.V.: Computational analysis of optical coherence tomography images for the detection of soft tissue sarcomas. In: Proceedings of SPIE 8580, 85800 T, 2, Dynamics and Fluctuations in Biomedical Photonics X (2013)
26. Wang, S., Liu, C.-H., Zakharov, V.P., Lazar, A.J., Pollock, R.E., Larin, K.V.: Three-dimensional computational analysis of optical coherence tomography images for the detection of soft tissue sarcomas. J. Biomed. Opt. **19**(2), 021102 (2014)
27. Zysk, A.M., Boppart, S.A.: Computational methods for analysis of human breast tumor tissue in optical coherence tomography images. J. Biomed. Opt. **11**(5), 054015 (2006)
28. Sullivan, A.C., Hunt, J.P., Oldenburg, A.L.: Fractal analysis for classification of breast carcinoma in Optical Coherence Tomography. J. Biomed. Opt. **16**(6), 066010 (2011)
29. Popescu, D.P., Flueraru, C., Mao, Y., Chang, S., Sowa, M.G.: Signal attenuation and box-counting fractal analysis of optical coherence tomography images of arterial tissue. Biomed. Opt. Express **1**(1), 268–277 (2010)
30. Mujat, M.R., Ferguson, D., Hammer, D.X., Gittins, C., Iftimia, N.: Automated algorithm for breast tissue differentiation in optical coherence tomography. J. Biomed. Opt. **14**(3), 034040 (2009)
31. Gossage, K.W., Tkaczyk, T.T., Rodriguez, J.J., Barton, J.K.: Texture analysis of Optical Coherence Tomography images: feasibility for tissue classification. J. Biomed. Opt. **8**(3), 570–575 (2003)
32. Gossage, K.W., Smith, C.M., Kanter, E.M., Hariri, L.P., Stone, A.L., Rodriguez, J.J., Williams, S.K., Barton, J.K.: Texture analysis of speckle in optical coherence tomography images of tissue phantoms. Phys. Med. Biol. **51**, 1563–1575 (2006)
33. Miettinen, M.M.: Diagnostic Soft Tissue Pathology. Churchill Livingstone (2003)
34. Miettinen, M.M.: Modern Soft Tissue Pathology, Tumors and Non-neoplastic Conditions. Cambridge University Press, Cambridge (2010)

Multi-modal Imaging for Shape Modelling of Dental Anatomies

Sandro Barone, Alessandro Paoli[✉], and Armando V. Razionale

Department of Civil and Industrial Engineering, University of Pisa,
Largo Lucio Lazzarino 1, 56126 Pisa, Italy
{s.barone,a.paoli,a.razionale}@ing.unipi.it

Abstract. In dentistry, standard radiographic imaging is a minimally invasive approach for anatomic tissue visualization and diagnostic assessment. However, this method does not provide 3D geometries of complete dental shapes, including crowns and roots, which are usually obtained by Computerized Tomography (CT) techniques. This paper describes a shape modelling process based on multi-modal imaging methodologies. In particular, 2D panoramic radiographs and 3D digital plaster casts, obtained by an optical scanner, are used to guide the creation of both shapes and orientations of complete teeth through the geometrical manipulation of general dental templates. The proposed methodology is independent on the tomographic device used to collect the panoramic radiograph.

Keywords: Multi-modal imaging · Dental shape modelling · PAN radiograph · Discrete Radon Transform

1   Introduction

Orthodontics is the branch of dentistry concerned with the study and treatment of irregular bites and deals with the practice of manipulating patient dentition in order to provide better functionalities and appearances.

Detection and correction of malocclusion problems caused by teeth irregularities and/or disproportionate jaw relationships represent the most critical aspects within an orthodontic diagnosis and treatment planning. The most common methodologies for non-surgical orthodontic treatments are based on the use of fixed appliances (dental brackets) or removable appliances (clear aligners) [1]. In clinical practice, the conventional approach to orthodontic diagnoses and treatment planning processes relies on the use of plaster models, which are manually analyzed and modified by clinicians in order to simulate and plan corrective interventions. These procedures however require labor intensive and time-consuming efforts, which are mainly restricted to highly experienced technicians. Recent progresses in three-dimensional surface scanning devices as well as CAD (Computer Aided Design)/CAM (Computer Aided Manufacturing) technologies have made feasible the complete planning process within virtual environments and its accurate transfer to the clinical field. In particular, orthodontic alignment procedures greatly benefit from the combined use of CAD/CAM methodologies which are used to

© Springer International Publishing Switzerland 2015
G. Plantier et al. (Eds.): BIOSTEC 2014, CCIS 511, pp. 135–145, 2015.
DOI: 10.1007/978-3-319-26129-4_9

produce custom tight-fitting devices worn by the patients [2]. In this context, the accurate and automatic reconstruction of individual tooth shapes obtained from digital 3D dental models is the key issue for planning customized treatment processes. Optical scanners may be used to digitize plaster models thus providing geometric representations of tooth crowns. However, even if both clear aligners and brackets accomplish the treatment plan only by acting onto the tooth crown surfaces, a correct orthodontic treatment should also take into account tooth roots in terms of position, shape and volume. In particular, position and volume of dental roots may cause dehiscence, gingival recession as well as root and bone resorption when teeth undergo movements during therapy. Cone beam computed tomography (CBCT) could provide comprehensive 3D tooth geometries. However, concerns about radiation doses absorbed by patients are raised. For this reason, the use of computed tomography as a routine in orthodontic dentistry is still a matter of discussion. Even if CBCT has greatly reduced the dose of absorbed x-rays, compared to traditional computed tomography (CT), it still produces a greater x-ray dose than a panoramic radiograph (PAN).

2-D panoramic radiographs are a routine approach in the field of dentistry since they represent an important source of information. In particular, they are able to inexpensively record the entire maxillomandibular region on a single image with low radiation exposure for the patient. However, they are also characterized by several limitations such as: lack of any 3D information, magnification factors which strongly vary within the image thus causing distortions, patient positioning which is very critical with regard to both sharpness and distortions. As a result, not only 3D measurements are impaired, but also reliable 2D dimensions cannot be retrieved.

The present paper is aimed at investigating the possibility to recover 3D geometry of individual teeth by customizing general templates over patient-specific dental anatomy. Information about patient anatomy is obtained by integrating the optical acquisition of plaster casts with 2D panoramic radiographs. Even if the reconstruction of patient-specific 3D dental information from 2D radiographs and casts represents a challenging issue, very few attempts have been made up to now within the scientific community [3, 4]. Moreover, these studies greatly rely on the knowledge of the specific tomographic device used to acquire the PAN image. The present study is focused on the formulation of a general solution, which could infer tooth roots shape without any assumption on the specific hardware as well as parameters used to collect patient data anatomy.

2   Materials and Methods

In this paper, complete 3D dental shapes are reconstructed by integrating template models with 3D crowns data deriving from the optical acquisition of plaster casts and 2D data deriving from PAN radiographic images.

An optical scanner based on a structured light stereo vision approach has been used to reconstruct both template dental models and patient's dental casts.

Panoramic radiographs have been captured by using a Planmeca ProMax unit (Planmeca Oy, Helsinki Finland); whose data are stored and processed in DICOM format.

2.1 3D Data Acquisition System

The optical scanner (Fig. 1) is composed of a monochrome digital CCD camera (1280 × 960 pixels) and a multimedia white light DLP projector (1024 × 768 pixels) which are used as active devices for a stereo triangulation process. In this paper, a multi-temporal *Gray Code Phase Shift Profilometry* (GCPSP) method is used for the 3D shape recovery through the projection of a sequence of black and white vertical fringes whose period is progressively halved [5]. The methodology is able to provide $n_p = l_h \times l_v$ measured points (where l_h is the horizontal resolution of the projector while l_v is the vertical resolution of the camera) with a spatial resolution of 0.1 mm and an overall accuracy of 0.01 mm.

Fig. 1. Optical scanner during the acquisition process of a tooth template.

The optical devices are integrated with two mechanical turntables (Fig. 1) which allow the automatic merging of different measurements collected from various directions conveniently selected.

2.2 Input Data

The methodology requires three different input data: (1) general dental CAD templates, (2) dental crowns shape and (3) a PAN image. Crowns shape and PAN image are patient-specific data while teeth templates can be obtained from existing libraries.

Dental CAD Templates. Teeth template models are composed of complete teeth crowns and roots and are placed in adequately shaped holes within transparent plastic soft tissue reproduction (Fig. 2). Teeth can be easily removed from their housing in order to allow full reconstructions through the 3D scanner without optical occlusions.

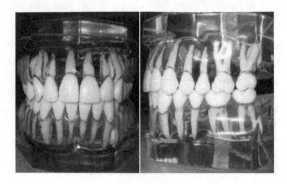

Fig. 2. Example of superior and inferior dental arch templates.

Patient Crowns Reconstruction. The patient dental crowns geometry can be acquired by scanning the plaster cast. Figure 3a shows the final digital reproduction of the patient tooth crowns with surrounding gingival tissue (digital mouth model) as obtained by merging twelve acquisitions of the superior plaster cast captured by different views. Tooth crown regions are segmented and disconnected from the oral soft tissue by exploiting the curvature of the digital mouth model. This model contains ridges and margin lines, which highlight the boundaries between different teeth, and between teeth and soft tissue. Regions with abrupt shape variations can be outlined by using curvature information [5]. Segmented crown shapes are finally closed by using computer-based filling tools (Fig. 3b).

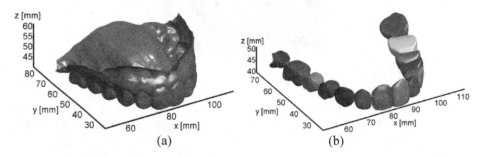

Fig. 3. (a) Reconstruction of the superior plaster cast as obtained by the optical scanner and (b) segmented patient crowns geometries.

Panoramic Radiograph. Dental panoramic systems provide comprehensive and detailed views of the patient maxillo-mandibular region by reproducing both dental arches on a single image film (Fig. 4).

A panoramic radiograph is acquired by simultaneously rotating the x-ray tube and the film around a single point or axis (*rotation center*). This process, which is known as tomography, allows the sharp imaging of the body regions disposed within a 3D horseshoe shaped volume (*focal trough* or *image layer*) while blurring superimposed structures from other layers. The rotation center changes as the film and x-ray tube are rotated around the patient's head. Location and number of rotation centers influence both size

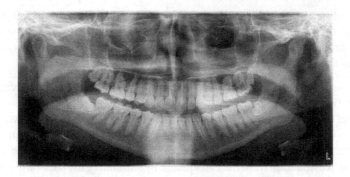

Fig. 4. Example of panoramic (PAN) radiograph.

and shape of the focal through which is therefore designed by manufacturers in order to accommodate the average jaw.

2.3   Methodology

The proposed methodology is based on scaling the tooth CAD template models accordingly to the information included within the patient segmented tooth crowns shape and the PAN image.

Segmented crown models are used to determine the axis of each patient tooth. Teeth templates are then linearly scaled by using non-uniform scale factors along three different dimensions (Fig. 5). In particular, the tooth width (taken along the mesiodistal line) and the tooth depth (taken along the vestibulo-lingual direction) values are directly determined from the patient crown geometries. The tooth height (taken along the vertical direction of the panoramic radiograph) is rather estimated by using the PAN image.

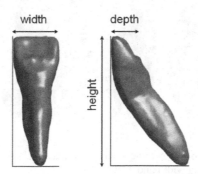

Fig. 5. Tooth dimensions used to scale CAD templates.

The height estimation process, which represents the core of the proposed method, is based on the reconstruction of a synthetic PAN image from the 3D patient crowns geometries. A panoramic radiograph essentially represents the sum of x-ray attenuation along each ray transmitted from the source to the film [6]. The attenuation is due to the x-ray absorption by tissues along the ray. For this reason, it is possible to emulate a panoramic radiograph by taking 2D projections through a data volume. In this paper, the *Discrete Radon Transform* (DRT) is used to calculate finite pixel intensity sums along rays normal to a curve, which approximate the medial axis of the crowns arch.

The whole methodology can be summarized in the following steps:

- Uniform scaling of the complete dental template arch by using the patient digitized cast (Fig. 6a);
- Alignment of each tooth template on the corresponding patient tooth crown geometry in order to determine the orientation and position with respect to the bone structure;
- Non-uniform scaling by using the tooth width and depth values (Fig. 6b);
- Tooth height estimation from the PAN image by simulating the panoramic radiograph process through the *Discrete Radon Transform* applied on the reconstructed patient crowns model.

The first three steps are quite straightforward and can be accomplished by using any CAD software. The last step is fully detailed in the following section.

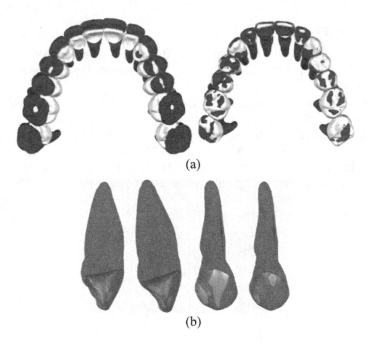

(a)

(b)

Fig. 6. (a) Uniform scaling of the complete dental template arch and (b) two examples of non-uniform tooth scaling by using width and depth values.

Tooth Height Estimation. The 3D patient crowns model must be spatially oriented, by a rigid motion, in order to make its projection consistent with the corresponding crowns region in the PAN radiograph. A set of n corresponding markers $[P_{PAN}^i \equiv (x_{PAN}^i, y_{PAN}^i, z_{PAN}^i), P_{cr}^i \equiv (x_{cr}^i, y_{cr}^i, z_{cr}^i)]$ is interactively selected on crown regions of both PAN image and segmented crowns model. A rigid motion, applied to the 3D model and described by a rotation matrix (R) and a translation vector (T), is then determined by minimizing an objective function defined as:

$$f(R, T) = \sum_{i=1}^{n} \left\| \Delta z_{PAN}^i - \Delta z_{cr}^i \right\|^2 \tag{1}$$

This transformation guarantees the alignment between the 3D patient crowns model and the radiograph along the z-direction (Fig. 7). A further transformation is then required in order to project the 3D model onto the panoramic image. This process is accomplished by computing multiple parallel-beam projections, from different angles, using the DRT. In particular, a 2D image is firstly created by projecting the crowns model onto the X_{cr}-Y_{cr} plane (Fig. 8). A fourth order polynomial curve (γ) is then determined by interpolating the projection of the selected $P_{cr}^i \equiv \left(x_{cr}^i, y_{cr}^i \right)$ points.

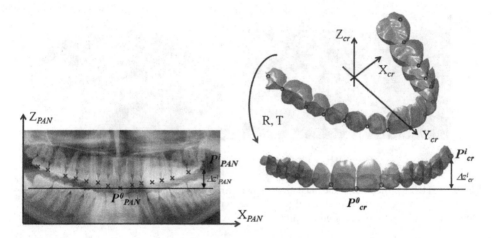

Fig. 7. Alignment between 3D crowns model and PAN image along the z-direction.

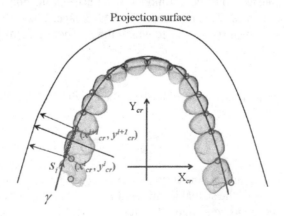

Fig. 8. Projection scheme of the 3D patient crowns model.

The 3D model is vertically sliced with the same vertical resolution of the PAN image. For each horizontal slice, crown contours are projected along the direction normal to γ in correspondence of each curve point by using the DRT (Fig. 8). The curve point sampling (s_i) is piecewisely estimated by matching the number of samples between two consecutive P^i_{cr} points with the number of pixels along the X_{PAN} direction between the corresponding P^i_{PAN} points. Figure 9a shows the DRT results for the projection of the crowns model illustrated in Fig. 3b, while Fig. 9b and c show its superimposition on the original PAN image.

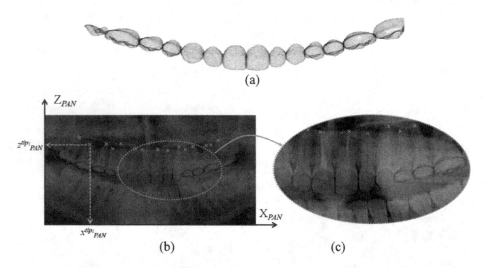

(a)

(b)   (c)

Fig. 9. (a) DRT projection of the 3D patient crowns model, (b) superimposition of the projection on the PAN image along with a detail (c). The crowns model projection is highlighted with a transparent cyan color (Color figure online).

Tooth heights are then extracted from the PAN image by the selection of root tips, which are back-projected onto the 3D model. This back-projection is performed by considering the coordinates of the root tip in the PAN image. The z-coordinate, up to a scale factor, is used to identify the slice to which the 3D root tip belongs:

$$z^{tip_i}_{cr} = scale_z \cdot z^{tip_i}_{PAN} \tag{2}$$

The x-coordinate is instead used to retrieve the curvilinear coordinate along the γ curve by:

$$s^{tip_i} = x^{tip_i}_{PAN} \tag{3}$$

The line normal to γ and passing from s^{tip_i} describes the projection ray through the root tip. It is then possible the spatial identification of a direction on which the 3D root tip must certainly lie (*constraint line*). The template tooth model, already scaled by considering width and depth values can then be finally scaled along the height direction in order to approach the above outlined constraint line. Clearly, an indetermination about the root inclination remains since the tooth root could be indifferently oriented to the buccal or lingual side of dentition. However, the preventive alignment of the tooth template on the patient crown model should guarantee the correct orientation of the final reconstructed tooth.

3   Results

The feasibility of the proposed methodology has been verified by reconstructing some teeth of a female patient superior dental arch. Figure 10 shows two views of the CAD templates aligned and scaled (using tooth width and depth values) on the crowns model, along with the directions on which respective root tips should lie. CAD templates are then further scaled along the tooth heights while tooth axes are oriented in order to intersect the respective constraint lines (Fig. 11). Crown geometries, acquired by the optical scanner, and root geometries, estimated by scaling CAD templates, are then merged together in order to create the final digital tooth model (Fig. 12).

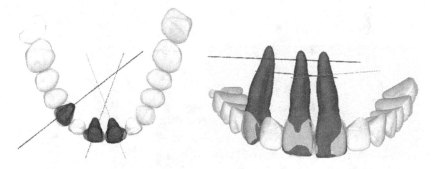

Fig. 10. CAD templates aligned and scaled on the crowns model with the respective constraint lines for root tips.

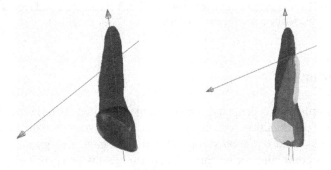

Fig. 11. Final height scaling and orientation (green model) of the tooth CAD template (blue model) (Color figure online).

Fig. 12. Merging process between crown geometry (gray model) and scaled tooth CAD template (green model) (Color figure online).

The reconstructed tooth shapes can be compared to those obtained by processing volumetric data from patient CBCT scans. In this case, segmented tooth geometries from CBCT data (Fig. 13a) can be used as ground truth to assess the accuracy of 3D models reconstructed by using minimally invasive imaging modalities (Fig. 13b, c).

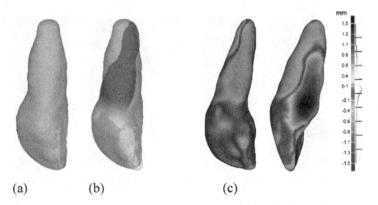

(a)   (b)   (c)

Fig. 13. (a) CBCT tooth ground truth, (b) overlapping between CBCT and reconstructed tooth model, (c) discrepancies between the two models.

4   Conclusions

In the field of orthodontic dentistry, one of the main challenges relies on the accurate determination of 3D dentition geometries by exposing the patient to the minimum radiation dose. In this context, the present paper outlines a methodology to infer 3D shape of tooth roots by combining the patient digital plaster cast with a panoramic radiograph. The method investigates the possibility to adapt general dental CAD templates over the real anatomy by exploiting geometrical information contained within the panoramic image and the digital plaster cast. The proposed modelling approach, which has showed encouraging preliminary results, allows a generalized formulation of the problem since assumptions about the tomographic device used for radiographic data capturing are not required.

Many are the variables involved in the adopted formulation. In particular, key issues are represented by the optimization of the γ curve, whose slope determines the orientation of root tip constraint lines, and the accurate evaluation of magnification factors along the z-direction of the PAN image.

These topics certainly require further research activities taking also into account, for example, additional information that could be extracted by supplementary lateral radiographs.

References

1. Kuncio, D., Maganzini, A., Shelton, C., Freeman, K.: Invisalign and traditional orthodontic treatment postretention outcomes compared using the American Board of Orthodontics Objective Grading System. Angle Orthod. **77**(5), 864–869 (2007)
2. Boyd, R.L.: Complex orthodontic treatment using a new protocol for the Invisalign appliance. J. Clin. Orthod. **41**(9), 525–547 (2007)
3. Pei, Y.R., Shi, F.H., Chen, H., Wei, J., Zha, H.B., Jiang, R.P., Xu, T.M.: Personalized tooth shape estimation from radiograph and cast. IEEE Trans. Bio-Med. Eng. **59**(9), 2400–2411 (2012)
4. Mazzotta, L., Cozzani, M., Razionale, A., Mutinelli, S., Castaldo, A., Silvestrini-Biavati, A.: From 2D to 3D: Construction of a 3D Parametric Model for Detection of Dental Roots Shape and Position from a Panoramic Radiograph - A Preliminary Report. Int. J. Dent. **2013**, 8 (2013)
5. Barone, S., Paoli, A., Razionale, A.: Creation of 3D multi-body orthodontic models by using independent imaging sensors. Sensors **13**(2), 2033–2050 (2013)
6. Tohnak, S., Mehnert, A., Crozier, S., Mahoney, M.: Synthesizing panoramic radiographs by unwrapping dental CT data. In: 28th Annual International Conference of the IEEE Engineering in Medicine and Biology Society, 1, pp. 3329–3332 (2006)

Bioinformatics Models, Methods and Algorithms

Formulation of Homeostasis by Realisability on Linear Temporal Logic

Sohei Ito[1]($\boxtimes$), Shigeki Hagihara[2], and Naoki Yonezaki[2]

[1] National Fisheries University, Shimonoseki, Japan
ito@fish-u.ac.jp
[2] Tokyo Institute of Technology, Tokyo, Japan

Abstract. Toward the system level understanding of the mechanisms contributing homeostasis in organisms, a computational framework to model a system and analyse its properties is indispensable. We propose a novel formalism to model and analyse homeostasis on gene networks. Since gene networks can be considered as *reactive systems* which respond to environmental input, we reduce the problem of analysing gene networks to that of verifying reactive system specifications. Based on this reduction, we formulate homeostasis as *realisability* of reactive system specifications. An advantage of this formulation is that we can consider any input sequence over time and any number of inputs, which are difficult to be captured by quantitative models. We demonstrate the usefulness and flexibility of our framework in analysing a number of small but tricky networks.

Keywords: Gene regulatory network · Systems biology · Homeostasis · Temporal logic · Realisability · Formal method

1 Introduction

Lack of quantitative information such as kinetic parameters or molecular concentrations about biological systems has been a problem in quantitative analysis. Even if we do not know such parameters, we can model and analyse them by using qualitative methods [5,6,9,18], which enables us to analyse qualitative properties of a system such as 'when this gene is expressed, that gene will be suppressed later'.

Ito et al. [13–15] proposed a method for analysing gene networks using linear temporal logic (LTL) [8], in which, a gene network is modelled as an LTL formula which specifies its possible behaviours. Their method for analysing gene networks is closely related with verification of *reactive system* specifications [1,4,12,19,20,23,25]. A reactive system is a system that responds to requests from an environment at an appropriate timing. Systems controlling an elevator or a vending machine are typical examples of reactive systems. Biological systems with external inputs or signals can be naturally considered as reactive systems.

© Springer International Publishing Switzerland 2015
G. Plantier et al. (Eds.): BIOSTEC 2014, CCIS 511, pp. 149–164, 2015.
DOI: 10.1007/978-3-319-26129-4_10

Realisability [1,20] is a desirable property of reactive system specifications which requires systems to behave according to a specification in reaction to any input from an environment. In terms of biological systems, this property means that a system behaves with satisfying a certain property (e.g. keeping a concentration within some range) in reaction to any input from an environment (e.g. for any stress or stimulation). That is to say, the system is *homeostatic* with respect to the property.

Using this correspondence, we formulate the notion of homeostasis by realisability of reactive systems. Our formulation captures homeostasis of not only logical structure of gene networks but also properties of any dynamic behaviours of networks. For example, we can analyse in our framework whether a given network maintains oscillation over time in response to any input sequence. This formulation yields not only a novel and simple characterisation of homeostasis but also provides a method to automatically check homeostasis of a system using realisability checkers [7,10,16,17]. Based on this formulation we analyse some homeostatic properties of a number of small but tricky gene networks.

This paper is organised as follows. Section 2 reviews the method for analysing gene networks using LTL [13–15] on that our work is grounded. In Sect. 3, we introduce the notion of realisability and formulate homeostasis by this notion. Based on this formulation, we show some example networks and analyse homeostatic properties of them in Sect. 4. We also discuss the scalability of our method in the same section. The final section offers conclusions and future directions.

2   Modelling Behaviours of Gene Networks in LTL

2.1   Linear Temporal Logic

First we introduce linear temporal logic.

If A is a finite set, A^ω denotes the set of all infinite sequences on A. The i-th element of $\sigma \in A^\omega$ is denoted by $\sigma[i]$. Let AP be a set of propositions. A *time structure* is a sequence $\sigma \in (2^{AP})^\omega$ where 2^{AP} is the powerset of AP. The formulae in LTL are defined as follows.

– $p \in AP$ is a formula.
– If ϕ and ψ are formulae, then $\neg\phi, \phi \wedge \psi, \phi \vee \psi$ and $\phi U \psi$ are also formulae.

We introduce the following abbreviations: $\bot \equiv p \wedge \neg p$ for some $p \in AP$, $\top \equiv \neg\bot$, $\phi \to \psi \equiv \neg\phi \vee \psi$, $\phi \leftrightarrow \psi \equiv (\phi \to \psi) \wedge (\psi \to \phi)$, $F\phi \equiv \top U\phi$, $G\phi \equiv \neg F\neg\phi$, and $\phi W \psi \equiv (\phi U \psi) \vee G\phi$. We assume that $\wedge, \vee$ and U binds more strongly than $\to$ and unary connectives binds more strongly than binary ones.

Now we define the formal semantics of LTL. Let σ be a time structure and ϕ be a formula. We write $\sigma \models \phi$ for ϕ is true in σ and we say σ *satisfies* ϕ. The satisfaction relation $\models$ is defined as follows.

$$\sigma \models p \quad \text{iff } p \in \sigma[0] \text{ for } p \in AP$$
$$\sigma \models \neg\phi \quad \text{iff } \sigma \not\models \phi$$
$$\sigma \models \phi \wedge \psi \text{ iff } \sigma \models \phi \text{ and } \sigma \models \psi$$
$$\sigma \models \phi \vee \psi \text{ iff } \sigma \models \phi \text{ or } \sigma \models \psi$$
$$\sigma \models \phi U \psi \text{ iff } (\exists i \geq 0)(\sigma^i \models \psi \text{ and } \forall j(0 \leq j < i)\sigma^j \models \phi)$$

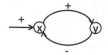

Fig. 1. An example network.

where $\sigma^i = \sigma[i]\sigma[i+1]\ldots$, i.e. the i-th suffix of σ. An LTL formula ϕ is *satisfiable* if there exists a time structure σ such that $\sigma \models \phi$.

Intuitively, $\neg\phi$ means 'ϕ is not true', $\phi \wedge \psi$ means 'both ϕ and ψ are true', $\phi \vee \psi$ means 'ϕ or ψ is true', and $\phi U \psi$ means 'ϕ continues to hold until ψ holds'. $\bot$ is a false proposition and $\top$ is a true proposition. $\phi \rightarrow \psi$ means 'if ϕ is true then ψ is true' and $\phi \leftrightarrow \psi$ means 'ϕ is true if and only if ψ is true'. $F\phi$ means 'ϕ holds at some future time', $G\phi$ means 'ϕ holds globally', $\phi W \psi$ is the 'weak until' operator in that ψ is not obliged to hold, in which case ϕ must always hold.

2.2 Specifying Possible Behaviours of Gene Networks in LTL

Now we review the method proposed in [13–15] to model behaviours of a given network in linear temporal logic, using an example network depicted in Fig. 1.

In this network gene x activates gene y and gene y inhibits gene x. Gene x has a positive environmental input. Let x_y be the threshold of gene x to activate gene y, y_x the threshold of gene y to inhibit gene x and e_x the threshold of the input to activate gene x. To specify possible behaviours of this network, we introduce the following propositions.

- on_x, on_y: whether gene x and y are ON respectively.
- x_y, y_x: whether gene x and y are expressed beyond the threshold x_y and y_x respectively[1].
- in_x: whether the input to x is ON.
- e_x: whether the positive input from the environment to x is beyond the threshold e_x.

The basic principles for characterising behaviours of a gene network are as follows:

- Genes are ON when their activators are expressed beyond some thresholds.
- Genes are OFF when their inhibitors are expressed beyond some thresholds.
- If genes are ON, the concentrations of their products increase.
- If genes are OFF, the concentrations of their products decrease.

We express these principles in LTL using the propositions introduced above.

---

[1] Although the same symbols (i.e. x_y and y_x) are used to represent both thresholds and propositions, we can clearly distinguish them from the context.

Genes' activation and inactivation. Gene y is positively regulated by gene x. Thus gene y is ON if gene x is expressed beyond the threshold x_y, which is the threshold of gene x to activate gene y. This can be described as

$$G(x_y \leftrightarrow on_y)$$

in LTL. Intuitively this formula says gene y is ON if, and only if, gene x is expressed beyond x_y due to positive regulation effect of gene x toward gene y. As for gene x, it is negatively regulated by gene y and has positive input from the environment. A condition for activation and inactivation of such multi-regulated genes depends on a function which merges the multiple effects. We assume that gene x is ON if gene y is not expressed beyond y_x and the input from the environment to gene x is beyond e_x; that is, the negative effect of gene y is not operating and the positive effect of the input is operating. Then this can be described as

$$G(e_x \wedge \neg y_x \rightarrow on_x).$$

This formula says that if the input level is beyond e_x (i.e. proposition e_x is true) and gene y is not expressed beyond y_x (i.e. proposition y_x is false; $\neg y_x$ is true), then gene x is ON (i.e. proposition on_x is true).

For the inactivation of gene x, we have choices to specify the rule. Let us assume that gene x is OFF when the input from the environment is under e_x and gene y is expressed beyond y_x, that is, the activation to gene x is not operating and the inhibition to gene x is operating, in which case gene x will surely be OFF. This is specified as

$$G(\neg e_x \wedge y_x \rightarrow \neg on_x). \tag{1}$$

For another choice, let us assume that the negative effect from gene y overpowers the positive input from the environment. Then we write

$$G(y_x \rightarrow \neg on_x), \tag{2}$$

which says that if the inhibition from gene y is operating, gene x becomes OFF regardless of the environmental input to gene x. Yet another choice is

$$G(\neg e_x \vee y_x \rightarrow \neg on_x) \tag{3}$$

which says that gene x is OFF when the positive input is not effective *or* negative regulation from gene y is effective. For example, although gene y is not expressed beyond the threshold y_x (i.e. the negative effect of gene y is not effective), gene x is OFF if the positive effect of the input is not effective.

We also have several options for the activation of gene x. The choice depends on a situation (or assumption) of a network under consideration.

Changes of expression levels of genes over time. If gene x is ON, it begins to be expressed and in some future it will reach the threshold for gene y unless gene x becomes OFF. This can be described as

$$G(on_x \rightarrow F(x_y \vee \neg on_x)).$$

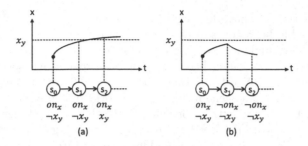

Fig. 2. If gene x is ON, (a) the expression level of gene x is over x_y, or (b) gene x becomes OFF before gene x reaches x_y, where $s_0s_1s_2$c is a time structure.

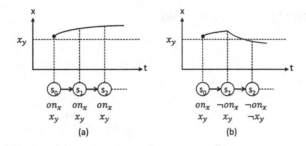

Fig. 3. If gene x is ON and the current expression level is over x_y, (a) the expression level of gene x keeps over x_y, or (b) gene x becomes OFF and as a result the expression level may fall below x_y.

This formula means 'if gene x is ON, in some future the expression level of gene x will be beyond x_y, or otherwise gene x will become off'. This situation is depicted in Fig. 2. If gene x is ON and expressed beyond x_y, it keeps the level until gene x is OFF. This can be described as

$$G(on_x \wedge x_y \rightarrow x_y W \neg on_x).$$

This formula means 'if gene x is ON and the current expression level of gene x is over x_y, gene x keeps its level until gene x becomes OFF, or otherwise gene x keeps its level always'. This situation is depicted in Fig. 3.

If gene x is OFF, its product decreases due to degradation. Thus if gene x is OFF and the current expression level of x is over x_y, it will fall below x_y in some future unless x becomes ON again. This can be specified as

$$G(\neg on_x \rightarrow F(\neg x_y \vee on_x)).$$

If the expression level of gene x is under x_y and x is OFF then it keeps the level (i.e. it does not increase and exceed x_y) until x is ON. This can be specified as

$$G(\neg on_x \wedge \neg x_y \rightarrow \neg x_y W on_x).$$

We have similar formulae for gene y and the input into gene x from the environment for increase and decrease of them.

The conjunction of above formulae (i.e. joining by ∧ operator) is the specification of possible behaviours of the network. In other words, time structures which satisfy the formula are possible behaviours of the network.

This method for modelling behaviours of gene regulatory networks can be contrasted to usual quantitative methods like ordinary differential equation models. We qualitatively model gene regulatory networks by temporal logic formulae instead of quantitative analytical formulae. Note that we have several possible temporal logic specifications for a single network depending on order of threshold values, functions for multi-regulations and how we capture increase and decrease of expression of genes. Interested reader may wish to consult [14,15] for detail.

3 Realisability and Homeostasis

In this section we discuss the connection between reactive systems and gene networks. Based on this connection, we formulate homeostasis of gene networks by realisability of reactive systems.

A *reactive system* is defined as a triple $\langle X, Y, r \rangle$, where X is a set of events caused by the environment, Y is a set of events caused by the system and $r : (2^X)^+ \to 2^Y$ is a reaction function. The set $(2^X)^+$ denotes the set of all finite sequences on subsets of X, that is to say, finite sequences on a set of environmental events. A reaction function determines how the system reacts to environmental input sequences. Reactive system is a natural formalisation of systems which appropriately respond to requests from the environment. Systems controlling vending machines, elevators, air traffic and nuclear power plants are examples of reactive systems. Gene networks which respond to inputs or stimulation from the environment such as glucose increase, change of temperature or blood pressure can also be considered as reactive systems.

A specification of a reactive system stipulates how it responds to inputs from the environment. For example, for a controller of an elevator system, a specification will be e.g. 'if the open button is pushed, the door opens' or 'if a call button of a certain floor is pushed, the lift will come to the floor'. It is important for a specification of a reactive system to satisfy *realisability* [1,20], which requires that there exists a reactive system such that for any environmental inputs of any timing, it produces system events (i.e. responds) so that it satisfies the specification.

To verify a reactive system specification, it should be described in a language with formal and rigorous semantics. Widespread research in specifying and developing reactive systems lead to the belief that temporal logic is the useful tool for reasoning them [1,3,20,24]. LTL is known to be one of many other formal languages suitable for this purpose and several realisability checkers of LTL are available [7,10,16,17].

Now we define the notion of realisability of LTL specifications. Let AP be a set of atomic propositions which is partitioned into X, a set of input propositions, and Y, a set of output propositions. X corresponds to input events and Y to output events. We denote a time structure σ on AP as $\langle x_0, y_0 \rangle \langle x_1, y_1 \rangle \ldots$ where

$x_i \subseteq X$, $y_i \subseteq Y$ and $\sigma[i] = x_i \cup y_i$. Let ϕ be an LTL specification. We say $\langle X, Y, \phi \rangle$ is *realisable* if there exists a reactive system $RS = \langle X, Y, r \rangle$ such that

$$\forall \tilde{x}. behave_{RS}(\tilde{x}) \models \phi,$$

where $\tilde{x} \in (2^X)^\omega$ and $behave_{RS}(\tilde{x})$ is the infinite behaviour determined by RS, that is,

$$behave_{RS}(\tilde{x}) = \langle x_0, y_0 \rangle \langle x_1, y_1 \rangle \ldots,$$

where $\tilde{x} = x_0 x_1 \ldots$ and $y_i = r(x_0 \ldots x_i)$.

Intuitively ϕ is realisable if for any sequence of input events there exists a system which produces output events such that its behaviour satisfies ϕ.

Example 1. Let $X = \{push_{open}, push_{close}\}$ and $Y = \{door_{open}\}$. The specification

$$G(push_{open} \to F door_{open})$$

is realisable since there is a reactive system $\langle X, Y, r \rangle$ with $r(\bar{x}a) = \{door_{open}\}$ where $\bar{x}$ is any finite sequence on 2^X and $a \supseteq \{push_{open}\}$. The specification

$$G((push_{open} \to F door_{open}) \wedge (push_{close} \to \neg door_{open}))$$

is not realisable since for input sequence $\{push_{open}, push_{close}\}^\omega$ there is no output sequence which satisfies the specification.

Realisability can be interpreted as the ability of a system to maintain its internal condition irrespective of environmental inputs. In the context of gene networks, realisability can be naturally interpreted as *homeostasis*. For example, a network for controlling glucose level responds to an environmental inputs such as glucose increase or decrease in a manner to maintain its glucose level within a normal range. In the framework described in Sect. 2, behaviour specifications of gene networks can be regarded as reactive system specifications. Based on this connection, we formulate homeostasis by realisability.

Let $\langle I, O, \phi \rangle$ be a behaviour specification of a gene network where I is the set of input propositions, O is the set of output propositions and ϕ is an LTL formula characterising possible behaviours of the network. Let ψ be a certain biological property of the network. A network property ψ is homeostatic in this network if for any input sequence $x_0 x_1 \ldots$ there exists a reaction function r such that the behaviour $\sigma = \langle x_0, r(x_0) \rangle \langle x_1, r(x_0, x_1) \rangle \ldots$ is a behaviour of the network (i.e. $\sigma \models \phi$) and σ also satisfies the property ψ (i.e. $\sigma \models \psi$). Thus we have the following simple definition of homeostasis:

Definition 1. *A property ψ is* homeostatic *with respect to a behaviour specification $\langle I, O, \phi \rangle$ if $\langle I, O, \phi \wedge \psi \rangle$ is realisable.*

In this definition we consider responses of a system not only to initial instantaneous inputs such as dose-response relationship but also to any input sequences (e.g. inputs are oscillating or sustained), which is difficult to be captured by ordinary differential models. Moreover, we have any number of environmental inputs thus we can consider homeostasis against compositive environmental inputs.

Based on the method described in Sect. 2 and this formulation, we can analyse homeostasis of gene networks using realisability checkers. In the next section, we demonstrate our method in analysing a number of small but tricky networks.

4   Demonstration: Analysis of Homeostasis for Example Networks

First we consider the network in Fig. 1 again. The network in Fig. 1 is expected to have a function that whenever gene x becomes ON, the expression of gene x will be suppressed afterward. This function maintains the expression level of gene x to its normal range (low level). Despite of the extreme situation that the input to gene x is always ON, the expression of gene x inevitably ceases due to the activation of gene y and its negative effect on gene x. Therefore this function is expected to be homeostatic. Now we formalise this verbal and informal reasoning with our framework. The property 'whenever gene x becomes ON, the expression of gene x will be suppressed afterward' is formally stated in LTL as:

$$G(on_x \rightarrow F(\neg on_x \wedge \neg x_y)). \tag{4}$$

This formula says that the property 'if on_x is true, it becomes false and gene x is suppressed below x_y in some future' always holds. We check whether this formula is realisable with respect to a behaviour specification introduced in Sect. 2.2. There are 6 propositions $on_x, on_y, x_y, y_x, in_x$ and e_x for this network. The partition of input propositions and output propositions are straightforward, that is, in_x is the only input proposition since the environment only controls the input to gene x. Other propositions represent internal states of the network. Note that the environment cannot directly control the proposition e_x, which represents whether the level of the input exceeds e_x. To exceed the level e_x, the environment needs to give the input for a certain duration.

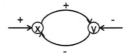

Fig. 4. The network in Fig. 1 with a negative input for gene y.

We had three options in the inactivation rule of gene x, i.e. formulae (1), (2) and (3). In all choices the property is realisable since even if in_x is always true, we need y_x being false for the activation of gene x due to the clause $G(e_x \wedge \neg y_x \rightarrow on_x)$. If in_x is always true, gene x will be expressed beyond x_y, and it induces y's expression. As a result, gene y can be expressed beyond y_x at which gene y inhibits gene x. Thus on_x may not be always true. If we replace the clause for the activation of gene x as $G(e_x \rightarrow on_x)$, which says if the input is effective gene

x must be ON regardless of the negative effect of gene y, then the property is not realisable.

For realisability checking, we used Lily[2] [16] which is a tool for checking realisability of LTL specifications. To use Lily, we specify input propositions, output propositions and an LTL formula. The result of checking (Yes or No) is output to command-line and if it is YES, it also outputs a state diagram.

Now we assume gene y accepts negative input from the environment (Fig. 4). We have the extra input proposition in_y and output proposition e_y. We describe the activation rule for gene y as follows in which gene y can be OFF by the negative input from the environment:

$$G(\neg e_y \wedge x_y \rightarrow on_y),$$
$$G(e_y \rightarrow \neg on_y).$$

Is the property (4) homeostatic with respect to this behaviour specification? The realisability checker answers 'No'. The reason is that if the input for gene y is always ON, gene y cannot be ON, therefore the negative effect from gene y to gene x cannot be effective. In this input scenario gene x cannot become OFF after gene x becomes ON.

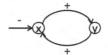

Fig. 5. A bistable switch.

Now we consider the next example depicted in Fig. 5. In this network we provide two thresholds y_x^0 and y_x^1 for gene y. The threshold y_x^0 is the level enough to activate gene x when the negative input from the environment is not effective. The threshold y_x^1 is the level enough to activate gene x regardless of negative effect from the environment, that is, y_x^1 is the threshold beyond which gene y overpowers the environmental input. The behaviour specification for this network will be somewhat complicated. First, we describe the fact that the threshold y_x^1 is greater than y_x^0, which is simply described as follows:

$$G(y_x^1 \rightarrow y_x^0),$$

which says 'if gene y is expressed beyond y_x^1, it is also beyond y_x^0 (since $y_x^1 > y_x^0$)'. Note that the proposition y_x^1 means 'gene y is expressed *beyond* the threshold y_x^1'. The activation rules and inactivation rules for gene x are as follows:

$$G(\neg y_x^0 \rightarrow \neg on_x), \tag{5}$$
$$G(e_x \wedge y_x^0 \wedge \neg y_x^1 \rightarrow \neg on_x), \tag{6}$$
$$G(\neg e_x \wedge y_x^0 \rightarrow on_x), \tag{7}$$
$$G(y_x^1 \rightarrow on_x). \tag{8}$$

[2] http://www.iaik.tugraz.at/content/research/design_verification/lily/.

Formula (5) says that if gene y is under y_x^0, gene x is OFF regardless of the environmental input. Formula (6) says that if gene y is in between y_x^0 and y_x^1 but the negative input is effective, gene x is OFF. Formula (7) says that gene x is ON when negative input is not effective and gene y is expressed over y_x^0. Formula (8) says that gene x is ON when gene y is just expressed over y_x^1.

The activation rule for gene y is simple:

$$G(x_y \leftrightarrow on_y)$$

The change of the expression level of gene y when it is ON are described as follows:

$$G(on_y \to F(y_x^0 \vee \neg on_y)), \tag{9}$$
$$G(on_y \wedge y_x^0 \to y_x^0 W \neg on_y), \tag{10}$$
$$G(on_y \wedge y_x^1 \to y_x^1 W \neg on_y). \tag{11}$$

Formula (9) says that if gene y is ON, it will reach the first threshold y_x^0 or otherwise it will become OFF. Formula (10) says that if gene y is ON and the current level is over the first threshold y_x^0, it will keep over y_x^0 (this means it can be expressed beyond y_x^1), or otherwise gene y becomes OFF. Formula (11) says that if gene y is ON and the current level is over the highest threshold y_x^1, it keeps y_x^1 or otherwise it will become OFF.

We have similar formulae for the change of the expression level of gene y when it is OFF.

$$G(\neg on_y \to F(\neg y_x^1 \vee on_y)),$$
$$G(\neg on_y \wedge \neg y_x^1 \to \neg y_x^1 W \ on_y),$$
$$G(\neg on_y \wedge \neg y_x^0 \to \neg y_x^0 W \ on_y).$$

For the change of the expression level of gene x and the environmental input we have similar formulae except they have only one threshold.

We check the bistability of the expression of gene x, that is to say, if gene x can always be ON or always be OFF. These properties are described as follows:

$$Gon_x, \tag{12}$$
$$G\neg on_x. \tag{13}$$

By using Lily, we checked that both properties are really homeostatic. Informal reasoning for the first property (12) is as follows. Suppose that the input sequence such that the negative input to x is always effective, which is the best choice for the environment to inactivate gene x. The system's response to satisfy the bistability is to start at a state in which both gene x and y are ON and gene x and gene y are expressed beyond x_y and y_x^1, respectively. Since gene y is expressed beyond y_x^1, gene x can continue to be ON regardless of negative input to x. The expression of gene y is supported by the positive effect from gene x. For the second property (13), we assume that the negative input is always ineffective.

The system's response is simply to start a state that both gene x and y is OFF and gene x and y are expressed below x_y and y_x^0, respectively. For x to be ON, we need y_x^0 being true but the system can control gene y to be OFF since gene x is OFF.

We expect both gene x and y are either ON or OFF simultaneously. This can be checked by the following properties:

$$Gon_x \wedge Gon_y, \tag{14}$$

$$G\neg on_x \wedge G\neg on_y. \tag{15}$$

Both properties are really homeostatic. Therefore gene x and gene y are 'interlocked' in a sense.

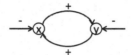

Fig. 6. A bistable switch with a negative input to gene y.

We further investigate this 'interlocking' property. Can gene x (and gene y) always be ON by its own? That is to say, are the following properties homeostatic?

$$Gon_x \wedge G\neg on_y, \tag{16}$$

$$G\neg on_x \wedge Gon_y. \tag{17}$$

The answers are 'No' for both properties. To keep gene x being ON gene x must be expressed beyond y_x and this prevents gene y to be always OFF. Thus the property (16) is not homeostatic. This property is even not satisfiable. That is to say, there is no input sequence to satisfy the property (16). Conversely, to keep gene y being ON gene x must be expressed beyond x_y and this prevents gene x to be always OFF. Thus the property (17) is not homeostatic and not satisfiable too.

Interestingly, provided gene y accepts a negative input from the environment (Fig. 6), the properties (12) and (13) are still homeostatic. Even if both negative inputs are always effective, each gene can be expressed thanks to the positive effect from the other gene. We confirmed the properties (12) and (13) are really homeostatic with respect to the following behaviour specification in which we have two thresholds for gene x (only activation rules for gene y are shown):

$$G(\neg x_y^0 \rightarrow \neg on_y),$$
$$G(e_y \wedge x_y^0 \wedge \neg x_y^1 \rightarrow \neg on_y),$$
$$G(\neg e_y \wedge x_y^0 \rightarrow on_y),$$
$$G(x_y^1 \rightarrow on_y).$$

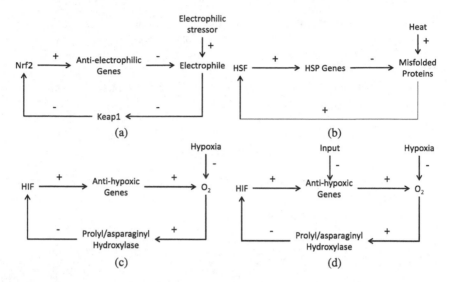

Fig. 7. Schematic representations of anti-stress gene regulatory networks that meditate (a) electrophilic stress response, (b) heat shock response, (c) hypoxic response and (d) the network (c) with hypothetic negative inputs.

Moreover, the properties (14) and (15) are still homeostatic. The properties (16) and (17) are also not homeostatic but are satisfiable in contrast to the previous case since if the environment appropriately controls the inputs, gene y can be ON and OFF alternately but keeps the expression level beyond y_x^0 so that gene x can be ON indefinitely.

The last examples are anti-stress networks [26] depicted in Fig. 7. The networks in Fig. 7 control the upper right objects to keep them within the tolerable ranges. Though these networks are schematic, we are just interested in the control mechanisms which contribute homeostasis against environmental stresses. Let us consider the network of Fig. 7(c). If the amount of O_2 becomes low, the network tries to recover the level of O_2. The property can be described as follows:

$$G(\neg o_2 \rightarrow F o_2)$$

In this formula we interpret proposition o_2 as 'the amount of O_2 is within the tolerable range' so $\neg o_2$ means it deviates the tolerable range. The behaviour specification of the network is obtained as usual[3]. We checked the property is really homeostatic.

Now we have a question: is this homeostatic function broken by the assumption that anti-hypoxic genes receive environmental negative input? (Fig. 7(d)) To check this hypothesis, we modified the behaviour specification in which anti-hypoxic genes receive a negative input from the environment. The activation rule for anti-hypoxic genes is modified considering the negative input. The result of

---

[3] We have 'on' propositions for each node and threshold propositions for each edge.

Table 1. Time of example analyses. Columns 'Input props' and 'Output props' respectively show the numbers of input propositions and output propositions in a formula. Column 'Size of formula' shows the size (length) of a formula (i.e. number of connectives and propositions). Here we chose formula (17) for both network Figs. 5 and 6.

Network	Input props	Output props	Size of formula	Realisable	Time(s)
Figure 1	1	5	164	YES	1.45
Figure 4	2	6	189	NO	13.47
Figure 5	1	6	185	NO	2.03
Figure 6	2	8	275	NO	201.35
Figure 7(a)	1	8	192	YES	260.10
Figure 7(b)	1	6	149	YES	2.55
Figure 7(c)	1	8	192	YES	226.73
Figure 7(d)	2	9	201	NO	226.43

realisability checking was 'No'. This analysis indicates that the homeostasis of this network may be broken by some environmental factor which hinders the operations of anti-hypoxic genes. Such analysis is difficult by observing dose-response relationship based on ordinary differential models.

The homeostatic properties for other two networks are similarly checked. Basic network topologies are almost the same and the modification of network specifications are minor.

As for the scalability of our method, since we use realisability checking whose complexity is double-exponential for the size of formula, we need some technique to mitigate this difficulty. For reference, we show the results of analyses we have employed in Table 1. These experiments are performed on a computer with Intel Core i7-3770 3.40 GHz CPU and 8 GB memory. The realisability checker used is Lily [16].

As for anti-stress networks (Fig. 7(a), (b), (c) and (d)), we used simplified specifications in that we do not introduce the threshold for inputs. This amounts to consider that the influence of input does not have time lag. The reason we simplified specifications is, if we do not so, the analyses take much time, e.g. the analysis of network in Fig. 7(c) takes about 47 min and network in Fig. 7(d) over 1 h (time out set by Lily). These results prompt us to develop an efficient way to analyse homeostasis in our framework.

5   Conclusion

In this paper we formulated the notion of homeostasis in gene regulatory networks by realisability in reactive systems. This formulation allows the automatic analysis of homeostasis of gene regulatory networks using realisability checkers. We analysed several networks with our method. In the analyses we can easily 'tweak' a network (such as appending extra-inputs from the environment) and observed whether the homeostatic properties can be maintained. Such flexibility in analysing networks is an advantage of our framework in the situation that we do not have the definite network topologies. To test several hypothetic networks, our method is more suitable than quantitative approaches using ordinary differential equation models.

There are several interesting future directions based on this work. First is to find more interesting applications in real biological examples. In association with this topic, we are interested in 'conditional' homeostasis which means that under certain constraints on input sequences, a property is homeostatic. This can be easily formulated as follows. Let I and O be the input and output propositions respectively. Let $\langle I, O, \phi \rangle$ be a behavioural specification and ψ be a property. Let σ be an assumption about input sequences e.g. 'inputs to gene x and gene y come infinitely often but not simultaneously'. Then the property ψ is *conditionally homeostatic* with respect to $\langle I, O, \phi \rangle$ under a condition σ if $\langle I, O, \sigma \rightarrow \phi \wedge \psi \rangle$ is realisable. The motivation of this definition is that in more realistic situation it is too strong to require a system to respond to *any* input sequence.

The next topic is to develop a method to suggest how we modify the model of a network when an expected or observed property is not homeostatic in a model. This problem is closely related to refinement of reactive system specifications [2,11]. We hope the techniques developed so far for verification of reactive systems can be imported to analysis of gene networks.

Another important future work is to develop a method to overcome high complexity in checking realisability of LTL formulae. The complexity of realisability checking is doubly exponential in the length of the given specification [20]. Thus it is intractable to directly apply our method to large networks. To circumvent this theoretical limitation we are interested in some approximate analysis method [14] or modular analysis method [13] in which a network is divided into several subnetworks and analyse them individually.

The last topic is to extend our method with some quantitative temporal logic (e.g. probability or real time) [21,22] to enable quantitative analysis.

References

1. Abadi, M., Lamport, L., Wolper, P.: Realizable and unrealizable specifications of reactive systems. In: Ronchi Della Rocca, S., Ausiello, G., Dezani-Ciancaglini, M. (eds.) ICALP 1989. LNCS, vol. 372, pp. 1–17. Springer, Heidelberg (1989)
2. Aoshima, T., Sakuma, K., Yonezaki, N.: An efficient verification procedure supporting evolution of reactive system specifications. In: Proceedings of the 4th International Workshop on Principles of Software Evolution, IWPSE 2001, pp. 182–185. ACM, New York (2001)
3. Barringer, H.: Up and down the temporal way. Comput. J. **30**(2), 134–148 (1987)
4. Barringer, H., Kuiper, R., Pnueli, A.: Now you may compose temporal logic specifications. In: Proceedings of the Sixteenth Annual ACM Symposium on Theory of Computing, STOC 1984, pp. 51–63. ACM, New York (1984)

5. Batt, G., Ropers, D., de Jong, H., Geiselmann, J., Mateescu, R., Page, M., Schneider, D.: Validation of qualitative models of genetic regulatory networks by model checking: analysis of the nutritional stress response in Escherichia coli. Bioinformatics **21**(suppl. 1), i19–i28 (2005)
6. Bernot, G., Comet, J., Richard, A., Guespin, J.: Application of formal methods to biological regulatory networks: extending Thomas' asynchronous logical approach with temporal logic. J. Theor. Biol. **229**(3), 339–347 (2004)
7. Bloem, R., Cimatti, A., Greimel, K., Hofferek, G., Könighofer, R., Roveri, M., Schuppan, V., Seeber, R.: RATSY – a new requirements analysis tool with synthesis. In: Touili, T., Cook, B., Jackson, P. (eds.) CAV 2010. LNCS, vol. 6174, pp. 425–429. Springer, Heidelberg (2010)
8. Emerson, E.A.: Temporal and modal logic. In: Handbook of Theoretical Computer Science. Formal Models and Sematics (B), vol. B, pp. 995–1072. MIT Press, Cambridge (1990)
9. Fages, F., Soliman, S., Chabrier-Rivier, N.: Modelling and querying interaction networks in the biochemical abstract machine BIOCHAM. J. Biol. Phys. Chem. **4**, 64–73 (2004)
10. Filiot, E., Jin, N., Raskin, J.-F.: An antichain algorithm for LTL realizability. In: Bouajjani, A., Maler, O. (eds.) CAV 2009. LNCS, vol. 5643, pp. 263–277. Springer, Heidelberg (2009)
11. Hagihara, S., Kitamura, Y., Shimakawa, M., Yonezaki, N.: Extracting environmental constraints to make reactive system specifications realizable. In: Proceedings of the 2009 16th Asia-Pacific Software Engineering Conference, APSEC 2009, pp. 61–68. IEEE Computer Society, Washington, DC (2009)
12. Hagihara, S., Yonezaki, N.: Completeness of verification methods for approaching to realizable reactive specifications. In: Completeness of Verification Methods for Approaching to Realizable Reactive Specifications, vol. 348, pp. 242–257 (2006)
13. Ito, S., Ichinose, T., Shimakawa, M., Izumi, N., Hagihara, S., Yonezaki, N.: Modular analysis of gene networks by linear temporal logic. J. Integr. Bioinform. **10**(2), 216 (2013)
14. Ito, S., Ichinose, T., Shimakawa, M., Izumi, N., Hagihara, S., Yonezaki, N.: Qualitative analysis of gene regulatory networks using network motifs. In: Proceedings of the 4th International Conference on Bioinformatics Models, Methods and Algorithms (BIOINFORMATICS 2013), pp. 15–24 (2013)
15. Ito, S., Izumi, N., Hagihara, S., Yonezaki, N.: Qualitative analysis of gene regulatory networks by satisfiability checking of linear temporal logic. In: Proceedings of the 10th IEEE International Conference on Bioinformatics & Bioengineering, pp. 232–237 (2010)
16. Jobstmann, B., Bloem, R.: Optimizations for LTL synthesis. In: Proceedings of the Formal Methods in Computer Aided Design, FMCAD 2006, pp. 117–124. IEEE Computer Society, Washington, DC (2006)
17. Jobstmann, B., Galler, S., Weiglhofer, M., Bloem, R.: Anzu: a tool for property synthesis. In: Damm, W., Hermanns, H. (eds.) CAV 2007. LNCS, vol. 4590, pp. 258–262. Springer, Heidelberg (2007)
18. de Jong, H., Geiselmann, J., Hernandez, G., Page, M.: Genetic network analyzer: qualitative simulation of genetic regulatory networks. Bioinformatics **19**(3), 336–344 (2003)
19. Mori, R., Yonezaki, N.: Several realizability concepts in reactive objects. In: Information Modeling and Knowledge Bases IV, pp. 407–424 (1993)

20. Pnueli, A., Rosner, R.: On the synthesis of a reactive module. In: POPL 1989: Proceedings of the 16th ACM SIGPLAN-SIGACT Symposium on Principles of Programming Languages, pp. 179–190. ACM, New York (1989)
21. Tomita, T., Hagihara, S., Yonezaki, N.: A probabilistic temporal logic with frequency operators and its model checking. In: INFINITY. EPTCS, pp. 79–93 (2011)
22. Tomita, T., Hiura, S., Hagihara, S., Yonezaki, N.: A temporal logic with mean-payoff constraints. In: Aoki, T., Taguchi, K. (eds.) ICFEM 2012. LNCS, vol. 7635, pp. 249–265. Springer, Heidelberg (2012)
23. Vanitha, V., Yamashita, K., Fukuzawa, K., Yonezaki., N.: A method for structural-isation of evolutional specifications of reactive systems. In: ICSE 2000, The Third International Workshop on Intelligent Software Engineering (WISE3), pp. 30–38 (2000)
24. Vardi, M.Y.: An automata-theoretic approach to fair realizability and synthesis. In: Wolper, P. (ed.) CAV 1995. LNCS, vol. 939, pp. 267–278. Springer, Heidelberg (1995)
25. Wong-Toi, H., Dill, D.L.: Synthesizing processes and schedulers from temporal specifications. In: Larsen, K.G., Skou, A. (eds.) CAV 1991. LNCS, vol. 575, pp. 272–281. Springer, Heidelberg (1992)
26. Zhang, Q., Andersen, M.E.: Dose response relationship in anti-stress gene regulatory networks. PLoS Comput. Biol. **3**(3), e24 (2007)

Interaction-Based Aggregation of mRNA and miRNA Expression Profiles to Differentiate Myelodysplastic Syndrome

Jiří Kléma[1](✉), Jan Zahálka[1], Michael Anděl[1], and Zdeněk Krejčík[2]

[1] Department of Computer Science and Engineering, Czech Technical University,
Technická 2, Prague, Czech Republic
{klema,andelmi2}@fel.cvut.cz, zahalka.j@gmail.com
http://ida.felk.cvut.cz
[2] Department of Molecular Genetics, Institute of Hematology and Blood Transfusion,
U Nemocnice, Prague, Czech Republic
zdenek.krejcik@uhkt.cz

Abstract. In this work we integrate conventional mRNA expression profiles with miRNA expressions using the knowledge of their validated or predicted interactions in order to improve class prediction in genetically determined diseases. The raw mRNA and miRNA expression features become enriched or replaced by new aggregated features that model the mRNA-miRNA interaction. The proposed subtractive integration method is directly motivated by the inhibition/degradation models of gene expression regulation. The method aggregates mRNA and miRNA expressions by subtracting a proportion of miRNA expression values from their respective target mRNAs. Further, its modification based on singular value decomposition that enables different subtractive weights for different miRNAs is introduced. Both the methods are used to model the outcome or development of myelodysplastic syndrome, a blood cell production disease often progressing to leukemia. The reached results demonstrate that the integration improves classification performance when dealing with mRNA and miRNA features of comparable significance. The proposed methods are available as a part of the web tool miXGENE.

Keywords: Gene expression · Machine learning · microRNA · Classification · Prior knowledge · Myelodysplastic syndrome

1 Introduction

Onset and progression of *myelodysplastic syndrome*, like other genetically determined diseases, depend on the overall activity of copious genes during their e xpression process. Current progress in microarray technologies [2] and RNA sequencing [19] enables affordable measurement of wide-scale gene activity, but only on the transcriptome level. Further levels of the gene expression (GE) process which prove disease, whether proteome or even phenome, are still difficult

© Springer International Publishing Switzerland 2015
G. Plantier et al. (Eds.): BIOSTEC 2014, CCIS 511, pp. 165–180, 2015.
DOI: 10.1007/978-3-319-26129-4_11

to capture. Henceforth, many natural learning tasks, such as disease diagnosis or classification, become non-trivial within current generation GE data. However, GE is a complex process with multiple phases, components, and regulatory mechanisms. Sensing GE at certain points of these phases and integrating the measurements with the aid of recent knowledge about subduing mechanisms may show the GE process in a broader, systematic view, and make the analysis comprehensible, robust and potentially more accurate.

The goal of our work is to integrate conventional GE data sources as mRNA profiles with microRNA measurements and to experimentally evaluate the merit of using the integrated data for class prediction. MicroRNAs (miRNAs) [16] serve as one component of complex machinery which eukaryotic organisms use to regulate gene expression and protein synthesis. Since their discovery, miRNA have shown to play crucial role in development and various pathologies [3,23]. They are short (21 nucleotides) noncoding RNA sequences which mediate post-transcriptional repression of mRNA via multiprotein complex called RISC complex (RNA-induced silencing complex) where miRNA serve as a template for recognizing complementary mRNA. The complementarity of miRNA-mRNA binding initiates one of the two possible mechanisms: the complete homology triggers degradation of target mRNA, whereas a partial complementarity leads to inhibition of translation of target mRNA. However, despite the progress in understanding the underlying mechanisms in recent years, the effects of miRNA on gene expression is still a developing field and many important facts about mechanism of action and possible interactions remain still unclear [7]. The level of expression of particular miRNAs can be measured by (e.g.) miRNA microarrays, analogically to well-known mRNA profiling. The resulting dataset, called the miRNA expression profile, contains, similarly to mRNA profiles, tissue samples as data instances; only this time the attributes are individual miRNA sequences. Integrating mRNA and miRNA data sources may provide a better picture about the true protein amount synthesized according to respective genes, regarding the mechanisms of disease occurrence.

We propose integration stemming from the knowledge which miRNA targets which mRNA. Target prediction is a topic of active research [26]. The most reliable form of target prediction is experimental *in vitro* validation. Complementary *in silico* target prediction offers more miRNA targets with a higher false detection rate. The predictive algorithms either work based directly on molecular biological theory, building the relationship based on miRNA/mRNA structure and properties [5,17], or be data-driven; i.e., determining targets empirically using statistical or machine learning methods on as much data as possible [14,29]. As an example of algorithms of the first class we should mention miRanda, as an extension of the Smith-Waterman algorithm [24], miRWalk [5] and TargetScan [17]; as to those of the second class refer to miRTarget2 [29] or PicTar 5 [14]. Target prediction algorithms usually output a score, which for a particular mRNA and a particular miRNA quantifies the strength of the belief that the two are truly related. While there is no guarantee that the results are truly correct, employing prediction algorithms on already existing gene/miRNA expression profiles is cheap and, with the possibility of thresholding the score, one can express confidence in the results, possibly eliminating fluke results.

Despite the above-mentioned problems in target prediction, the main challenge in mRNA and miRNA data integration is different. The relationship between miRNAs and mRNAs is many-to-many, a miRNA binds to different mRNAs, while an mRNA molecule hosts binding sites for different miRNAs. Moreover, the binding can be, and often is, imperfect; with a miRNA binding only partly to its target site. One miRNA can, in addition, potentially bind to multiple locations on one mRNA. Due to all of these aspects and the fact that the mRNA-miRNA interaction itself is far from being fully understood, mRNA-miRNA data integration is a non-trivial task. Simply merging mRNA and miRNA probesets [15] may increase current difficulties in GE classification, such as overfitting caused by the immense number of features. Hence, a smart method of reasonable integrating miRNA and mRNA features is desired.

The authors of [10] present an interesting tool for inferring a disease specific miRNA-gene regulatory network based on prior knowledge and user data (miRNA and mRNA profiles). However, this approach does not address the method of breaking down the large inferred network into smaller regulatory units, which are essential for subsequent classification. The method of *data specific* identification of miRNA-gene regulatory modules is proposed in [20,27], where the modules are searched as maximal bi-cliques or induced as decision rules respectively. But none of these methods offer an intuitive way to *express* the identified modules within the sample set. Contrariwise, [12] provides a black box integration procedure for several data sources like mRNAs, miRNAs, methylation data etc., with an immediate classification output. Nevertheless, this method contains no natural interpretation of the learned predictive models, which is unsuitable for an expert decision-making tool.

In this work, we propose a novel feature extraction and data integration method for the accurate and interpretable classification of biological samples based on their mRNA and miRNA expression profiles. The main idea is to use the knowledge of miRNA targets and better approximate the actual protein amount synthesized in the sample. The raw mRNA and miRNA expression features become enriched or replaced by new aggregated features that model the mRNA-miRNA regulation instead. The sample profile presumably gets closer to the phenotype being predicted. The proposed subtractive aggregation method directly implements a simple mRNA-miRNA interaction model in which mRNA expression is modified using the expression of its targeting miRNAs. A similar approach has already been demonstrated in [1], where we employ matrix factorization proposed in [31] instead. In comparison to the subtractive method under study, the matrix-factorization approach leaves room for developing features corresponding to larger functional co-modules, but it could overfit training data when dealing with a small number of samples.

The method widely used for analyzing associations between two heterogeneous genome-wide measurements acquired on the same cohort is canonical correlation analysis (CCA) [21,25,30]. CCA is applicable for mRNA and miRNA expression integration. However, CCA is based purely on mutual correlation between distinct feature sets and disregards prior knowledge of their interaction. It rather aims to describe or simplify the underlying data, while we focus on prediction of the decrease of respective protein level due to inhibition that does not primarily manifest in correlation. In [18] the authors model heterogeneous genomic data by the means of sparse regression. The method explains mRNA matrix through decomposition into miRNA expression, copy number value and DNA methylation matrices. It follows similar descriptive goals as CCA.

Incentive for our method design comes from probe sessions performed on patients with myelodysplastic syndrome (MDS). MDS is a heterogeneous group of clonal hematological diseases characterized by ineffective hematopoiesis originating from hemato-poietic stem cells [28]. Patients with MDS usually develop severe anemia (or other cytopenias) and require frequent blood transfusion. MDS is also characterized by a high risk of transformation into secondary acute myeloid leukemia, and thus could serve as a model for the research of leukemic transformation.

Of the different cytogenetic abnormalities found in MDS, deletion of the long arm of chromosome 5 (del(5q)) is the most common aberration. MDS with isolated del(5q) exhibits a distinct clinical profile and a favorable outcome. Lenalidomide is a relatively new and potent immunomodulatory drug for the treatment of patients with transfusion-dependent MDS with del(5q). It has pleiotropic biologic effects, including a selective cytotoxic effect on del(5q) myelodysplastic clones. As miRNAs serve as key regulators of many cellular processes including hematopoiesis, a number of miRNAs have been also implicated in the pathophysiology of MDS [4,22].

The paper is organized as follows. Section 2 describes the proposed subtractive method (SubAgg) including its SVD-based modification (SVDAgg) that enables different subtractive weights for different miRNAs. Section 3 describes the MDS domain, defines the learning tasks and summarizes the experimental protocol. Section 4 provides experimental results. Section 5 analyzes the intermediate results to deeper understand the functioning of the proposed methods. Section 6 concludes the paper.

2   Materials and Methods

This section covers the procedures proposed for the integration of mRNA and miRNA data. First, inputs required for correct functionality of the methods are defined in Sect. 2.1. Then dataset merge, a simple integration technique serving as a benchmark, is presented in Sect. 2.2. The new integration method, subtractive aggregation is presented in Sect. 2.3. In Sect. 2.4 we introduce another integration method that can be perceived as an extension of subtractive aggregation that learns subtractive weights for different miRNAs by the singular value decomposition. Section 2.5 gives more details about availability of the proposed methods.

2.1 Inputs

The integration method requires two datasets; one containing mRNA measurements, and one containing miRNA measurements. Those two datasets must be matched; i.e., both must contain samples taken from the same patients and the same tissue types.

Let $\mathcal{G} = \{g_1, ..., g_n\}$ be the genes, $\mathcal{R} = \{r_1, ..., r_m\}$ be the miRNAs and $\mathcal{S} = \{S_1, ..., S_s\}$ be the interrogated samples (tissues, patients, experiments). Then $x^G : \mathcal{G} \times \mathcal{S} \to \mathbb{R}$ is the amount of respective mRNA measured by mRNA chip in particular samples, and $x^\mu : \mathcal{R} \times \mathcal{S} \to \mathbb{R}$ is the expression profile of known miRNA sequences; i.e., the amount of respective molecules measured by the miRNA chip within the samples.

For further reference, the mRNA dataset will be denoted as an $s \times n$ matrix $\mathbf{X}^G$, with s samples and n genes. Similarly, the miRNA dataset will be referred to as an $s \times m$ matrix $\mathbf{X}^\mu$, with m miRNAs. Henceforth, column vectors of the two data matrices, $\left\{\mathbf{x}_i^G\right\}_{i=1}^n$ and $\left\{\mathbf{x}_i^\mu\right\}_{i=1}^m$, represent measured expression of particular genes and miRNAs, respectively.

The integration method requires information pertaining to which miRNA targets which mRNA. The known miRNA-gene control system is represented by binary relation $\mathcal{T} \subset \mathcal{R} \times \mathcal{G}$.

2.2 Dataset Merge

The most straightforward method of obtaining integrated mRNA and miRNA data is merging the two respective datasets. This method, as mentioned above, was presented by [15] and is included in our experimental evaluation as a benchmark. The resulting *merged* dataset simply contains column-wise concatenated mRNA and miRNA data matrices. The advantages of this integration approach are no required prior knowledge of targets and computational efficiency. Excluding prior knowledge of targets, however, means that the target relationships are to be induced empirically by the classifier itself. The question remains as to whether the classifier is capable of doing that. Also, this approach increases the already-high number of features.

2.3 Subtractive Aggregation (SubAgg)

Due to the fact, that many aspects of miRNA-mRNA interactions are not yet fully understood and remain unclear, we were forced to involve several simplifying assumptions as follows: (1) miRNA effect is strictly subtractive, (2) the measured miRNA amount is proportionally distributed among its targets, and (3) the mRNA inhibition rate is proportional to the amount of available targeting miRNA.

The method aggregates mRNA and miRNA values by subtracting a proportion of miRNA expression values from their respective target mRNAs. At the same time, it minimizes the number of parameters needed to be learned to 1.

This characteristic complies with the inconvenient sample set size and the feature set size rate.

Each gene, or rather its mRNA transcript, $g \in \mathcal{G}$ has a defined set of miRNAs which target it, $\mathcal{R}_g \subset \mathcal{R}$. Conversely, each miRNA $r \in \mathcal{R}$ has a defined set of mRNAs which it targets, $\mathcal{G}_r \subset \mathcal{G}$. Let, x_g^G be the amount of mRNA measured for respective gene g in an arbitrary tissue sample and x_r^μ be the amount of particular miRNA r measured in an arbitrary sample. Let p_r be the proportion of the amount of $r \in \mathcal{R}_g$ used to regulate the expression of gene g and σ be a coefficient representing the strength of the inhibition of mRNA by miRNAs. Since the process is considered to be strictly subtractive, the aggregated value representing the inhibited mRNA of gene g, denoted x_g^{sub} would be obtained by subtracting as follows:

$$x_g^{\text{sub}} = X_g^G - \sigma \sum_{r \in \mathcal{R}_g} p_r x_r^\mu. \tag{1}$$

This equation takes an above-mentioned simplified view of inhibition of the gene by all targeting miRNAs. Hence, proportion p_r is defined as a ratio of X_g^G to the sum of levels of all targeted mRNAs. The inhibition equation is then expanded:

$$x_g^{\text{sub}} = X_g^G - \frac{c}{|\mathcal{R}_g|} \sum_{r \in \mathcal{R}_g} \frac{X_g^G}{\sum_{t \in \mathcal{G}_r} X_t^G} x_r^\mu. \tag{2}$$

Further, the parameter σ has been expanded in (2). The strength of inhibition is an unknown value, but needs to be somehow represented nonetheless. In this method, it is modeled as the product of a real parameter c and a normalizer defined as $1/|\mathcal{R}_g|$. The real parameter c represents the unknown strength of the relationship and its values are subject to experimentation. Intuitively, the larger c is, the more prominent the miRNA data are (larger c amplifies the inhibition). The c parameter can be set uniformly for all genes, or alternatively, different c values may be employed for different mRNAs. Concerning the limited sample sets and the risk of overfitting, we worked with the uniform c for all mRNAs. Its setting is further discussed in the experimental part of the paper.

It is possible to obtain the overall data matrix $\mathbf{X}^{\text{sub}}$ of inhibited mRNA by iteratively updating the submatrix $X_{1\ldots s, \mathcal{G}_r}^G$, thus calculating all x_g^{sub} in Eq. 2 pertaining to one miRNA and all samples in one step. Henceforth, the implementation of (2) is iterated over particular miRNAs, as there are far fewer miRNAs than mRNAs:

$$X_{1\ldots s, \mathcal{G}_r}^{\text{sub}} = X_{1\ldots s, \mathcal{G}_r}^G - c\Delta(\mathbf{u})\Delta(\mathbf{x}_r)\Delta(\mathbf{s})^{-1} X_{1\ldots s, \mathcal{G}_r}^G, \tag{3}$$

where $\Delta(\mathbf{v})$ denotes a diagonal matrix, whose (i, i)-th item is equal to the i-th value of a vector $\mathbf{v}$; $\mathbf{u}$ is a vector containing the number of targeting miRNAs for each mRNA and $\mathbf{s} = X_{1\ldots s, \mathcal{G}_r}^G \mathbf{1}_{|\mathcal{G}_r|}$ is a vector of mRNA value sums pertaining to miRNA targets in all samples.

2.4 SVD-based Aggregation (SVDAgg)

The aim of an integration method in general, is to reduce the mRNA vectors and their respective targeting miRNAs into one aggregated feature. A common known method which reduces data, preserving as much useful information contained in the non-reduced vectors as possible, is *Singular value decomposition* (SVD) [6].

The second method we propose, SVD-based aggregation, is based on the idea that targeting miRNAs of each gene can be represented in one dimensional basis space. So, for each gene g, the expression data submatrix $\mathbf{X}^{\mu}_{1...s,\mathcal{R}_g}$, referring to the respective targeting miRNAs, is projected into its first singular vector:

$$\mathbf{x}^{\mu,\text{svd}} = \mathbf{X}^{\mu}_{1...s,\mathcal{R}_g} \mathbf{V}_{1...|R_g|,1}, \tag{4}$$

where $\mathbf{V}$ is the singular vector matrix of targeting miRNAs. Vector $\mathbf{x}^{\mu,\text{svd}}$, the new representative of targeting miRNAs, is then joined to the respective mRNA vector, and reduced into one dimensional space again:

$$\mathbf{x}^{G,\text{svd}}_g = \left[\mathbf{x}^G_g, \mathbf{x}^{\mu,\text{svd}}\right] \mathbf{U}_{1...2,1}, \tag{5}$$

where $\mathbf{U}_{1...2,1}$ is the first singular vector of the two concatenated vectors.

The new feature $\mathbf{x}^{G,\text{svd}}_g$, a virtual profile comprising the gene and its miR-NAs, is computed in two steps. The reason for not aggregating the mRNA vector and respective miRNA vectors together at the same time follows. Such an alternative approach gives almost all the power to the miRNAs, and since they would constitute a majority of vectors, SVD would have a tendency to disregard the information contained in the mRNA, which is necessary to avoid. Moreover, this effect would increase with the increasing number of targeting miRNAs.

2.5 SubAgg and SVDAgg Availability

Both SubAgg and SVDAgg are available as a part of the web tool miXGENE (http://mixgene.felk.cvut.cz/) [9]. miXGENE is a workflow management system dedicated for machine learning from heterogeneous gene expression data using prior knowledge. The main idea is to facilitate development of predictive phenotype models that do not merely capture the transcriptional phase of gene expression quantified by the amount of mRNA, but analyze them concurrently with miRNA and epigenetic data to explain unexpected transcriptional irregularities. This paper evaluates the proposed methods in the MDS domain, miXGENE website presents another case study that employs these (and other) methods in the domain of germ cell tumor classification.

3 Experiments

This section describes the MDS domain, defines the learning tasks, and summarizes the experimental protocol.

3.1 Datasets

The data were acquired in collaboration with the Institute of Hematology and Blood Transfusion in Prague. Illumina miRNA (Human v2 MicroRNA Expression Profiling Kit, Illumina, San Diego, USA) and mRNA (HumanRef-8 v3 and HumanHT-12 v4 Expression BeadChips, Illumina) expression profiling were used to investigate the effect of lenalidomide treatment on miRNA and mRNA expression in bone marrow (BM) CD34+ progenitor cells and peripheral blood (PB) CD14+ monocytes. Quantile normalization was performed independently for both the expression sets, the datasets were scaled to have the identical median of 1 then. The mRNA dataset has 16,666 attributes representing the GE level through the amount of corresponding mRNA measured, while the miRNA dataset has 1,146 attributes representing the expression level of particular miR-NAs. The measurements were conducted on 75 tissue samples categorized according to the following conditions: (1) tissue type: peripheral blood monocytes vs. bone marrow cells, (2) presence of MDS and del(5q), (3) lenalidomide treatment stage: before treatment (BT) vs. during treatment (DT). Henceforth, the samples can be broken into 10 categories. The categories, along with the actual number of samples, are shown in Table 1:

Table 1. The overview of MDS classes.

PB	Healthy		10
	5q-	BT	9
		DT	13
	non 5q-	BT	4
		DT	5
BM	Healthy		10
	5q-	BT	11
		DT	5
	non 5q-	BT	6
		DT	2

The domain experts defined 16 binary classification tasks with a clear diagnostic and therapeutic motivation. There are 8 tasks for each tissue type, the tissue types are encoded in the task names, while the numbers of samples are shown in parentheses. The afflicted group comprises all MDS patients regardless their treatment status.

1. **PB1**: healthy (10) × afflicted (31),
2. **BM1**: healthy (10) × afflicted (24),
3. **PB2**: healthy (10) × BT (13),
4. **BM2**: healthy (10) × BT (17),
5. **PB3**: healthy (10) × BT with del(5q) (9),

6. **BM3**: healthy (10) × BT with del(5q) (11),
7. **PB4**: healthy (10) × DT (18),
8. **BM4**: healthy (10) × DT (7),
9. **PB5**: afflicted: del(5q) (9) × non del(5q) (22),
10. **BM5**: afflicted: del(5q) (8) × non del(5q) (16).
11. **PB6**: healthy (10) × DT del(5q) (13),
12. **BM6**: healthy (10) × DT del(5q) (5),
13. **PB7**: healthy (10) × BT non del(5q) (4),
14. **BM7**: healthy (10) × BT non del(5q)(6),
15. **PB8**: del(5q): BT (9) × DT (13),
16. **BM8**: del(5q): BT (11) × DT (5).

3.2   Prior Knowledge

Considering prior knowledge, we downloaded the interactions between genes and miRNAs from publicly available databases. TarBase 6.0, strives to encompass as many miRNA-mRNA validated targeting relations scattered in literature as possible. The database, maintained by DIANA Lab, was built utilizing text-mining-assisted literature curation – literature covering the discovery of new target relationships were downloaded in XML format from MedLine, processed using text mining and the resulting candidates for addition to the database were reviewed before the actual entry by the curators (DIANA Lab personnel). Its respective target matrix, filtered so as to contain solely human data, covers 228 miRNAs, 11,996 mRNAs and 20,107 target relationships between them. When selecting only the mRNAs and miRNAs available in the actual chip probesets and carefully translating and unifying miRNA identifiers using miRBase [13], the TarBase covers 179 miRNAs, 8,188 mRNAs and contains 14,404 target relationships.

The miRWalk database [5], comprises both validated and predicted targets. In our experiments, only the predicted target database is used; the entries in the validated target database are already included in TarBase 6.0. Since, according to the authors, no target prediction algorithm consistently achieves better results than the others, the predicted target database includes not only targets obtained using the eponymous miRWalk algorithm, but also targets provided by other prediction algorithms. Our experiments use five of them, which are outlined in Sect. 1. The predicted targets dataset used in the experiments was obtained from miRWalk by merging the results of multiple queries on the mRNA targets of canonically-named miRNAs present in the experimental miRNA dataset. Each query consisted of up to 20 miRNAs (limit imposed by the miRWalk site), each query was restricted to targets in the 3' UTR region with p-value less or equal to 0.01. The resulting dataset obtained contains 392 miRNAs, 14,550 mRNAs and 89,402 unique predicted human miRNA-mRNA target relationships. 389 miRNAs, 12,847 mRNAs and 79,014 relationships were applicable in terms of our actual mRNA and miRNA probesets.

The merged target dataset concatenates both the above-mentioned resources. It is further referred to as the extended predicted database and contains 93,325 target relationships.

3.3   Experimental Protocol

The main aim of the experiments is to verify whether the features, extracted by prior knowledge, can improve classification quality. Since we deal with classes of different sizes, we use the Mathews correlation coefficient as a balanced quality measure. It returns a value between -1 and +1, +1 represents a perfect match between annotation and prediction. We employ three benchmarking feature sets to tackle this issue. The first contains mRNA profiles only, the second takes purely miRNA profiles, and the third concatenates them as described in Sect. 2.2. The knowledge-based feature sets denoted as SubAgg and SVDAgg take the merged feature set and concatenate it with the aggregated features obtained in 2, 3 and 5 respectively.

We used 5 times repeated stratified 5-fold cross validation to assess the performance of the proposed methods as well as their benchmarking counterparts. The whole learning workflow was implemented in R environment.

SVDAgg has no parameters, SubAgg has the inhibition strength parameter c that needs to be optimized. The most direct way is to set it to 1 relying purely on mRNA and miRNA expression normalization. However, the absolute mRNA and miRNA expression values can hardly be directly matched. Moreover, the relative predictive power of mRNA and miRNA feature sets varies for different tasks. That is why we tuned the optimal value of c in terms of internal cross-validation. The parameter values 10^k, k$\in${-2,1,0,1,2} were concerned, the best value was taken in each experimental setting and fold uniformly for all mRNAs.

In order to keep a reasonable number of features, to minimize overfitting, and maintain the constant number of features across different feature sets in terms of one learning task, we applied the well-known feature selection method SVM-RFE [8]. In each of the learning tasks, the size of the reduced feature set was established as follows. We found the number of active mRNAs and the number of active miRNAs, and took their minimum. This value served as the target feature set size for mRNA, miRNA, merged and both subtractive classifiers.

We deal with 8 binary MDS tasks defined in Sect. 3.1. At the same time, we have two distinct target relations (validated and extended) as described in the previous section. These target relations have different domains and ranges, the domain and range of the validated target relation make subsets of their extended counterparts. As the aggregated features concern purely the domain miRNAs and the range mRNAs we filter out the rest of mRNA and miRNA profiles from the benchmarking datasets as well. This is done in order to make the comparison of classifier performance on benchmarking datasets more relatable and better identify the potential asset of the target relationships. The absolute score is not important, the main issue is the relative comparison in terms of a single learning task. In this way, 64 different experimental settings originate (2 tissue types × 8 task definitions × 2 target relations × 2 classification algorithms). The settings are independent between tissue types, however, they deal with overlapping sample and feature sets within the same tissue type.

We employed two diverse classification algorithms to avoid the dependence of experimental results on a specific choice of learning method. Support Vector

Machine (SVM) with a linear kernel and the regularization parameter $C = 1$ was taken as the first option. SVM prevails in predictive modeling of gene expression data and is usually associated with high resistance to noise in data. C setting proves robust even when learning with many relevant features [11]. Naïve Bayes is a simple and interpretable classifier.

4 Results

The individual feature sets were tested and compared under all the experimental settings defined above. The results reached are available in Table 2; the table summarizes the results achieved by the two classification algorithms.

   The following direct observations can be drawn from the result tables. There are settings that can be perfectly solved by either the mRNA or miRNA profiles. Then, there are settings with incomparable score reached with the mRNA and miRNA feature set. Naturally, these settings are not suitable for any integration including the concatenation as this integration can hardly outperform the better of the raw feature sets. These settings can be a priori identified and omitted from the integration procedure, or the procedure can be parametrized in such a way that the inferior dataset has no influence on the final feature set (e.g., c parameter in SubAgg is set to 0).

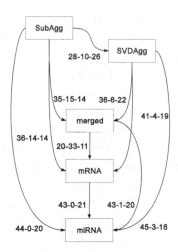

Fig. 1. Pair-wise classification comparison graph. The nodes represent particular feature sets, an edge from node a to node b, annotated as x-y-z means that method a outperforms method b in x experiments, in y ties and in z losses.

On the other hand, when dealing with mRNA and miRNA profiles of comparable predictive power, the integration improves classification performance. In general, the knowledge-based methods outperform their concatenation benchmark. As already mentioned, we deal with dependent tasks and settings while traditional

176   J. Kléma et al.

Table 2. Classification performance of two learners in terms of MCC. *Relat.* stands for the target relation type (*val* means validated and *ext* extended), mR for mRNA, miR for miRNA, mer for merged, Sub stands for SubAgg and SVD for SVDAgg. The last row gives average ranking of each feature set; the lower the rank, the better.

Task	Relat.	SVM					Naïve Bayes				
		mR	miR	mer	Sub	SVD	mR	miR	mer	Sub	SVD
PB1	ext	0.96	0.65	0.96	1.00	1.00	0.84	0.73	0.84	0.84	0.80
PB2	ext	0.98	0.88	0.98	0.98	1.00	0.87	0.85	0.88	0.85	0.90
PB3	ext	0.98	0.81	0.98	1.00	1.00	0.85	0.69	0.85	0.90	0.88
PB4	ext	1.00	0.80	1.00	1.00	0.97	0.83	0.73	0.83	0.84	0.82
PB5	ext	0.86	0.97	0.89	0.89	0.94	0.79	0.98	0.86	0.88	0.86
PB6	ext	1.00	0.82	1.00	1.00	0.98	0.82	0.77	0.82	0.88	0.97
PB7	ext	0.76	0.86	0.72	0.83	0.79	0.65	0.79	0.65	0.65	0.97
PB8	ext	0.62	0.49	0.56	0.56	0.64	0.32	0.43	0.30	0.31	0.52
BM1	ext	0.97	0.92	0.99	0.96	0.96	0.93	0.92	0.94	0.93	0.94
BM2	ext	0.91	0.95	0.91	0.94	0.93	0.95	0.95	0.95	0.87	0.98
BM3	ext	0.94	0.98	0.94	0.94	0.94	0.91	0.98	0.91	0.91	0.96
BM4	ext	0.98	0.88	0.98	0.95	0.84	0.98	0.79	0.98	0.98	0.79
BM5	ext	0.73	0.91	0.77	0.82	0.91	0.59	0.87	0.73	0.71	0.75
BM6	ext	0.88	0.85	0.88	0.91	0.85	1.00	0.71	0.80	0.94	0.71
BM7	ext	0.95	0.97	0.97	0.95	1.00	0.87	0.87	0.87	0.87	0.90
BM8	ext	0.57	0.54	0.54	0.43	0.45	0.32	0.40	0.38	0.41	0.27
PB1	val	0.96	0.78	0.96	1.00	0.99	0.85	0.55	0.85	0.88	0.88
PB2	val	0.98	0.83	0.98	0.98	0.92	0.83	0.63	0.85	0.87	0.97
PB3	val	0.94	0.77	0.98	1.00	0.96	0.88	0.64	0.88	1.00	1.00
PB4	val	1.00	0.76	1.00	1.00	0.97	0.86	0.69	0.86	0.88	0.84
PB5	val	0.86	0.94	0.89	0.89	0.89	0.84	0.94	0.84	0.88	0.86
PB6	val	1.00	0.90	1.00	1.00	1.00	0.81	0.63	0.81	0.90	0.91
PB7	val	0.72	-0.16	0.65	0.79	0.61	0.86	0.83	0.90	0.97	0.83
PB8	val	0.62	0.51	0.62	0.52	0.59	0.32	0.21	0.30	0.36	0.52
BM1	val	0.97	0.85	0.99	0.96	0.99	0.89	0.89	0.92	0.93	0.93
BM2	val	0.90	0.87	0.91	0.95	0.95	0.95	0.83	0.95	0.90	0.95
BM3	val	0.96	0.91	0.96	1.00	0.98	0.91	0.91	0.91	0.89	0.94
BM4	val	0.98	0.86	0.98	0.95	1.00	0.95	0.88	0.95	0.98	0.91
BM5	val	0.69	0.89	0.73	0.77	0.72	0.58	0.82	0.69	0.73	0.69
BM6	val	0.91	0.80	0.88	0.91	0.94	0.94	0.85	0.88	0.97	0.91
BM7	val	0.92	0.90	0.97	0.97	0.95	0.87	0.87	0.87	0.87	0.87
BM8	val	0.57	0.61	0.50	0.54	0.57	0.15	0.20	0.19	0.27	0.36
Average rank		3.08	3.81	2.88	2.56	2.67	3.41	3.72	3.17	2.36	2.34

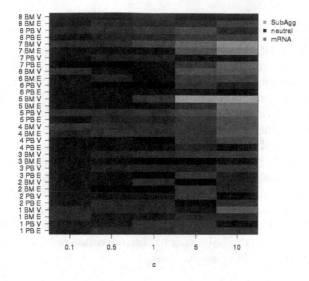

Fig. 2. The heat map illustrating the role of original mRNA profiles and their subtracted counterparts within SubAgg. The experiments in which SVM-RFE preferred the aggregated to original features are shown in green. In the red-colored experiments, the original features prevailed. The heat map concerns top 100 features. The lightest green color observed in the map corresponds to the distribution of 33 original versus 67 aggregated features, the lightest red stands for 54 original and 46 aggregated features(Color figure online).

hypothesis testing asks for independence. That is why we cannot apply Wilcoxon, Friedman, or other classical tests. Instead, the methods are sorted and ranked according to their pair-wise comparison in each of the particular settings; Fig. 1 provides an overall comparison graph and the last row of result tables gives the ranks averaged across all the settings. The comparison suggests that the knowledge-based feature sets dominate the rest of the feature pool.

Another useful comparison measure is the overall number of occurrences, denoted as *synergies*, in which the knowledge based features outperform both raw feature sets. The presented results show 31 and 26 synergies occurred in the case of SubAgg and SVDAgg methods respectively; only 10 synergies can be observed in the case of the benchmark integration. In the other words, when dealing with settings that cannot be perfectly solved by the original features, the knowledge based integration helps.

SVM turns out to be a better choice than naïve Bayes. Let us stress that the choice of target type (validated, extended) may seem to largely affect classification quality; however, the main reason for this difference lies in the filtering mentioned in Sect. 3.3. The validated and extended runs cannot be directly compared (validated clearly worse than extended). The relative comparison between the merged and the other knowledge-based methods suggests that when including the predicted targets into the aggregation, no clear improvement can be observed.

5   Discussion

In order to understand in more depth the functioning of the proposed methods we analyzed the intermediate results in the auxiliary experiment that avoided the internal cross-validation. In particular, we focused on the role of c value (changed in smaller steps) in combination with the relationship between mRNA and miRNA predictive strengths. The individual feature sets were tested and compared under all the experimental settings defined above.

The following list of conclusions can be drawn from the heat map in Fig. 2 and the results in Table 2. Firstly, the aggregated profiles tend to replace the original mRNA ones in the tasks with predictive miRNA features. Secondly, the predictive strength of some mRNA profiles still deteriorates. General replacement of all the original mRNA profiles cannot be recommended. Thirdly, feature selection that leads to the enrichment of the original set of features is preferable. It may serve as a tool for automatic balancing of the individual feature classes based on their predictive strength in the given task.

6   Conclusions

Molecular classification of biological samples based on their expression profiles represents a natural task. However, the task proved conceptually difficult due to the inconvenient rate of the sample and feature set sizes and complexity and heterogeneity of the expression process. These characteristics often cause overfitting. Classifiers do not sufficiently generalize; instead of revealing the underlying relationships, they capture perturbations in training data. This problem can be minimized by regularization; i.e., introduction of additional knowledge. The regularized models should be more comprehensible and potentially more accurate than standard models based solely on a large amount of raw measurements.

The integration of heterogeneous measurements and prior knowledge is nontrivial, though. In this paper we proposed the subtractive method that aggregates mRNA and miRNA values by subtracting a proportion of miRNA expression values from their respective target mRNAs. The method simplifies the mRNA-miRNA interaction and minimizes the number of parameters needed to be learned to 1. We also proposed another integration method that can be perceived as an extension that enables different subtractive weights for different miRNAs; the weights are learned by SVD.

In this work we classified myelodysplastic syndrome patients under 64 experimental settings. We compared five types of feature sets. Two of them represented raw homogeneous expression measurements (mRNa and miRNA profiles), the third implemented their straightforward concatenation, and the last two resulted from SubAgg and SVDAgg integration. The comparison suggests that the knowledge-based feature sets dominate the rest of the feature pool, and the features resulting from the mRNA-miRNA target relation can improve classification performance.

There is still a lot of future work. More problem domains need to be considered. The prior knowledge should be extended to cover the gene regulatory network (the protein-protein interactions, interactions between genes, and their transcription factors). Another challenge is to employ epigenetic data, namely DNA methylation. Concerning the algorithmic issues, we intend to develop another parameter-free integration method where the prior knowledge controls pseudorandom construction of weak classifiers vaguely corresponding to the individual biological processes. The weak classifiers will later be merged into an ensemble classifier.

Acknowledgements. This research was supported by the grants NT14539 and NT1 4377 of the Ministry of Health of the Czech Republic.

References

1. Anděl, M., Kléma, J., Krejčík, Z.: Integrating mRNA and miRNA expressions with interaction knowledge to predict myelodysplastic syndrome. In: Information Technologies - Applications and Theory, Workshop on Bioinformatics in Genomics and Proteomics, ITAT 2013, pp. 48–55 (2013)
2. Brewster, J.L., Beason, K.B., Eckdahl, T.T., et al.: The microarray revolution: perspectives from educators. Biochem. Mol. Biol. Educ. **32**(4), 217–227 (2004)
3. Croce, C.M.: Causes and consequences of microRNA dysregulation in cancer. Nat. Rev. Genet. **10**(10), 704–714 (2009)
4. Merkerova, M.D., Krejcik, Z., Votavova, H., et al.: Distinctive microRNA expression profiles in CD34+ bone marrow cells from patients with myelodysplastic syndrome. Eur. J. Hum. Genet. **19**(3), 313–319 (2011)
5. Dweep, H., Sticht, C., Pandey, P., et al.: miRWalk - database: prediction of possible miRNA binding sites by "walking" the genes of three genomes. J. Biomed. Inform. **44**(5), 839–847 (2011)
6. Eckart, C., Young, G.: The approximation of one matrix by another of lower rank. Psychometrika **1**, 211–8 (1936)
7. Fabian, M.R., Sonenberg, N.: The mechanics of miRNA-mediated gene silencing: a look under the hood of miRISC. Nat. Struct. Mol. Biol. **19**(6), 586–593 (2012)
8. Guyon, I., Weston, J., Barnhill, S., Vapnik, V.: Gene selection for cancer classification using support vector machines. Mach. Learn. **46**, 389–422 (2002)
9. Holec, M., Gologuzov, V., Kléma, J.: miXGENE tool for learning from heterogeneous gene expression data using prior knowledge. In: Proceedings of the 27th IEEE International Symposium on Computer-Based Medical Systems 2014 (2014) (to appear)
10. Huang, G.T., Athanassiou, C., Benos, P.V.: mirConnX: condition-specific mRNA-microRNA network integrator. Nucleic Acids Res. **39**, W416–W423 (2011). Web Server issue
11. Joachims, T.: Text categorization with support vector machines: learning with many relevant features. In: Nédellec, C., Rouveirol, C. (eds.) ECML 1998. LNCS, vol. 1398. Springer, Heidelberg (1998)
12. Kim, D., Shin, H., Song, Y.S., et al.: Synergistic effect of different levels of genomic data for cancer clinical outcome prediction. J. Biomed. Inform. **45**(6), 1191–1198 (2012)
13. Kozomara, A., Griffiths-Jones, S.: miRBase: integrating microRNA annotation and deep-sequencing data. Nucleic Acids Res. **39**, 152–157 (2011). Database-Issue

14. Krek, A., Grün, D., Poy, M.N., et al.: Combinatorial microRNA target predictions. Nat. Genet. **37**(5), 495–500 (2005)
15. Lanza, G., Ferracin, M., Gafà, R., et al.: mRNA/microRNA gene expression profile in microsatellite unstable colorectal cancer. Mol. Cancer **6**, 54 (2007)
16. Lee, R.C., Feinbaum, R.L., Ambros, V.: The C. elegans heterochronic gene lin-4 encodes small RNAs with antisense complementarity to lin-14. Cell **75**(5), 843–854 (1993)
17. Lewis, B.P., Shih, I.H.H., et al.: Prediction of mammalian microRNA targets. Cell **115**(7), 787–798 (2003)
18. Li, W., Zhang, S.H., et al.: Identifying multi-layer gene regulatory modules from multi-dimensional genomic data. Bioinformatics **28**(19), 2458–66 (2012)
19. Morin, R., Bainbridge, M., Fejes, A., et al.: Profiling the HeLa S3 transcriptome using randomly primed cDNA and massively parallel short-read sequencing. BioTechniques **45**(1), 81–94 (2008)
20. Peng, X., Li, Y., Walters, K.A., et al.: Computational identification of hepatitis C virus associated microRNA-mRNA regulatory modules in human livers. BMC Genomics **10**(1), 373 (2009)
21. Pollack, J.R., Sørlie, T., Perou, C.M., et al.: Microarray analysis reveals a major direct role of DNA copy number alteration in the transcriptional program of human breast tumors. Proc. Natl. Acad. Sci. USA **99**(20), 12963–12968 (2002)
22. Rhyasen, G.W., Starczynowski, D.T.: Deregulation of microRNAs in myelodysplastic syndrome. Leukemia **26**(1), 13–22 (2012)
23. Sayed, D., Abdellatif, M.: MicroRNAs in development and disease. Physiol. Rev. **91**(3), 827–887 (2011)
24. Smith, T.F., Waterman, M.S.: Identification of common molecular subsequences. J. Mol. Biol. **147**(1), 195–197 (1981)
25. Stranger, B.E., Forrest, M.S., Dunning, M., et al.: Relative impact of nucleotide and copy number variation on gene expression phenotypes. Science **315**(5813), 848–853 (2007)
26. Tan Gana, N.H., Victoriano, A.F., Okamoto, T.: Evaluation of online miRNA resources for biomedical applications. Genes Cells **17**(1), 11–27 (2012)
27. Tran, D.H., Satou, K., Ho, T.B.: Finding microRNA regulatory modules in human genome using rule induction. BMC Bioinform. **9**(12), S5 (2008)
28. Vašíková, A., Běličková, M., Budinská, E., et al.: A distinct expression of various gene subsets in cd34+ cells from patients with early and advanced myelodysplastic syndrome. Leuk. Res. **34**(12), 1566–1572 (2010)
29. Wang, X., Naqa, I.M.E.: Prediction of both conserved and nonconserved microRNA targets in animals. Bioinformatics **24**(3), 325–332 (2008)
30. Witten, D.M., Tibshirani, R.J.: Extensions of sparse canonical correlation analysis with applications to genomic data. Stat. Appl. Genet. Mol. Biol. **8**(1), 28 (2009)
31. Zhang, S.H., Li, Q., et al.: A novel computational framework for simultaneous integration of multiple types of genomic data to identify microRNA-gene regulatory modules. Bioinformatics **27**(13), 401–409 (2011)

A Computational Pipeline to Identify New Potential Regulatory Motifs in Melanoma Progression

Gianfranco Politano[1], Alfredo Benso[1,4], Stefano Di Carlo[1],
Francesca Orso[2,3], Alessandro Savino[1(✉)], and Daniela Taverna[2,3]

[1] Politecnico di Torino, Corso Duca Degli Abruzzi 24, 10129 Torino, Italy
{alfredo.benso,stefano.dicarlo,gianfranco.politano,
alessandro.savino}@polito.it
[2] Molecular Biotechnology Center (MBC), Torino, Italy
{francesca.orso,daniela.taverna}@unito.it
[3] Center for Molecular Systems Biology, Università di Torino, Torino, Italy
[4] Consorzio Interuniversitario Nazionale per L'Informatica, Verres (AO), Italy

Abstract. Molecular biology experiments allow to obtain reliable data about the expression of different classes of molecules involved in several cellular processes. This information is mostly static and does not give much clue about the causal relationships (i.e., regulation) among the different molecules. A typical scenario is the presence of a set of modulated mRNAs (up or down regulated) along with an over expression of one or more small non-coding RNAs molecules like miRNAs. To computationally identify the presence of transcriptional or post-transcriptional regulatory modules between one or more miRNAs and a set of target modulated genes, we propose a computational pipeline designed to integrate data from multiple online data repositories. The pipeline produces a set of three types of putative regulatory motifs involving coding genes, intronic miRNAs, and transcription factors. We used this pipeline to analyze the results of a set of expression experiments on a melanoma cell line that showed an over expression of miR-214 along with the modulation of a set of 73 other genes. The results suggest the presence of 27 putative regulatory modules involving *miR-214*, *NFKB1*, *SREBPF2*, *miR-33a* and 9 out of the 73 *miR-214* modulated genes (*ALCAM*, *POSTN*, *TFAP2A*, *ADAM9*, *NCAM1*, *SEMA3A*, *PVRL2*, *JAG1*, *EGFR1*). As a preliminary experimental validation we focused on 9 out of the 27 identified regulatory modules that involve *miR-33a* and *SREBF2*. The results confirm the importance of the predictions obtained with the presented computational approach.

Keywords: microRNA · miR-214 · Melanomas · Biological pathways · Gene regulation · Post-transcriptional regulation

1 Introduction

Expression experiments on melanomas cell lines (as well as on tumor cell lines in general) often reveal an aberrant expression of coding and non-coding molecules,

G. Plantier et al. (Eds.): BIOSTEC 2014, CCIS 511, pp. 181–194, 2015.
DOI: 10.1007/978-3-319-26129-4_12

such as microRNAs (miRNAs). miRNAs are 20 to 24 nucleotides long non-coding RNAs involved in the post-transcriptional down-regulation of protein-coding genes through imperfect base pairing with their target mRNAs. The peculiar characteristic of miRNAs is their ability to simultaneously affect the expression of several genes. Consequently, miRNAs have been implicated as possible key factors in several diseases [1–5]. Referring to melanomas, miRNAs such as *let-7a/b*, *miR-23a/b*, *miR-148*, *miR-155*, *miR-182*, *miR-200c*, *miR-211*, *miR-214* and *miR-221/222* have been found to be differentially expressed in benign melanocytes versus melanoma cell lines or in benign melanocytic lesions versus melanomas in human samples. Targets of some of the above listed miR-NAs are well-known melanoma-associated genes, like the oncogene *NRAS*, the microphthalmia-associated transcription factor (*MITF*), the receptor tyrosine kinase *c-KIT*, and the *AP-2* transcription factor (*TFAP2*). *miR-214* is known to control in vitro tumor cell movement and survival to anoikis, as well as in vivo malignant cell extravasation from blood vessels and lung metastasis formation. [6,7] show that *miR-214*, the product of an intron of the *Dynamin-3* gene on human chromosome 1, coordinates melanoma metastasis formation by modulating the expression of over 70 different genes, including two activating protein transcription factors (*TFAP2A* and *TFAP2C*) and the adhesion molecule *ALCAM*. The alteration in the expression level of some of these genes leads to important downstream effects on a number of key processes such as apoptosis, proliferation, migration and invasion. The static information about genes and miRNAs expression is very important but not sufficient to precisely understand the regulatory dynamics that cause the aberrant phenotype. A significant help in this direction can be provided by Systems Biology approaches designed to integrate the huge amount of biological data available online to infer possible regulatory motifs involving the molecules of interest highlighted by microarray expression experiments. In this work, we present a computational pipeline that allows biologists to identify the presence of different classes of regulatory modules between a miRNA and a set of target genes of interest. We used the pipeline to analyze the possible relationships between *miR-214* and the set of 73 proteins found modulated in melanoma expression experiments.

The pipeline is designed to automatically verify the presence of three different classes of regulatory modules, all characterized by an interplay between Transcription Factors (TFs) and miRNAs. Several studies (such as [8–11]) suggested that this type of interaction is particularly critical in cellular regulation during tumor genesis. The regulatory motifs that are analyzed (see Fig. 1) are:

1. Type-0 (direct interactions), where the *miRNA* directly down-regulates one of the target proteins;
2. Type-1 (one-level indirect interactions), where the *miRNA* down-regulates a Transcription Factor, which eventually regulates one of the targets genes;
3. Type-2 (two-levels indirect interactions), where the *miRNA* targets a Transcription Factor (TF). The TF is then regulating a gene, which hosts an intragenic miRNA that acts as down-regulator of one of the target proteins.

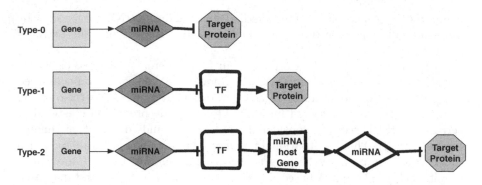

Fig. 1. Three classes of regulatory modules involving *miR-214* investigated in this paper: Type-0 (direct interactions), Type-1 (one-level indirect interactions), and Type-2 (two-levels indirect interactions).

Whereas detecting Type-0 and Type-1 interactions is a quite simple data integration task, the identification of all the necessary evidence to infer the existence of Type-2 interactions requires a more complex data integration process. Obviously, these three motifs are not the only interactions that may be investigated. Deeper interactions could be addresses, for example three-level integrations that involve two TFs before reaching a target protein (*miR-214* → TF1 → TF2 → Target Protein). At this stage these complex interactions have not been considered due to the difficulty to experimentally validate their existence.

2 Methods

The computational analysis to investigate the set of previously defined regulatory motifs (Fig. 1) requires the successful integration of heterogeneous data sources. This section describes, at first, the set of public repositories included in the flow and, eventually, the computational pipeline that integrates and elaborates these data sources to search for the three classes of interactions.

2.1 Data Sources

The following public repositories represent the main sources of information in our computational process:

– **microRNA.org** database [12] is used to search for miRNA target genes. MicroRNA.org uses the miRanda algorithm [13] for target predictions. The algorithm computes optimal sequence complementarity between a miRNA and a mRNA using a weighted dynamic programming algorithm. The overall database consists of 16,228,619 predicted miRNA target sites in 34,911 distinct 3'UTR from isoforms of 19,898 human genes. Each prediction is associated to a prediction confidence score, the mirSVR score. The mirSVR is a real number computed by a machine learning method for ranking miRNA target sites by a down-regulation score [14]. The lower (negative) is the score, the better is the prediction.

microRNA.org provides data organized in four different packages: (1) Good mirSVR score, Conserved miRNA, (2) Good mirSVR score, Non-conserved miRNA, (3) Non-good mirSVR score, Conserved miRNA, (4) Non-good mirSVR score, Non-conserved miRNA. They include prediction clustered in terms of mirSVR score (if considered good or not) and conservation (highly, low conserved). The computational pipeline includes a single database, which unifies the four packages keeping all source information intact (stored in a specific field). The mirSVR score information is still included and it is used in order to be able to refine queries by filtering unreliable predictions out.

– **Transcription Factor Encyclopedia** (TFE) [15] and **Targetmine** [16,17] have been used to identify genes TFs. TFE provides details of Transcription Factor binding sites in close collaboration with Pazar, a public database of TFs and regulatory sequence information. While TFE includes both Upstream and Downstream Transcription Factors, Targetmine contains only Upstream Transcription Factors. For each gene, the database retrieves all upstream regulatory genes from the AMADEUS [18,19] and ORegAnno compiled TF-Target gene sets. AMADEUS contains TF and miRNA target sets in *human*, *mouse*, *D. Melanogaster*, and *C. Elegans*, collected from the literature. For each TF, it is reported its set of targets, given as a list of Ensembl gene IDs.

– **e-Utils programming utilities** [20] and **Mirbase.org** [21,22] allow for retrieving coordinates of precursor miRNAs and genes. miRBase is a searchable database of miRNA sequences and annotations already published. About 94.5 % of the available mature miRNA sequences considered in this database have experimental evidence, thus representing a reliable source of information. Each miRNA entry in miRBase is annotated with the information on the location that is exploited to identify the host genes.

2.2   Computational Pipeline

The full pipeline has been developed in PHP language and coupled with a MySQL database, which mirrors an optimized subset of data coming from multiple online repositories. To better explain its functioning, we will refer to an actual analysis that we performed to search for *miR-214* mediated interactions. Nevertheless, the same analysis could be executed with any other miRNA and a set of target genes of interest. Figure 2 highlights the implemented computational pipeline. The modulator miRNA is *miR-214* and the list of target genes of interest is the set of 73 protein-coding genes reported in Table 1. Previous microarray experiments, [6], suggest that *miR-214* modulates (directly or indirectly) these proteins. In Fig. 2 these genes are named "Target Proteins".

The computational pipeline is organized into four main data integrations steps, which search for Type-0, Type-1 and Type-2 interactions.

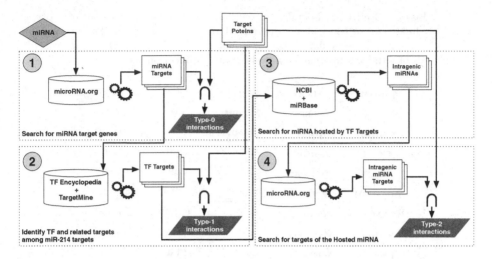

Fig. 2. The four steps pipeline to investigate the presence of transcriptional or post-transcriptional regulatory pathways.

Step 1 - Detection of Type-0 Interactions. Target genes that are directly regulated by *miR-214* represent the way Type-0 interactions manifest. microRNA.org provides information on *miR-214* direct targets. Thus, we queried microRNA.org database to search for all direct targets predicted for *miR-214*. Due to the computational approach used by microRNA.org to predict miRNA targets, query results can be affected by false positives. By restricting the query to "Good mirSVR score, Conserved miRNA" and to the "Good mirSVR score, Non-conserved miRNA" packages, a more reliable set of computed targets is represented, positively affecting all investigation results. Once the set of computed targets is retrieved, we applied a further filtering process according to the mirSVR score. In [14], the mirSVR score is shown to be meaningful with a cut-off of at most -0.1, derived from the empirical distribution of the extent of target down-regulation (measured as log-fold change) that is expected given a mirSVR score. For scores closer to zero, according to [23], the probability of meaningful down-regulation decreases while the number of predictions drastically increases. The final selection is done with mirSVR < -0.3 in order to guarantee high reliable predictions.

Once the set of *miR-214* targets is defined, the pipeline intersects the full list with the set of 73 Target Proteins, in order to highlight the Type-0 interactions.

Step 2 - Detection of Type-1 Interactions. Step 2 starts from the full list of *miR-214* targets already computed during Step 1. In fact, the Type-1 interactions require to identify, among the miRNA targets, those that are also TFs for other genes.

The search is done for each *miR-214* target both in Transcription Factor Encyclopedia (TFE) and TargetMine. The output is the subset of *miR-214*

Table 1. List of 73 miR-214 modulated genes. In green and bold the set of proteins that result linked to miR-214 in the discovered regulatory modules. The arrow indicates if the gene was up regulated (up) or down regulated (down) in the microarray experiments; in red, proteins that do not show any connection.

↑ADAM9	↓JAM3	↑THY1	CD44	ENG
↑ALCAM	↓LRP6	↑TIMP3	CD9	EPCAM
↓BMPR1B	↑MET	ADAM15	CDH1	ERBB2
↓CD40	↑MMP2	ADAM8	CDH11	ERBB3
↑CD99	↑NCAM1	APP	CDH2	EREG
↑CEACAM1	↓POSTN	ARHGAP12	CDH4	F2
↓CEACAM5	↑PVRL2	BCAM	CDHR5	FCER2
↓EGFR	↓SEMA3A	BSG	CLU	FLT1
↑HBEGF	↓TFAP2A	CD36	CTSD	HRG
↓JAG1	↓TFAP2C	CD40LG	CX3CL1	ICAM2
IL1R2	LCN2	TIMP1	IL8	LGALS3BP
TIMP2	ITGA3	MITF	VCAM1	ITGA6
PAK2	ITGAV	PODXL	ITGB1	PODXL2
ITGB3	PTEN	JAM1	PVR	JAM2
SELE	KDR	TGFBI		

targets that are also TFs. For each of them, the related targets are extracted and added to a list of TF Targets. The pipeline identifies all possible Type-1 interactions by intersecting this list with the set of 73 targets.

For each TF-target interaction we also check if TFE stores the information about its sign (either activation or inhibition). When available, this information is useful to have a better insight about the behavior of the regulatory module. When the sign of the TF is positive (i.e., activation), an increase in the expression of the miRNA is expected to lead to the down-regulation of its target. On the other side, a negative sign (i.e., inhibition) will result in a correlated expression between the miRNA and its target. If this prediction correlates with the actual expression of the target genes, this information can provide an additional biological validation of the detected motif.

Step 3 - Intragenic miRNAs Enrichment. The last two steps of the proposed computational pipeline are used to identify Type-2 interactions, which represent the most complex considered motif.

The previous list of TF Targets needs to be enriched in order to identify a set of candidate intragenic miRNAs. Intragenic miRNAs represent around 50 % of the mammalian miRNAs. Most of these intragenic miRNAs are located within introns of protein coding genes (miRNA host genes) and are referred to as intronic miRNAs, whereas the remaining miRNAs are overlapping with exons of their host genes and are thus called exonic miRNAs. Moreover, the majority of intragenic miRNAs are sense strand located, while only a very small portion is anti-sense strand located.

In this step we want to consider both intronic and exonic miRNAs, either sense or anti-sense strand located. For each TF Target, its set of candidate intragenic miRNAs are retrieved by querying the miRBase database. To be able to correctly identify intragenic miRNAs of a given host gene, we use e-Utils to resolve the genomic coordinates of the gene. The gene coordinates are the right input to query the miRBase database for all miRNAs whose coordinates are embraced within the ones of the gene. The enrichment feeds the last step to complete the detection of Type-2 interactions.

Step 4 - Detection of Type-2 Interactions. So far, the *miR-214* has been correlated to all its targets annotated as TF (Step 2), and each of them with the intragenic miRNAs (if any) of the gene they target (Step 3). At this point, with the full list of possibly modulated Intragenic miRNAs, the pipeline searches microRNA.org to obtain the list of their possible targets. By filtering this list and keeping only those targets that correspond to one of the 73 genes of interest, the identification of the possible Type-2 interactions is complete.

2.3   Data Integration: Strategy, Bottlenecks, and Optimizations

To efficiently design and reliably use the data produced by a pipeline of this type, it is necessary to take into account a number of bottlenecks related to data integration and data reliability.

Given the large amount of available databases and their lack of standardization rules for data consistency and interoperability [24], the integration of information coming from heterogeneous sources is often a hard and computationally expensive task. To properly collect all necessary data, the pipeline needs to interrogate multiple databases; to integrate this data, it is necessary that records that refer to the same item (gene, TF, miRNA) share a set of compatible fields that can be used to link these data together. Field compatibility is based upon synonyms and accession numbers across the selected information sources. In this work we specifically designed and populated a custom database to cope with this issue. The database contains information from the databases listed in Sect. 2.1 enhanced with additional meta data. Such meta data are collections of accession numbers (i.e., Entrez GeneID and approved common symbol) whose presence simplifies SQL join operations on queries, both reducing the execution time and avoiding either multiple or ambiguous matches.

From a data reliability point of view, the type of analysis and data integration performed by this pipeline is obviously affected by the reliability of the source data extracted by online repositories. In the presented pipeline there are two critical and potential sources of unreliable or inconsistent data:

1. miRNA and TF targets;
2. TF regulatory action (enhancer, silencer).

In the case of miRNA and TF targets, it is well known how the different algorithms used to predict the potential targets can lead to very different results. In this case a possible way to approach the problem is to define a policy to merge similar information coming from different sources (i.e., TF targets extracted from TFE and TargetMine, or miRNA targets extracted from different databases). In this scenario, the pipeline has to handle potential incoherencies between the data sources. In the presented work, to handle TF targets we designed two user selectable policies: (i) intersection or (ii) union of data. In the first case, only targets available in both databases are selected, whereas in the second case the union of the targets of both databases is considered. Both policies offer drawbacks and advantages: while intersecting data among sources seems to offer a more reliable dataset but possibly a higher rate of false negatives, merging multiple sources results in a larger dataset is able to provide a more investigative outcome but also a higher probability of false positives. Although more keen to false-positives, the results presented in this work are obtained using the "union of data" policy in order to include in the analysis as much information as possible. This choice is also justified by the high level of abstraction of the proposed work together with its aim to infer putative relations that will have in any case to be biologically validated.

For what concerns miRNA targets, the presented pipeline uses, for now, a single database (microRNA.org) and relies on a mirSVR score threshold to exclude target predictions that are too unreliable. Nevertheless, also in this case an approach based on data integration policies could be more reliable and give the user more flexibility. In particular the availability of several different miRNA target prediction databases could allow the implementation of a "majority voting" policy, where a target is considered reliable if it appears in at least k *out of* n databases, with k selectable by the user.

Finally, concerning TF regulatory action, it is important to point out here that the proposed computational analysis cannot always predict the sign of the resulting differential expression (up or down regulation). In fact, following the Type-2 regulatory chain, if *miR-214* is silenced the expression of the target protein is very likely inhibited. If, instead, *miR-214* is over expressed, the regulatory module "removes" the inhibition and allows the target gene expression to possibly change. The only realistic way to experimentally verify the presence of the Type-2 regulatory module is to correlate the over expression of *miR-214* with the under expression of the cascade TF $\rightarrow$ gene $\rightarrow$ miRNA that follows *miR-214* (see Fig. 1). This is obviously true unless the transcription factor acts as a repressor of its own target, which is statistically unlikely to happen. The type of regulatory action of a TF on its target is an information that is very difficult to find and the only database that offers it (Transcription Factor Encyclopedia) does so for a very limited set of TFs. Consequently, for most motifs it is not possible to predict if its presence is or not compatible with the actual modulation detected in the expression experiment.

2.4 Biological Methods

Computational predictions have been validated against the following biological setup.

Cell Culture. MA-2 cells were provided by R.O. Hynes [25] and maintained as described in [6].

Transient Transfections of Pre-miRs. To obtain transient *miR-214* over expression, cells were plated in 6-well plates at 30–50% confluency and transfected 24 h later using RNAiFect (QIAGEN, Stanford, CA) reagent, according to manufacturers instructions, with 75 nM Pre-miR™ miRNA Precursor Molecules-Negative Control (a non-specific sequence) or Pre-miR-214.

RNA Isolation and qRT-PCR for MiRNA or mRNA Detection. Total RNA was isolated from cells using TRIzol® Reagent (Invitrogen Life Technologies, Carlsbad, CA). qRT-PCRs for miR detection were performed with TaqMan® MicroRNA Assays *hsa-miR-33a* assay ID 002306, U6 snRNA assay ID001973 (all from Applied Biosystems, Foster City, CA) on 10 ng total RNA according to the manufacturer's instructions. For mRNA detection, 1 ug of DNAse-treated RNA (DNA-free™ kit, Ambion, Austin, TX) was retrotranscribed with RETROscript™ reagents (Ambion, Austin, TX) and qRT-PCRs were carried out using SREBPF2 gene-specific primers (FW:gccctggaagtgac agagag, RV: tgctttcccagggagtga) and the Probe #21 of the Universal Probe Library (Roche, Mannheim, GmbH) using a 7900HT Fast Real Time PCR System. The expression of the U6 small nucleolar RNA or of 18 S was quantitatively normalized for miR or mRNA detection, respectively. The relative expression levels between samples were calculated using the comparative delta CT (threshold cycle number) method (2-DDCT) with a control sample as the reference point [26].

3   Results and Discussion

Using the presented pipeline we were able to identify 27 Type-2 interactions. No Type-0 or Type-1 have been identified. All results have been manually verified and the pipeline has also been tested against very simple (and known) examples of each interaction type. Thus, the fact that no Type-0 and Type-1 interactions were found has no biological meaning: it only shows that, in the available database, there is no evidence of their presence.

The 27 Type-2 interactions target 22 out of the 73 considered *miR-214* potential interacting proteins, which have been marked in green in Table 1. The full list of the 27 identified regulatory modules is shown in Table 2.

From our predictions, *miR-214* influences two transcription factors: *NFKB1* and *TP53* (average mirSVR = −0.4). Seven of the genes regulated by these two

190   G. Politano et al.

Table 2. The 27 Type-2 regulatory modules related to miR-214 as obtained by the pipeline after data scraping. The set of final targets (surface protein in the table) is limited to the 73 genes listed in Table 1. The first 9 modules have been experimentally validated.

miR_214	mirSVR	TF	miRNA_Host	Intragenic_miRNA	Surface Protein	mirSVR
miR-214	-0.4056	NFKB1	SREBF2	hsa-mir-33a	ALCAM	-0.504
miR-214	-0.4056	NFKB1	SREBF2	hsa-mir-33a	POSTN	-0.9944
miR-214	-0.4056	NFKB1	SREBF2	hsa-mir-33a	TFAP2A	-1.3043
miR-214	-0.4056	NFKB1	SREBF2	hsa-mir-33a	ADAM9	-0.8819
miR-214	-0.4056	NFKB1	SREBF2	hsa-mir-33a	NCAM1	-1.1293
miR-214	-0.4056	NFKB1	SREBF2	hsa-mir-33a	SEMA3A	-1.0884
miR-214	-0.4056	NFKB1	SREBF2	hsa-mir-33a	PVRL2	-0.3633
miR-214	-0.4056	NFKB1	SREBF2	hsa-mir-33a	JAG1	-0.7951
miR-214	-0.4056	NFKB1	SREBF2	hsa-mir-33a	EGFR	-0.5771
miR-214	-0.4056	NFKB1	SVIL	hsa-mir-604	MMP2	-0.5526
miR-214	-0.4056	NFKB1	SVIL	hsa-mir-604	CEACAM5	-0.6373
miR-214	-0.4056	NFKB1	C11orf10	hsa-mir-611	THY1	-0.3774
miR-214	-0.4056	NFKB1	C11orf10	hsa-mir-611	NCAM1	-0.4402
miR-214	-0.4056	NFKB1	APOLD1	hsa-mir-613	MET	-0.8579
miR-214	-0.4056	NFKB1	APOLD1	hsa-mir-613	ALCAM	-0.5254
miR-214	-0.4056	NFKB1	APOLD1	hsa-mir-613	TIMP3	-0.582
miR-214	-0.4056	NFKB1	APOLD1	hsa-mir-613	CEACAM1	-0.9242
miR-214	-0.4056	NFKB1	APOLD1	hsa-mir-613	BMPR1B	-0.7156
miR-214	-0.4056	NFKB1	APOLD1	hsa-mir-613	TFAP2C	-0.6921
miR-214	-0.4056	NFKB1	APOLD1	hsa-mir-613	JAG1	-0.4012
miR-214	-0.4056	NFKB1	NFATC2	hsa-mir-3194	CD99	-0.8366
miR-214	-0.4056	NFKB1	NFATC2	hsa-mir-3194	CD40	-0.7136
miR-214	-0.3966	TP53	GDF15	hsa-mir-3189	JAM3	-0.8858
miR-214	-0.3966	TP53	GDF15	hsa-mir-3189	PVRL2	-0.5146
miR-214	-0.3966	TP53	GDF15	hsa-mir-3189	HBEGF	-0.3806
miR-214	-0.3966	TP53	GDF15	hsa-mir-3189	LRP6	-0.6945
miR-214	-0.3966	TP53	BBC3	hsa-mir-3191	HBEGF	-0.8502

TFs were identified as host genes for miRNAs targeting at least one of the 73 *miR-214* modulated proteins: *APOLD1*, *BBC3*, *C11orf10*, *GDF15*, *NFATC2*, *SREBF2*, and *SVIL*. The hosted miRNAs are: *hsa-mir-33a*, *hsa-mir-604*, *hsa-mir-611*, *hsa-mir-613*, *hsa-mir-3189*, *hsa-mir-3191*, and *hsa-mir-3194*. The average mirSVR score is significantly low (average mirSVR < -0.71). The high significance of the mirSVR scores, resulting from interactions between the intragenic miRNAs and their target proteins, is particularly evident for *TFAP2A*, which outperforms the others with a mirSVR score of -1.3043.

In this work, as a preliminary experimental validation, we focused our attention on the first 9 identified regulatory modules involving *miR-214*, *NFKB1*, *SREBF2*, *miR-33a* and 9 of the 73 considered proteins (*ALCAM*, *POSTN*, *TFAP2A*, *ADAM9*, *NCAM1*, *SEMA3A*, *PVRL2*, *JAG1* and *EGFR1*). We evaluated *miR-33a* and *SREBPF2* expression levels following *miR-214* over expression in MA-2 melanoma cells and we observed a decrease in *miR-33a* and *SREBF2* expression as shown in Fig. 3.

The observed co-regulation of *miR-33a* and *SREBPF2* is in agreement with literature data published in [27], thus supporting our computational predictions. The down-regulation of *miR-33a* following *miR-214* over expression could contribute to *miR-214*-mediated cell invasion, in fact it has been demonstrated that an enforced expression of *miR-33a* inhibits the motility of lung cancer cells [28].

This regulatory module resulted to be very interesting also because *SREBPF2* and *miR-33a* act in concert to control cholesterol homeostasis [27].

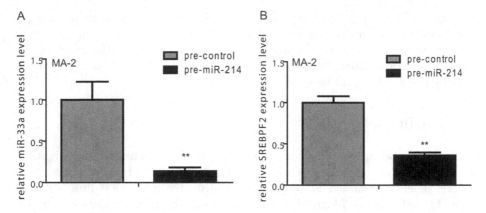

Fig. 3. miR-33a, and SREBPF2 expression modulations. (A) miR-33a expression levels tested by qRT-PCR in the MA-2 melanoma cell line following transfection with miR-214 precursors or their negative controls (pre-miR-214 or control). (B) SREBPF2 mRNA expression levels were evaluated in MA-2 cells by Real Time PCR analysis 72 h following transient transfection with miR-214 precursors or their negative controls (pre-miR-214 or control). Results are shown as fold changes (meanSD of triplicates) relative to controls, normalized on U6 RNA level and 18S, respectively. All experiments performed in our work were tested for miR-214 modulations; representative results are shown here.

In fact, *SREBPF2* controls the expression of many cholesterogenic and lipogenic genes, such as low-density lipoprotein (*LDL*) receptor, 3-hydroxy-3-methylglutaryl coenzyme A reductase, and fatty acid synthase. Instead, *miR-33a* targets the adenosine triphosphate-binding cassette A1 (*ABCA1*) cholesterol transporter, a key mediator of intracellular cholesterol efflux from liver to apolipoprotein A-I (*apoA-I*) to obtain high-density lipoprotein (*HDL*). Considering that the lipogenic pathway is a metabolic hallmark of cancer cells, these preliminary data suggest a potential role of *miR-214* in this aspect of cancer formation and progression. Our hypothesis is further supported by experimental results [6], obtained from microarray analysis in a context of *miR-214* over expression.

We applied an Ingenuity Functional Analysis in order to look for molecular and cellular functions associations within the almost 500 differentially expressed genes detected by microarray analysis comparing cells over expressing *miR-214* versus controls. The Ingenuity Pathways Knowledge Base (http://www.ingenuity.com/) is currently the world largest database of knowledge on biological networks, with annotations performed by experts. The significance value obtained with the Functional Analysis for a dataset is a measure of the likelihood that the association between a set of Functional Analysis molecules in our experiment and a given process or pathway is due to random chance. The p-value is calculated using the right-tailed Fisher Exact Test and it considers both the number of functional analysis molecules that participate in that function and the total number of molecules that are known to be associated with that function in the Ingenuity Knowledge Base. In our case, the most significant functions associated to our dataset resulted to be Cellular Assembly and Organization (7.08E-04 ÷ 3.95E-02, 25 molecules) and Lipid Metabolism (9.54E-04 ÷ 4.23E-02, 18 molecules).

4   Conclusions

In this paper we presented a computational pipeline created for investigating possible regulatory pathways between a miRNA and a set of target genes.

The pipeline identified 27 putative regulatory pathways that link together *miR-214* and a set of 73 proteins already annotated as co-regulated with the miRNA in melanomas. A preliminary experimental validation performed on 9 out of the 27 pathways provided interesting insights about the regulatory mechanisms involving *miR-214* in the considered disease. The analysis suggests the involvement of *miR-214* in metabolic pathways that could control metastatization. Moreover, the study highlights the relevance of specific *miR-214* modulated genes, such as *ALCAM*, *HBEGF*, *JAG1*, *NCAM1*, and *PVRL2*, which correspond to surface proteins redundantly regulated by multiple pathways. Further laboratory experiments are under way to complete the validations of the full set of identified regulatory modules. Nevertheless, the preliminary results presented in this work already represent a significant achievement that seems to confirm the overall quality of the predictions obtained with the proposed computational approach.

Acknowledgements. This work has been partially supported by grants from Regione Valle d'Aosta (for the project: "Open Health Care Network Analysis" - CUP B15G1300 0010006), from the Italian Ministry of Education, University.

References

1. Beezhold, K.J., Castranova, V., Chen, F.: Review microprocessor of microRNAs: regulation and potential for therapeutic intervention. Molecular Cancer 9 (2010)
2. Tu, K., Yu, H., Hua, Y.J., Li, Y.Y., Liu, L., Xie, L., Li, Y.X.: Combinatorial network of primary and secondary microrna-driven regulatory mechanisms. Nucleic Acids Res. **37**, 5969–5980 (2009)
3. Benso, A., Di Carlo, S., Politano, G., Savino, A.: A new mirna motif protects pathways' expression in gene regulatory networks. In: Proceedings IWBBIO 2013: International Work-Conference on Bioinformatics and Biomedical Engineering, pp. 377–384 (2013)
4. Di Carlo, S., Politano, G., Savino, A., Benso, A.: A systematic analysis of a miRNA inter-pathway regulatory motif. J. Clin. Biol. **3**, 20 (2013)
5. Yuan, X., Liu, C., Yang, P., He, S., Liao, Q., Kang, S., Zhao, Y.: Clustered microRNAs' coordination in regulating protein-protein interaction network. BMC Syst. Biol. **3**(1), 65 (2009)

6. Penna, E., Orso, F., Cimino, D., Tenaglia, E., Lembo, A., Quaglino, E., Poliseno, L., Haimovic, A., Osella-Abate, S., De Pittà, C., et al.: microRNA-214 contributes to melanoma tumour progression through suppression of TFAP2C. EMBO J. **30**, 1990–2007 (2011)
7. Penna, E., Orso, F., Cimino, D., Vercellino, I., Grassi, E., Quaglino, E., Turco, E., Taverna, D.: miR-214 coordinates melanoma progression by upregulating ALCAM through TFAP2 and miR-148b downmodulation. Cancer Res. **73**, 4098–4111 (2013)
8. Zhao, M., Sun, J., Zhao, Z.: Synergetic regulatory networks mediated by oncogene-driven micrornas and transcription factors in serous ovarian cancer. Mol. BioSyst. **9**, 3187–3198 (2013)
9. Peng, C., Wang, M., Shen, Y., Feng, H., Li, A.: Reconstruction and analysis of transcription factor-miRNA co-regulatory feed-forward loops in human cancers using filter-wrapper feature selection. PLoS ONE **8**, e78197 (2013)
10. Delfino, K.R., Rodriguez-Zas, S.L.: Transcription factor-microRNA-target gene networks associated with ovarian cancer survival and recurrence. PLoS ONE **8**, e58608 (2013)
11. Iwama, H.: Coordinated networks of micrornas and transcription factors with evolutionary perspectives. In: Schmitz, U., Wolkenhauer, O., Vera, J. (eds.) MicroRNA Cancer Regulation. Advances in Experimental Medicine and Biology, vol. 774, pp. 169–187. Springer, Netherlands (2013)
12. Betel, D., Wilson, M., Gabow, A., Marks, D.S., Sander, C.: The microRNA.org resource: targets and expression. Nucleic Acids Res. **36**, D149–D153 (2008)
13. John, B., Enright, A., Aravin, A., Tuschl, T., Sander, C., Marks, D.: Human microRNA targets. PLoS Biol. **2**, e363 (2004)
14. Betel, D., Koppal, A., Agius, P., Sander, C., Leslie, C.: Comprehensive modeling of microRNA targets predicts functional non-conserved and non-canonical sites. Genome Biol. **11**, R90 (2010)
15. Wasserman Lab: Transcription factor encyclopedia (TFe) (2012). http://www.cisreg.ca/cgi-bin/tfe/home.pl
16. The Mizuguchi Laboratory: Targetmine (2013). http://targetmine.nibio.go.jp/
17. Chen, Y.A., Tripathi, L.P., Mizuguchi, K.: Targetmine, an integrated data warehouse for candidate gene prioritisation and target discovery. PLoS ONE **6**, e17844 (2011)
18. Linhart, C., Halperin, Y., Shamir, R.: Amadeus (2013). http://acgt.cs.tau.ac.il/amadeus/download.html
19. Linhart, C., Halperin, Y., Shamir, R.: Transcription factor and microRNA motif discovery: the amadeus platform and a compendium of metazoan target sets. Genome Res. **18**, 1180–1189 (2008)
20. NCBI: Entrez programming utilities help (2013). http://www.ncbi.nlm.nih.gov/books/NBK25501
21. mirbase.org: mirbase.org (2013). http://www.mirbase.org
22. Griffiths-Jones, S., Grocock, R.J., Van Dongen, S., Bateman, A., Enright, A.J.: miRBase: microRNA sequences, targets and gene nomenclature. Nucleic Acids Res. **34**, D140–D144 (2006)
23. MicroRNA.org: Microrna.org - release notes (2013). http://www.microrna.org/microrna/releaseNotes.do

24. Gaudet, P., Bairoch, A., Field, D., Sansone, S. A., Taylor, C., Attwood, T. K., Bateman, A., Blake, J. A., Bult, C. J., Cherry, J.M., Chisholm, R.L., Cochrane, G., Cook, C.E., Eppig, J. T., Galperin, M. Y., Gentleman, R., Goble, C. A., Gojobori, T., Hancock, J. M., Howe, D. G., Imanishi, T., Kelso, J., Landsman, D., Lewis, S. E., Karsch Mizrachi, I., Orchard, S., Ouellette, B. F., Ranganathan, S., Richardson, L., Rocca-Serra, P., Schofield, P. N., Smedley, D., Southan, C., Tan, T. W., Tatusova, T., Whetzel, P. L., White, O., Yamasaki, C., on behalf of the BioDBCore working group: Towards biodbcore: a community-defined information specification for biological databases. Database 2011 (2011)
25. Xu, L., Shen, S.S., Hoshida, Y., Subramanian, A., Ross, K., Brunet, J.P., Wagner, S.N., Ramaswamy, S., Mesirov, J.P., Hynes, R.O.: Gene expression changes in an animal melanoma model correlate with aggressiveness of human melanoma metastases. Mol. Cancer Res. **6**, 760–769 (2008)
26. Bookout, A.L., Mangelsdorf, D.J.: Quantitative real-time PCR protocol for analysis of nuclear receptor signaling pathways. Nucl. Recept. Signal. **1**, e012 (2003)
27. Najafi-Shoushtari, S.H., Kristo, F., Li, Y., Shioda, T., Cohen, D.E., Gerszten, R.E., Näär, A.M.: MicroRNA-33 and the SREBP host genes cooperate to control cholesterol homeostasis. Science **328**, 1566–1569 (2010)
28. Rice, S.J., Lai, S.C., Wood, L.W., Helsley, K.R., Runkle, E.A., Winslow, M.M., Mu, D.: MicroRNA-33a mediates the regulation of high mobility group at-hook 2 gene (HMGA2) by thyroid transcription factor 1 (TTF-1/NK2-1). J. Biol. Chem. **288**, 16348–16360 (2013)

Empirical Study of Domain Adaptation Algorithms on the Task of Splice Site Prediction

Nic Herndon$^{(\boxtimes)}$ and Doina Caragea

Kansas State University, 234 Nichols Hall, Manhattan, KS 66506, USA
{nherndon,dcaragea}@ksu.edu

Abstract. Many biological problems that rely on machine learning do not have enough labeled data to use a classic classifier. To address this, we propose two domain adaptation algorithms, derived from the multinomial naïve Bayes classifier, that leverage the large corpus of labeled data from a similar, well-studied organism (the source domain), in conjunction with the unlabeled and some labeled data from an organism of interest (the target domain). When evaluated on the splice site prediction, a difficult and essential step in gene prediction, they correctly classified instances with highest average area under precision-recall curve (auPRC) values between 18.46 % and 78.01 %. We show that the algorithms learned meaningful patterns by evaluating them on shuffled instances and labels. Then we used one of the algorithms in an ensemble setting and produced even better results when there is not much labeled data or the domains are distantly related.

Keywords: Domain adaptation · Naïve Bayes · Splice site prediction · Unbalanced data · Ensemble learning

1 Introduction

In recent years, a rapid increase in the volume of digital data generated has been observed. The total volume of digital data was estimated to double every 18 months [1]. Contributing to this growth, in the field of biology, large amounts of raw genomic data are generated with next generation sequencing technologies, as well as data derived from primary sequences. The availability and scale of the biological data creates great opportunities in terms of medical, agricultural, and environmental discoveries, to name just a few.

A stepping stone towards advancements in such fields is the identification of genes in a genome. Accurate identification of genes in eukaryotic genomes depends heavily on the ability to accurately determine the location of the splice sites [2,3], the sections in the DNA that separate exons from introns in a gene. In addition to identifying the gene structure, the splice sites also determine the amino acid composition of the proteins encoded by the genes. Therefore, considering that the content of a protein plays a major role with respect to its function, the splice site prediction is a crucial task in identifying genes and ultimately the functions of their products.

© Springer International Publishing Switzerland 2015
G. Plantier et al. (Eds.): BIOSTEC 2014, CCIS 511, pp. 195–211, 2015.
DOI: 10.1007/978-3-319-26129-4_13

To identify genes, we need to identify two types of splice sites: donor and acceptor. The donor splice site indicates where an exon ends and an intron begins, while the acceptor splice site indicates where an intron ends, and an exon begins. Virtually most donor sites are the GT dimer and most acceptor sites are the AG dimer, with very few exceptions. However, not every GT or AG dimer is a splice site. In fact, only a very small percent of them are (possibly less than one percent), which makes the task of identifying splice sites very difficult.

Nonetheless, this problem can be addressed with supervised machine learning algorithms, which have been successfully used for many biological problems, including gene prediction. For example, hidden Markov models have been used for *ab initio* gene predictions, while support vector machines (SVMs) have been used successfully for problems such as identification of translation initiation sites [4,5], labeling gene expression profiles as malign or benign [6], *ab initio* gene prediction [2], and protein function prediction [7].

However, one drawback of these algorithms is that they require large amounts of labeled data to learn accurate classifiers, and many times labeled data is not available for a problem of interest, yet it is available for a different but related problem. For example, in biology, a newly sequenced organism is generally scarce in labeled data, while a well-studied model organism is rich in labeled data. However, the use of the classifier learned for the related problem to classify unlabeled data for the problem of interest does not always produce accurate predictions. A better alternative is to learn a classifier in a domain adaptation (DA) framework. In this setting, the large corpus of labeled data from the related, well studied organism is used in conjunction with available labeled and unlabeled data from the new organism.

Towards this goal we propose two similar algorithms, described in Sect. 3.2. Each algorithm trains a classifier in this configuration, by using the large volume of labeled data from a well studied organism, also called source domain, some labeled data, and any available unlabeled data from the new organism, called target domain. Once learned, the classifier can be used to classify new data from the new organism. These algorithms are similar to the algorithms in [8–10], which produced good results on the tasks of sentiment analysis and protein localization, respectively.

The algorithm proposed by [9] uses a weighted multinomial Naïve Bayes classifier combined with the iterative approach of the Expectation-Maximization algorithm and self-training [11–13]. It iterates until the probabilities in the expectation step converge. In the maximization step, the prior probabilities and the likelihood are estimated using a weighted combination between the labeled data from the source domain and the labeled and unlabeled data from the target domain. In the expectation step, the conditional class probabilities for each instance in the target unlabeled dataset are estimated with the probability values from the maximization step using Bayes theorem. After each iteration, a number of instances from the unlabeled target dataset are considered labeled and moved to the labeled dataset. The number of these instances is proportional to the prior probability, with a minimum of one instance selected from each class. For example, if the target labeled dataset had 1 % positive instances and 99 % negative instances, and at each iteration we select 10 instances, one instance with the top positive probability and nine instances with the top negative probabilities would be selected. In addition, the weight is shifted from the source data to the target data with each iteration.

While this algorithm worked well on the task of protein localization prediction, it did not perform well on the task of splice site prediction [9]. To improve its classification accuracy on the splice site prediction task, we made the following four changes to the original algorithm:

- We normalized the counts used in computing the prior probability and the likelihood, and assigned different weights to the target labeled and unlabeled datasets.
- We used mutual information to rank the features instead of the marginal probabilities of the features.
- We assigned different weights to the features instead of selecting which features to use from the source domain.
- We used features that took into consideration the location of nucleotides instead of counts of 8-mers generated with a sliding window.

With these changes, the two algorithms produced promising results, shown in Sect. 4, when evaluated on the splice site prediction problem with the data presented in Sect. 3.4. They increased the highest average auPRC from about 1 % in [9] all the way up to 78.01 %. In addition, to further asses the relevance of these algorithms, we evaluated them on the same data sets but with the data and labels shuffled and show that in this case the algorithms were unable to learn any patterns, leading to poor results. And, in a final assesment trying to address the highly unbalanced data sets, we evaluated how the proposed method performed when we split the training data into class-balanced data sets and used ensemble learning on these data sets.

2 Related Work

Although there are good domain adaptation results for text classification, even with algorithms that make significant simplifying assumptions, such as naïve Bayes, there are only a few domain adaptation algorithms designed for biological problems. For example, for text classification, [14] proposed an algorithm that combined naïve Bayes (NB) with expectation-maximization (EM) for classifying text documents into top categories. This algorithm performed better than SVM and naïve Bayes classifiers. Two other domain adaptation algorithms that used a combination of NB with EM are [8,10]. The latter produced promising results on the task of sentiment analysis of text documents, and the former had good results on the task of protein localization.

However, to the best of our knowledge, for the task of splice site prediction, up until now, most of the work used supervised learning classifiers with only two exceptions in the domain adaptation setting: one that analyzed several support vector machine algorithms [15], and another that used naïve Bayes [9]. The latter did not obtain good results, primarily due to the features generated. It used the number of occurrences in each instance of the k-mers of length eight generated with a sliding window.

One major challenge with the task of splice site prediction is the fact that the classes are highly imbalanced, with only about one percent positive instances. This challenge is common to other problems as well, such as intrusion detection, medical diagnosis, risk management, and text classification, to name a few. The proposed solutions address this problem at the algorithmic level or at the data level through resampling. For an overview of solutions to imbalanced data sets see [16,17].

For supervised learning, [18] evaluated the discriminating power of each position in the DNA sequence around the splice site using the chi-square test. With this information, they proposed a SVM algorithm with an RBF kernel function that used a combination of scaled component features, nucleotide frequencies at conserved sites, and correlative information of two sites, to train a classifier for the human genome. Other supervised learning approaches based on SVM include [19–21], while [22] proposed an algorithm for splice site prediction based on Bayesian networks, [23] proposed a hidden Markov model algorithm, and [24] proposed a method using Bahadur expansion truncated at the second order. However, supervised learning algorithms typically require a large volume of labeled data to train a classifier.

For domain adaptation setting, [15] obtained good results using an SVM classifier with weighted degree kernel [25], especially when the source and target domains were not close. However, the complexity of SVM classifiers increases with the number of training instances and the number of features, when training the classifier [15]. Besides, the classification results are not as easy to interpret as the results of probabilistic classifiers, such as naïve Bayes, to gain insight into the problem studied. For example, [15] further processed the results to analyze the relevant biological features. In addition, their algorithms did not use the large volume of target unlabeled data. Although unlabeled, this dataset could improve the classification accuracy of the algorithm.

3 Methodology

The two algorithms we propose are based on the algorithm presented in our previous work [9]. That algorithm modified the algorithm introduced by [10], by using self-training and the labeled data from the target domain, to make it suitable for biological sequences. To highlight the changes made to our previous algorithm we'll describe it here.

3.1 Naïve Bayes DA for Biological Sequences

The algorithm in [9] was designed to use labeled data from the source domain in conjunction with labeled and unlabeled data from the target domain to build a classifier to be used on the target domain. It is an iterative algorithm that uses a combination of the Bayes' theorem with the simplifying assumption that features are independent, and the expectation-maximization algorithm [26], to estimate the posterior probability as proportional to the product of the prior and the likelihood:

$$P(c_k \mid d_i) \propto P(c_k) \prod_{t=1}^{|V|} [P(w_t \mid c_k)]^{N_{t,i}^T} \tag{1}$$

where the prior is defined as

$$P(c_k) = \frac{(1-\lambda) \sum_{i'=1}^{|D_S|} P(c_k \mid d_{i'}) + \lambda \sum_{i''=1}^{|D_T|} P(c_k \mid d_{i''})}{(1-\lambda)|D_S| + \lambda|D_T|} \tag{2}$$

and the likelihood is defined as

$$P(w_t \mid c_k) = \frac{(1-\lambda)(\eta_t N_{t,k}^S) + \lambda N_{t,k}^T + 1}{(1-\lambda) \sum_{t=1}^{|V|} \eta_t N_{t,k}^S + \lambda \sum_{t=1}^{|V|} N_{t,k}^T + |V|} \tag{3}$$

where λ is a weight factor between the two domains:

$$\lambda = \min\{\delta \cdot \tau, 1\}$$

τ is the iteration number, with a value of 0 for the first iteration, and $\delta \in (0,1)$ is a constant that indicates how fast the weight of the source domain decreases while the weight of the target domain increases; c_k stands for class k, d_i for document i, and w_t for feature t.

Let $|D_x|$ be the number of instances in x domain ($x \in \{S, T\}$, where S denotes the source domain and T denotes the target domain), $|V|$ be the number of features, and $N_{t,k}^x$ be the number of times feature w_t occurs in x domain in instances labeled with class c_k:

$$N_{t,k}^x = \sum_{i=1}^{|D_x|} N_{t,i}^x P(c_k \mid d_i)$$

where $N_{t,i}^x$ is the number of occurrences in x domain of feature w_t in instance d_i.

$$\eta_t = \begin{cases} 1, & \text{if feature } w_t \text{ is generalizable.} \\ 0, & \text{otherwise.} \end{cases}$$

To determine which features from the source domain are generalizable to the target domain, i.e., generalizable from use on D_S to use on D_T, they are ranked with the following measure and only the top ranking features are kept:

$$f(w_t) = \log \frac{P_S(w_t) \cdot P_T(w_t)}{|P_S(w_t) - P_T(w_t)| + \alpha} \tag{4}$$

where α is a small constant used to prevent division by 0 when $P_S = P_T$, and P_x is the probability of feature w_t in x domain, i.e., the number of times feature w_t occurs in x domain divided by the sum over all features of the times each feature occurs in x domain.

This algorithm iterates until convergence. In the maximization step, the prior and likelihood are simultaneously estimated using (2) and (3), respectively, while in the expectation step the posterior for the target unlabeled instances is estimated using (1). In addition to expectation-maximization, this algorithm uses self-training, i.e., at each iteration it selects a number of unlabeled instances with highest class probabilities, proportional to the class distribution, and considers them to be labeled during subsequent iterations, while the remaining unlabeled instances are assigned "soft labels" for the following iteration. By soft labels we mean that the class probability distribution is used instead of the class label. For the target domain, during the first iteration, only the instances from the labeled dataset are used, and for the rest of the iterations, instances from both labeled and unlabeled datasets are used.

Note that when the algorithm goes through just one iteration and $\lambda = 1$, the prior and likelihood, (2) and (3), respectively, reduce to

$$P(c_k) = \frac{\sum_{i=1}^{|D_S|} P(c_k \mid d_i)}{|D_S|} \tag{5}$$

and

$$P(w_t \mid c_k) = \frac{\sum_{i=1}^{|D_S|} N_{t,i} P(c_k \mid d_i) + 1}{\sum_{t=1}^{|V|} \sum_{i=1}^{|D_S|} N_{t,i} P(c_k \mid d_i) + |V|} \tag{6}$$

respectively, which are the equations for the multinomial naïve Bayes classifier[27] trained and tested with data from the same domain.

Although the algorithm in [9] performed well on the task of protein localization with maximum auPRC values between 73.98 % and 96.14 %, on the task of splice site prediction, the algorithm performed poorly. The splice site prediction is a more difficult task because, in general, less than 1 % of the AG dimer occurrences in a genome correspond to splice sites. To simulate this proportion, unbalanced datasets were used, with each containing only about 1 % positive instances. Therefore, the training sets for splice site prediction were much more unbalanced than the training sets for the protein localization, leading to worse performance for the former.

3.2   Our Approach

One major drawback of the algorithm in [9] is that it assigns low weight to the target data (through λ in (2) and (3)), including the labeled data, during the first iterations. This biases the classifier towards the source domain. However, it is not effective to only assign a different weight to the target labeled data in (2) and (3). This is because we'd like to use much more labeled instances from the source domain, as well as much more unlabeled instances from the target domain, i.e., $|D_S| \gg |D_{TL}|$ and $|D_{TU}| \gg |D_{TL}|$, where subscripts S, TL, and TU stand for source data, target labeled data, and target unlabeled data, respectively. This would cause the sums and counts for the target labeled data in these two equations to be much smaller than their counterparts for the source data and target unlabeled data, rendering the weight assignment useless. Instead, we need to also normalize these values, or better yet, use their probabilities. Thus, we estimate the prior as

$$P(c_k) = \beta P_{TL}(c_k) + (1 - \beta)[(1 - \lambda)P_S(c_k) + \lambda P_{TU}(c_k)] \tag{7}$$

and the likelihood as

$$
\begin{aligned}
P(w_t \mid c_k) = {} & \beta P_{TL}(w_t \mid c_k) \\
& + (1 - \beta)[(1 - \lambda)P_S(w_t \mid c_k) + \lambda P_{TU}(w_t \mid c_k)]
\end{aligned}
\tag{8}
$$

where $\beta \in (0, 1)$ is a constant weight, and λ is defined the same as in our previous approach.

In addition to using different formulas for prior and likelihood, we made a second change to the algorithm in [9]. We replaced the measure for ranking the features, (4), with the following measure in (9). We made this change because the goal of ranking the features is to select top features in terms of their correlation with the class, or assign them different weights: higher weights to features that are more correlated with the class, and lower weights to the features that are less correlated with the class. Therefore, the mutual information [28] of each feature is a more appropriate measure in determining the correlation of the feature with the class rather than the marginal probability of the feature. With this new formula, the features are ranked better based on their generalizability between the source and target domains:

$$f(w_t) = \frac{I_S(w_t; c_k) \cdot I_{TL}(w_t; c_k)}{|I_S(w_t; c_k) - I_{TL}(w_t; c_k)| + \alpha} \tag{9}$$

where

$$I_x(w_t, c_k) = \sum_{t=1}^{|V|} \sum_{k=1}^{|C|} P_x(w_t, c_k) \log \frac{P_x(w_t, c_k)}{P_x(w_t)P_x(c_k)}$$

is the mutual information between feature w_t and class c_k, and $x \in \{S, TL\}$. The numerator in (9) ranks higher the features that have higher mutual information in their domains, while the denominator ranks higher the features that have closer mutual information between the domains, as shown in Fig. 1.

If instead of ranking the features and using the top ranked features, we want to assign different weights to the features and higher values for $f(w_t)$ we can compute them with the formula in (10). For the splice site prediction problem

Algorithm 1. DA with feature selection.

1: *Select the features to be used from the source domain* using (9).
2: Simultaneously compute the prior and likelihood, 7 and 8, respectively. Note that for the source domain all labeled instances are used *but only with the features selected in step 1*, while for the target domain only labeled instances are used with all their features.
3: Assign labels to the unlabeled instances from the target domain using (1). Note that we use self-training, i.e., a number of instances, proportional to the class distribution, with the highest class probability are considered to be labeled in subsequent iterations.
4: **while** labels assigned to unlabeled data change **do**
5:    **M-step**: Same as step 2, except that we also use the instances in the target unlabeled dataset; for this dataset we use the class for the self-trained instances, and the class distribution for the rest of the instances.
6:    **E-step**: Same as step 3.
7: **end while**
8: Use classifier on new target instances.

Algorithm 2. DA with weighted features.

1: *Assign different weights to the features of the source dataset* using either (9) or (10).
2: Simultaneously compute the prior and likelihood, 7 and 8, respectively. Note that for the source domain all labeled instances are used *but the features are assigned different weights in step 1*, while for the target domain only labeled instances are used with all their features.
3: Assign labels to the unlabeled instances from the target domain using (1). Note that we use self-training, i.e., a number of instances, proportional to the class distribution, with the highest class probability are considered to be labeled in subsequent iterations.
4: **while** labels assigned to unlabeled data change **do**
5:    **M-step**: Same as step 2, except that we also use the instances in the target unlabeled dataset; for this dataset we use the class for the self-trained instances, and the class distribution for the rest of the instances.
6:    **E-step**: Same as step 3.
7: **end while**
8: Use classifier on new target instances.

---

using the data from Sect. 3.4 with the features described in Sect. 3.5, most of the mutual information values are close to zero since each feature contributes very little to the classification of an instance. Therefore, the nominator is about one order of magnitude smaller than the denominator in (9), while in (10), the numerator and denominator have the same order of magnitude, resulting in higher values for $f(w_t)$.

The final change we made to [9] was the use of different features, as explained in Sect. 3.5.

$$f(w_t) = \frac{I_S(w_t; c_k) \cdot I_{TL}(w_t; c_k)}{(I_S(w_t; c_k) - I_{TL}(w_t; c_k))^2 + \alpha} \tag{10}$$

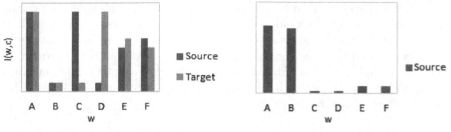

(a) Mutual information of features in both domains   (b) Rank of features in the source domain

Fig. 1. Ranking features in the source domain based on mutual information using (9), showcasing the different combinations of mutual information of the features in the two domains. Feature A has high mutual information in both domains, as shown on the left of subfigure (a), resulting in large numerator and small denominator, and thus a high rank, as shown on the left of subfigure (b). Feature B has low mutual information in both domains, resulting in small denominator, and thus a high rank, but lower than the rank of A since the numerator for A is greater than the numerator for B. Features C and D have high mutual information in one domain and low mutual information in the other domain, resulting in large denominator, and thus a low rank. Features E and F have close mutual information between the two domains, resulting in small denominator, and thus a higher rank than Features C and D. Note that features A through F are hypothetical features meant to show the ranking of the real features based on the patterns in the data.

Thus, the two algorithms we propose use the source labeled data and the target labeled and unlabeled data to train a classifier for the target domain. For the source domain, Algorithm 1 selects generalizable features to be used, while Algorithm 2 assigns different weights to the features. These differences are highlighted in steps 1 and 2 of the algorithms by using italics text. The algorithms iterate until convergence. In the first iteration, they use only the source and target labeled data to calculate and assign the posterior probabilities for the unlabeled data. Proportional to the prior probability distribution, the algorithms select top instances from the target unlabeled dataset and consider them to be labeled for subsequent iterations. For the rest of the iterations, the algorithms use the source labeled data, the target labeled data, and the target unlabeled data to build a classifier for labeling the remaining unlabeled instances. For the target unlabeled data, the algorithms use the probability distributions, while for the other datasets, source and target labeled, they use the labels of each instance when computing the prior probabilities and the likelihood.

3.3   Ensemble Learning

Considering that the data is highly imbalanced, we also evaluated our method in an ensemble learning setting. We used an approach similar to [29]. We sampled with no replacement the labeled data, from the source and target domains, and generated 99 balanced subsets, with 50 % positive instances and 50 %

negative instances each. That is, in each balanced subset we included all positive instances from the initial set along with 1 % negative instances sampled with no replacement from the initial set. We generated 99 balanced subsets because the ratio of positive to negative instances is 1 to 99. Then we ran our algorithm 99 times with sampled labeled data and whole unlabeled and test data sets from the target domain. When predicting the class for a new instance we used the majority voting of the 99 algorithms.

3.4 Data Set

To evaluate these algorithms, we used a dataset for which [9] did not perform well. Specifically, we used the splice site dataset[1], first introduced in [15], which contains DNA sequences of 141 nucleotides long, with the AG dimer at sixty-first position in the sequence, and a class label that indicates whether this dimer is an acceptor splice site or not. Although this dataset contains only acceptor splice sites, classifying donor splice sites can be done similarly to classifying the acceptor splice sites. This dataset contains sequences from one source organism, *C.elegans*, and four target organisms at increasing evolutionary distance, namely, *C.remanei*, *P.pacificus*, *D.melanogaster*, and *A.thaliana*. The source organism contains a set of 100,000 instances, while each target organism has three folds each of 1,000, 2,500, 6,500, 16,000, 25,000, 40,000, and 100,000 instances, respectively, that can be used for training, as well as three corresponding folds of 20,000 instances to be used for testing. In each file, there are 1 % positive instances with small variations (variance is 0.01) and the remaining instances are negative.

3.5 Data Preparation and Experimental Setup

To limit the number of experiments, we used the following datasets, as shown in Fig. 2:

- The 100,000 *C.elegans* instances as source labeled data used for training.
- Only the sets with 2,500, 6,500, 16,000, and 40,000 instances as labeled target data used for training.
- The set with 100,000 instances as unlabeled target data used for training.
- The corresponding folds of the 20,000 instances as target data used for testing.

We represent each sequence as a combination of features that indicate the nucleotide and "codon" present at each location, i.e., 1-mer and 3-mer, respectively. We chose these features to create a balanced combination of simple features, 1-mers, and features that capture some correlation between the nucleotides, 3-mers, while keeping the number of features small. For example, a sequence starting with AAGATTCGC... and class −1 would be represented in WEKA ARFF[2] format as:

---

[1] Downloaded from ftp://ftp.tuebingen.mpg.de/fml/cwidmer/
[2] WEKA Attribute-Relation File Format (ARFF) is described at http://www.cs.waikato.ac.nz/ml/weka/arff.html.

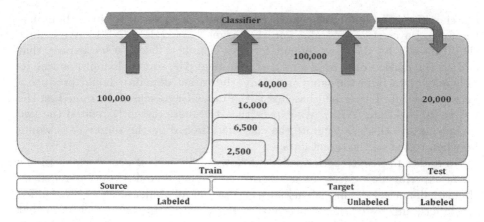

Fig. 2. Experimental setup. Our algorithms are trained on the source labeled dataset (100,000 instances) from *C.elegans*, target labeled instances (2,500, 6,500, 16,000, or 40,000 instances) and target unlabeled instances (100,000 instances) from *C.remanei*, *P.pacificus*, *D.melanogaster*, or *A.thaliana*. Once trained, each algorithm is tested on 20,000 instances from the corresponding target dataset.

```
@RELATION remanei_2500_0
@ATTRIBUTE NUCLEOTIDE1 {A,C,G,T}
  ⋮
@ATTRIBUTE NUCLEOTIDE141 {A,C,G,T}
@ATTRIBUTE CODON1 {AAA,AAC,...,TTT}
  ⋮
@ATTRIBUTE CODON139 {AAA,AAC,...,TTT}
@ATTRIBUTE cls {1,-1}
@DATA
A,A,G,A,T,T,C,G,C,...,AAG,AGA,GAT,...,-1
```

In our previous work we used a bag-of-words approach with k-mers of length eight. This led to 4^8 or 65,535 *sparse features*. Moreover, those features did not take into consideration the location of each nucleotide. We believe that the major improvement we achieved with our current algorithms is due to the features used, fewer and more location-aware.

We ran two different experiments with a grid search for the optimal values for $\beta \in \{0.1, 0.2, \ldots, 0.9\}$ and $\delta \in \{0.01, 0.02, \ldots, 0.08, 0.09, 0.1, 0.2\}$. In the first one, we used only the nucleotides as features, while in the second one we used nucleotides and codons as features. In both settings, we ran Algorithm 1 to select the features in the source domain (A1), Algorithm 2 using (9) to weigh the features in the source domain (A2E9), and Algorithm 2 with (10) to weigh the source domain features (A2E10).

To ensure that our results are unbiased, we repeated the experiments three times with different training and test splits. We used two baselines to evaluate our classifiers. The first baseline, which we expect to represent the lower bound, is the naïve Bayes classifier trained on the target labeled data (NBT). We believe that this will be the lower bound for our classifiers because we expect that adding the labeled data from a related organism (the source domain) as well as unlabeled data from the same organism (the target domain) should produce a better classifier. The second baseline is the naïve Bayes classifier trained on the source labeled data (NBS). We expect that the more distantly related the two organisms are, the less accurate the classifier trained on the source data would be when tested on the target data.

In addition, we also performed the following two sets of experiments. In one set of experiments we tested our algorithms versus the null hypothesis, i.e., we shuffled all nucleotides in each instance except the AG dimer at index 61, shuffled the labels, and tested our algorithms on these data sets. The intuition is that in this case our algorithms should not learn any meaningful patterns, and thus obtain poor results when evaluated on the test data. In another set of experiments we evaluated how our best-overall algorithm performed in an ensemble setting with balanced data. For this, we split the source and target labeled data into 99 class-balanced data sets, and used an ensemble of Algorithm 2 classifiers to predict the labels for the test data. We ran these experiments to investigate if this approach is better suited for highly unbalanced data sets.

Our goal with this experimental setup was to determine how each of the following influence the performance of the classifier:

1. Features used, i.e., nucleotides, or nucleotides and codons.
2. Algorithm used, i.e., Algorithm 1 which keeps only the generalizable features in the source domain, or Algorithm 2 which assigns different weights to the features in the source domain.
3. Number of target labeled instances used in training the classifier, i.e., 2,500, 6,500, 16,000, or 40,000.
4. The evolutionary distance between the source and target domains.
5. Using source labeled data and target unlabeled data when training the classifier.
6. Using real data versus shuffled data with shuffled labels.
7. Using unbalanced data sets or creating balanced subsets and using an ensemble of classifiers.

4   Results

Since the class distribution is highly skewed, with each set containing about 1 % positive instances and the rest, about 99 %, negative instances, we used the area under the precision-recall curve as a metric for evaluating the accuracy of our classifiers, and we present the highest auPRCs averaged over the three folds.

In Table 1, we list the auPRC for our classifiers, for the best overall algorithm in [15], SVM$_{S,T}$, and for the classifiers in [9]. Although our results are not as good as the ones in [15], they are greatly improved compared to the results in [9], yet not as much as we expected. Even though they are not as good as the SVM$_{S,T}$

in [15], the algorithms we proposed could be superior in some contexts. In addition, as we mentioned in Sect. 2, the complexity of SVM classifiers increases with the number of training instances and the number of features, when training the classifier [15]. While their algorithms "required an equivalent of about 1,500 days of computing time on state-of-the-art CPU cores" to tune their parameters [15], and additional computing to analyze the biological features, our algorithms required the equivalent of only 300 days of computing, and the results can be easily interpreted. Furthermore, we believe that our algorithms are interesting from a theoretical perspective and it is useful to know how well they perform.

Based on these results, we make the following observations:

1. Using a classifier with both the nucleotides and the codons as features performs better as the size of the target labeled data increases, while a classifier using only the nucleotides as features performs better with smaller target labeled datasets. We believe that this is due to the fact that the codon features are sparser than the nucleotide features, and when there is little target labeled data the classifier does not have enough data to separate the positive from the negative instances.

2. The A1 classifier, based on Algorithm 1, and the A2E10 classifier, based on Algorithm 2, performed better than the other classifier, A2E9, each producing the best results in two and five cases, respectively, when the data was not balanced. As mentioned above, the sparsity of the data affects the performance of the classifiers based on the features used. The A1 classifier, which uses only the nucleotides as features, performs better when the target labeled dataset is small, and also when the two organisms are more distantly related, while the A2E10 classifier, which uses the nucleotides and codons as features, performs better when the target labeled dataset is larger and the two organisms are more closely related.

3. The general trend in all classifiers is that they perform better as more target labeled instances are used for training. This conforms with our intuition that using a small dataset for training does not produce a good classifier.

4. For the non-ensemble classifiers, as the evolutionary distance increases between the source and target organisms, the performance of our classifiers decreases. When the source and target organisms are closely related, as is the case with *C.elegans* and *C.remanei*, the large volume of labeled source data significantly contributes to generating a better classifier, compared to training a classifier on the target data alone. However, as the source and target organisms diverge, the source data contributes less to the classifier.

5. Using the source data and the target unlabeled data in addition to the target labeled data improves the performance of the classifier (e.g., A1 v. NBT for *A.thaliana* with 2,500 instances) compared to training a classifier on just the target labeled data. The improvement occurs even with larger datasets, although less substantial than with smaller datasets.

6. When we shuffled the data and the labels the algorithms were unable to learn meaningful patterns, resulting in poor classifiers with auPRC values less than one percent.

Table 1. auPRC values for four target organisms based on the number of labeled target instances used for training: 2,500, 6,500, 16,000, and 40,000. The tables on the left show the values for the algorithms trained with nucleotide features, while the tables on the right show the values for the algorithms trained with nucleotide and codons features. The highest values for our algorithms are displayed in bold font. NBT and NBS are baseline naïve Bayes classifiers trained on target labeled and source data, respectively, for the two algorithms we proposed, Algorithm 1 (A1) and Algorithm 2 (A2E9 and A2E10 when using (9) and (10), respectively). We show for comparison with our algorithms the values for the best overall algorithm in [15], SVM$_{S,T}$, and the values in [9], ANB. Note that these values are shown in each table even though the SVM$_{S,T}$ and ANB algorithms used different features. S1 and S2 are the results when we shuffled the data and the labels, and we used Algorithms 1 and 2 with (10), respectively. BAL are the results of the ensemble of classifiers with Algorithm 2 using (10).

	2,500	6,500	16,000	40,000
S1	0.90	0.94	0.90	0.92
S2	0.90	0.94	0.90	0.93
ANB	1.13	1.13	1.13	1.10
NBT	23.42	45.44	54.57	59.68
BAL	**60.24**	62.27	63.36	63.83
A1	59.18	63.10	63.95	63.80
A2E9	35.03	46.08	54.89	59.73
A2E10	48.92	60.83	63.06	63.59
NBS		63.77		
SVM	77.06	77.80	77.89	79.02

C.remanei (nucleotides)

	2,500	6,500	16,000	40,000
S1	0.91	0.93	0.90	0.93
S2	0.91	0.93	0.91	0.93
ANB	1.13	1.13	1.13	1.10
NBT	22.94	58.39	68.40	75.75
BAL	25.83	26.08	25.99	26.09
A1	45.29	**72.00**	74.83	77.07
A2E9	24.96	61.45	69.11	75.91
A2E10	49.22	70.23	**75.43**	**78.01**
NBS		77.67		
SVM	77.06	77.80	77.89	79.02

C.remanei (nucleotides+codons)

	2,500	6,500	16,000	40,000
S1	0.91	0.92	0.96	0.91
S2	0.90	0.91	0.95	0.92
ANB	1.00	0.97	1.07	1.10
NBT	19.22	37.33	45.33	52.84
BAL	**50.60**	50.64	52.34	55.16
A1	45.32	49.82	52.09	54.62
A2E9	19.85	37.51	45.64	52.91
A2E10	37.20	48.71	52.31	55.62
NBS		49.12		
SVM	64.72	66.39	68.44	71.00

P.pacificus (nucleotides)

	2,500	6,500	16,000	40,000
S1	0.92	0.89	0.95	0.91
S2	0.90	0.89	0.94	0.91
ANB	1.00	0.97	1.07	1.10
NBT	26.39	48.54	59.29	68.78
BAL	23.48	23.61	23.38	23.72
A1	20.21	53.29	62.33	69.88
A2E9	20.16	43.95	57.44	65.80
A2E10	20.19	**57.21**	**65.99**	**70.94**
NBS		67.10		
SVM	64.72	66.39	68.44	71.00

P.pacificus (nucleotides+codons)

	2,500	6,500	16,000	40,000
S1	0.89	0.93	0.96	0.91
S2	0.89	0.91	0.96	0.90
ANB	1.07	1.13	1.07	1.03
NBT	14.90	26.05	35.21	39.42
BAL	33.57	36.87	39.35	39.62
A1	33.31	36.43	40.32	42.37
A2E9	16.27	26.21	35.12	39.16
A2E10	22.86	32.92	36.95	37.55
NBS		31.23		
SVM	40.80	37.87	52.33	58.17

D.melanogaster (nucleotides)

	2,500	6,500	16,000	40,000
S1	0.89	0.88	0.94	0.92
S2	0.89	0.89	0.94	0.89
ANB	1.07	1.13	1.07	1.03
NBT	13.87	25.00	35.28	45.85
BAL	**42.12**	**42.71**	**45.85**	47.42
A1	25.83	32.58	39.10	**47.49**
A2E9	15.03	26.45	34.73	42.90
A2E10	22.53	29.47	36.18	42.92
NBS		34.09		
SVM	40.80	37.87	52.33	58.17

D.melanogaster (nucleotides+codons)

	2,500	6,500	16,000	40,000
S1	0.97	0.91	0.96	0.91
S2	0.92	0.90	0.96	0.91
ANB	1.20	1.17	1.20	1.17
NBT	7.20	17.90	28.10	34.82
BAL	28.41	30.98	33.50	36.70
A1	18.46	25.04	31.47	36.95
A2E9	8.42	18.39	28.22	34.79
A2E10	13.61	22.28	29.05	34.66
NBS		11.97		
SVM	24.21	27.30	38.49	49.75

A.thaliana (nucleotides)

	2,500	6,500	16,000	40,000
S1	0.91	0.90	0.97	0.90
S2	0.91	0.90	0.96	0.90
ANB	1.20	1.17	1.20	1.17
NBT	3.10	8.76	28.11	40.92
BAL	**39.80**	**42.27**	**43.10**	**45.05**
A1	3.99	13.96	33.62	43.20
A2E9	2.65	8.72	29.39	40.35
A2E10	3.64	10.00	30.85	40.40
NBS		13.98		
SVM	24.21	27.30	38.49	49.75

A.thaliana (nucleotides+codons)

7. When there are only a few labeled instances in the target domain (e.g., 2,500 instances), splitting the data into class-balanced data sets and using an ensemble of classifiers performs better than using the unbalanced data sets. In addition, when the source and target domains are distantly related, the ensemble classifier outperforms even the SVM classifier (e.g., when the target domain is *D.melanogaster* and there are 2,500 or 6,500 labeled instances in the target domain, or for *A.thaliana* with 2,500 to 16,000 labeled instances in the target domain).

5   Conclusions and Future Work

In this paper, we presented two similar domain adaptation algorithms based on the naïve Bayes classifier. They are based on the algorithms in [8,9], to which we made four changes. The first change was to use probabilities for computing the prior and likelihood, (7) and (8), instead of the counts in (2) and (3). The second change was to use mutual information in (9) and (10) instead of marginal probabilities to rank the features in (4). The third change was to assign different weights to the features from the source domain instead of selecting the features to use during training. The final change was to use fewer but more informative features. We used nucleotides and codons features that are aware of their location in the DNA sequence instead of 8-mers generated with a sliding window approach.

With these changes, we significantly improved the classification performance as compared to [9]. In addition, empirical results on the splice site prediction task support our hypothesis that augmenting a small labeled dataset with a large labeled dataset from a close domain and unlabeled data from the same domain improves the performance of the classifier. This is especially the case when we have small amounts of labeled data but the same trend occurs for larger labeled datasets as well.

In future work, we would like to more thoroughly evaluate the predictive power of the features we used, and to investigate whether other algorithms might produce better results when used with these features. To evaluate the features, we plan to use data from other organisms. To identify algorithms that might potentially be better for the task addressed in this work, we will investigate other transfer learning algorithms and compare their results with the results of our proposed algorithms.

In addition, we would like to further improve our classifier and are considering three approaches. In one approach, besides the features derived from biological background, nucleotides (1-mers) and codons (3-mers), we would like to use additional features, such as 2-mers, for example, or biological features, such as pyrimidine-rich motifs around the acceptor splice site. In another approach, we would like to address the splice site prediction problem as an anomaly detection problem, since the preponderance of positive instances is so small. And in the final approach, we would like to balance the ratio of positive and negative instances going as far as using only the positive labeled data for training.

Acknowledgements. Supported in part by the Kansas INBRE, P20 GM103418. The computing for this project was performed on the Beocat Research Cluster at Kansas State University, which is funded in part by NSF grants CNS-1006860, EPS-1006860, EPS-0919443, and MRI-1126709.

References

1. Gantz, J.H., Reinsel, D., Chute, C., Schlinchting, W., McArthur, J., Minton, S., Xheneti, I., Toncheva, A., Manfrediz, A.: The Expanding Digital Universe (2007)
2. Bernal, A., Crammer, K., Hatzigeorgiou, A., Pereira, F.: Global discriminative learning for higher-accuracy computational gene prediction. PLoS Comput. Biol. **3**, e54 (2007)
3. Rätsch, G., Sonnenburg, S., Srinivasan, J., Witte, H., Müller, K.R., Sommer, R., Schölkopf, B.: Improving the C. elegans genome annotation using machine learning. PLoS Comput. Biol. **3**, e20 (2007)
4. Müller, K.R., Mika, S., Rätsch, G., Tsuda, S., Schölkopf, B.: An introduction to kernel-based learning algorithms. IEEE Trans. Neural Networks **12**, 181–202 (2001)
5. Zien, A., Rätsch, G., Mika, S., Schölkopf, B., Lengauer, T., Müller, K.R.: Engineering support vector machine kernels that recognize translation initiation sites. Bioinformatics **16**, 799–807 (2000)
6. Noble, W.S.: What is a support vector machine? Nat. Biotech. **24**, 1565–1567 (2006)
7. Brown, M.P.S., Grundy, W.N., Lin, D., Cristianini, N., Sugnet, C., Furey, T.S., Ares, J.M., Haussler, D.: Knowledge-based analysis of microarray gene expression data using support vector machines. PNAS **97**, 262–267 (2000)
8. Herndon, N., Caragea, D.: Naïve Bayes domain adaptation for biological sequences. In: Proceedings of the 4th International Conference on Bioinformatics Models, Methods and Algorithms, BIOINFORMATICS 2013, pp. 62–70 (2013)
9. Herndon, N., Caragea, D.: Predicting protein localization using a domain adaptation approach. In: Fernández-Chimeno, M., Fernandes, P.L., Alvarez, S., Stacey, D., Solé-Casals, J., Fred, A., Gamboa, H. (eds.) Biomedical Engineering Systems and Technologies. CCIS, pp. 191–206. Springer, Heidelberg (2013)
10. Tan, S., Cheng, X., Wang, Y., Xu, H.: Adapting Naive Bayes to domain adaptation for sentiment analysis. In: Boughanem, M., Berrut, C., Mothe, J., Soule-Dupuy, C. (eds.) ECIR 2009. LNCS, vol. 5478, pp. 337–349. Springer, Heidelberg (2009)
11. Maeireizo, B., Litman, D., Hwa, R.: Co-training for predicting emotions with spoken dialogue data. In: Proceedings of the ACL 2004 on Interactive poster and demonstration sessions. ACLdemo 2004. Association for Computational Linguistics, Stroudsburg (2004)
12. Riloff, E., Wiebe, J., Wilson, T.: Learning subjective nouns using extraction pattern bootstrapping. In: Proceedings of the seventh conference on Natural language learning at HLT-NAACL 2003, CONLL 2003, vol. 4, pp. 25–32. Association for Computational Linguistics, Stroudsburg (2003)
13. Yarowsky, D.: Unsupervised word sense disambiguation rivaling supervised methods. In: Proceedings of the 33rd annual meeting on Association for Computational Linguistics, ACL 1995, pp. 189–196. Association for Computational Linguistics, Stroudsburg (1995)
14. Dai, W., Xue, G., Yang, Q., Yu, Y.: Transferring Naïve Bayes classifiers for text classification. In: Proceedings of the 22nd AAAI Conference on Artificial Intelligence (2007)

15. Schweikert, G., Widmer, C., Schölkopf, B., Rätsch, G.: An empirical analysis of domain adaptation algorithms for genomic sequence analysis. In: NIPS 2008, pp. 1433–1440 (2008)
16. Chawla, N.V., Japkowicz, N., Kotcz, A.: Editorial: special issue on learning from imbalanced data sets. SIGKDD Explor. Newsl. **6**, 1–6 (2004)
17. He, H., Garcia, E.: Learning from imbalanced data. IEEE Trans. Knowl. Data Eng. **21**, 1263–1284 (2009)
18. Li, J., Wang, L., Wang, H., Bai, L., Yuan, Z.: High-accuracy splice site prediction based on sequence component and position features. Genet. Mol. Res. **11**, 3431–3451 (2012)
19. Baten, A., Chang, B., Halgamuge, S., Li, J.: Splice site identification using probabilistic parameters and SVM classification. BMC Bioinform. **7**(Suppl 5), S15 (2006)
20. Sonnenburg, S., Schweikert, G., Philips, P., Behr, J., Rätsch, G.: Accurate splice site prediction using support vector machines. BMC Bioinform. **8**, 1–16 (2007)
21. Zhang, Y., Chu, C.H., Chen, Y., Zha, H., Ji, X.: Splice site prediction using support vector machines with a Bayes kernel. Expert Syst. Appl. **30**, 73–81 (2006)
22. Cai, D., Delcher, A., Kao, B., Kasif, S.: Modeling splice sites with Bayes networks. Bioinformatics **16**, 152–158 (2000)
23. Baten, A.K.M.A., Halgamuge, S.K., Chang, B., Wickramarachchi, N.: Biological sequence data preprocessing for classification: a case study in splice site identification. In: Liu, D., Fei, S., Hou, Z., Zhang, H., Sun, C. (eds.) ISNN 2007, Part II. LNCS, vol. 4492, pp. 1221–1230. Springer, Heidelberg (2007)
24. Arita, M., Tsuda, K., Asai, K.: Modeling splicing sites with pairwise correlations. Bioinformatics **18**(suppl 2), S27–S34 (2002)
25. Rätsch, G., Sonnenburg, S.: Accurate Splice Site Prediction for Caenorhabditis Elegans. In: Kernel Methods in Computational Biology. MIT Press series on Computational Molecular Biology. MIT Press (2004) 277–298
26. Dempster, A.P., Laird, N.M., Rubin, D.B.: Maximum likelihood from incomplete data via the EM algorithm. J. Roy. Stat. Soc.: Ser. B (Methodol.) **39**, 1–38 (1977)
27. Mccallum, A., Nigam, K.: A comparison of event models for Naïve Bayes text classification. In: AAAI-1998 Workshop on 'Learning for Text Categorization' (1998)
28. Shannon, C.E.: A mathematical theory of communication. Bell Syst. Tech. J. **27**(379–423), 623–656 (1948)
29. Breiman, L.: Bagging predictors. Mach. Learn. **24**, 123–140 (1996)

Automatic Extraction of Highly Predictive Sequence Features that Incorporate Contiguity and Mutation

Hao Wan, Carolina Ruiz$^{(\boxtimes)}$, and Joseph Beck

Department of Computer Science,
Worcester Polytechnic Institute, Worcester, MA, USA
{hale, ruiz, josephbeck}@wpi.edu

Abstract. This paper investigates the problem of extracting sequence features that can be useful in the construction of prediction models. The method introduced in this paper generates such features by considering contiguous subsequences and their mutations, and by selecting those candidate features that have a strong association with the classification target according to the Gini index. Experimental results on three genetic data sets provide evidence of the superiority of this method over other sequence feature generation methods from the literature, especially in domains where presence, not specific location, of features within a sequence is pertinent for classification.

Keywords: Sequence feature generation · Sequence classification · Mutation

1 Introduction

Supervised sequence classification deals with the problem of learning models from labelled sequences. The resulting models are used to assign appropriate class labels to unlabelled sequences. Sequence classification methods can be used for example to predict whether a segment of DNA is in the promoter region of a gene or not.

In general, sequence classification is more difficult than classification of tabular data, mainly because of two reasons: in sequence classification it is unclear what features should be used to characterize a given data set of sequences (e.g., Fig. 1 shows three different candidate features whose presence, or lack of, in a sequence can be used to characterize the sequence); and the number of such potential features is very large. For instance, given a set of sequences of maximum length l, over an alphabet B, where $|B| = d$ symbols are considered as features, then there are d^k potential features. Furthermore, the number of potential features will grow exponentially as the length of subsequences under consideration increases, up to d^l. Thus, determining which features to use to characterize a set of sequences is a crucial problem in sequence classification.

A mutation is a change in an element of a sequence. Like in DNA sequences, this could result from unrepaired damage to DNA or from an error during sequence replication. We are interested in whether these changes affect the sequences' function. Thus, we focus on generating features that represent mutation patterns in the sequences, and on selecting the generated features that are most suitable for classification.

© Springer International Publishing Switzerland 2015
G. Plantier et al. (Eds.): BIOSTEC 2014, CCIS 511, pp. 212–229, 2015.
DOI: 10.1007/978-3-319-26129-4_14

AACTAGCCTAACCGGTACTGCAGTA

(a)   (b)   (c)

Fig. 1. Example of different candidate features for a given sequence. (a) 4-grams: AACT; (b) 4-grams with one gap of size 2; (c) 2-gapped pair (TA,2).

More specifically, our proposed Mutated Subsequence Generation (MGS) algorithm generates features from sequence data by regarding contiguous subsequences and mutated subsequences as potential features. It first generates all contiguous subsequences of a fixed length from the sequences in the data set. Then it checks whether or not each pair of candidate feature subsequences that differ in only one position should be joined into a mutated subsequence. The join is performed if the resulting joint mutated subsequence has a stronger association with the target class than the two subsequences in the candidate pair do. If that is the case, the algorithm keeps the joint subsequence and removes the two subsequences in the candidate pair from consideration. Otherwise, the algorithm keeps the candidate pair instead of the joint mutated subsequence. After all the generated candidate pairs of all lengths have been checked, a new data set is constructed containing the target class and the generated features.

The features in the resulting data set represent (possibly mutated) segments of the original sequences that have a strong connection with the sequences' function. We then build classification models over the new data set that are able to predict the function (i.e., class value) of novel sequences.

The contributions of this paper are the introduction of a new feature generation method based on mutated subsequences for sequence classification, and a comparison of the performance of our algorithm with that of other feature generation algorithms.

2   Related Work

Feature-based classification algorithms transform sequences into features for use in sequence classification [1]. A number of feature-based classification algorithms have been proposed in the literature. For example, k-grams are substrings of k consecutive characters, where k is fixed beforehand, see Fig. 1(a). Damashek used k-grams to calculate the similarity between text documents during text categorization [2]. In [3], the authors vary k-grams by adding gap constraints into the features, see Fig. 1(b). Another kind of feature, k-gapped pair, is a pair of characters with constraint $l_2 - l_1 = k$, where k is a constant, and l_1 and l_2 are the locations in the sequence where the characters in the pair occur, see Fig. 1(c). The k-gapped pair method is used to generate features for Support Vector Machines in [4, 5]. In contrast with our method, features generated by their approach cannot represent mutations in the sequences.

Another method is mismatch string kernel [6]. It constructs a (k, m) – mismatch tree for each pair of sequences to extract k-mer features with at most m mismatches. It then uses these features to compute the string kernel function. A similarity between this method and our method is that both generate features that are subsequences with mutations. However, there are three major differences between them:

1. In mismatch string kernel, the features are generated from pairs of sequences and used to update the kernel matrix. In contrast, our MSG method generates features from the entire set of data sequences in order to transform the sequence data set into a feature vector data set.
2. In the process of computing candidate mutated subsequences, our MSG method does not only consider mutations in the subsequences, but also takes into account correlations between these mutated subsequences and the target classes. In contrast, the mismatch string kernel method disregards the latter part.
3. The mismatch string kernel method can be used in Support Vector Machines and other distance based classifiers for sequence data. Our MSG approach is more general as it transforms the sequences into feature vectors. In other words, the data set that results from the MSG transformation can be used with any classifier defined on feature vectors.

3   Background

3.1   Feature Selection

Feature selection methods focus on selecting the most relevant features of a data set. These methods can help prediction models in three main aspects: improving the prediction accuracy; reducing the cost of building the models; and making the models, and even the data, more understandable.

To obtain the features that maximize classification performance every possible feature set should be considered. However, exhaustively searching all the feature sets is known to be an NP-hard problem [7]. Thus, a number of feature selection algorithms has been developed based on greedy search methods like best-first and hill climbing (see [8]). These greedy algorithms use three main search strategies: *forward selection*, which starts with a null feature set and at each step adds the most important unselected feature (according to a specific metric); *backward deletion*, which starts at the full feature set and at each step removes an unimportant feature; and *bi-directional selection*, which also starts at a null feature set and at each step applies forward selection and then backward deletion until a stable feature set is reached.

3.2   CFS Evaluation

The Correlation-based Feature Selection (CFS) algorithm introduces a heuristic function for evaluating the association between a set of features and the target classes. It selects a set of features that are highly correlated with the target classes, yet uncorrelated with each other. This method was introduced in [9]. In this paper, we use this algorithm to select a subset of the generated features that is expected to have high classification performance.

3.3   Gini Index

In decision tree induction, the Gini index is used to measure the degree of impurity of the target class in the instances grouped by a set of features [10]. Similarly in our research, the Gini index is used to measure the strength of the association between candidate features and the target class. Specifically, we use it during feature generation to determine whether or not to replace a pair of subsequences (candidate features) that differ just in one position with their joined mutated subsequence. Details are described in Sect. 4.1. The Gini index of a data set is defined as: $1 - \sum_{c \in C} p(c)^2$, where C is the set of all target classes, and p denotes probability (estimated as frequency in the data set). Given a discrete feature (or attribute) X, the data set can be partitioned into disjoint groups by the different values of X. The Gini index can be used to calculate the impurity of the target class in each of these groups. Then, the association between X and the target class can be regarded as the weighted average of the impurity in each of the groups: $Gini(X) = \sum_{x \in X} p(x) * (1 - \sum_{c \in C} p(c|x)^2)$.

4   Our MSG Algorithm

4.1   Feature Generation

Our Mutated Subsequence Generation (MSG) algorithm belongs in the category of feature-based sequence classification according to the classification methods described in [1]. The MSG algorithm transforms the original sequences into contiguous subsequences and mutated subsequences, which are used as candidate features in the construction of classification models.

The MSG algorithm generates features of different lengths according to two user-defined input parameters: the minimum length l_{min} and the maximum length l_{max}. It first generates the candidate subsequences of length l_{min}, then length $l_{min} + 1$, and all the way to the subsequences of length l_{max}. Then it takes the union of all the generated subsequences of different lengths. Finally, the MSG algorithm constructs a new data set containing a Boolean feature for each generated subsequence (see Table 2 for an example). Each sequence in the original data set is represented as a vector of 0's and 1's in the new data set, where a feature's entry in this vector is 1 if the feature's corresponding subsequence is present in the sequence, and 0 if not.

The feature generation process consists of five main steps described below. Figure 2 shows an example of the transformation of a sequence data set into the subsequence features, from Step a to Step d.

(a) The MSG algorithm generates all the contiguous subsequences of a specific length from each original sequence;
(b) It divides the contiguous subsequences into n categories according to which class they are most frequent in, where n is the number of different classes;
(c) For each category, it generates mutated subsequences based on the Gini index;
(d) It combines together all the features from each category;

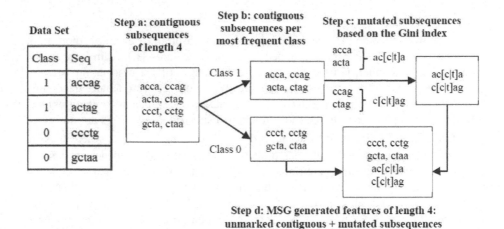

Fig. 2. Illustration of the mutated subsequence generation process from Step (a) to Step (d). In this example, features of length 4, including mutated subsequences and contiguous subsequences, are obtained. Here, the Gini index of each of the mutated subsequences ac[c|t]a and c[c|t]ag is better than the Gini indexes of its forming contiguous subsequences.

(e) It repeats Step *a* to Step *d* with a different length, until features of all the lengths in the user defined range are generated. Then it combines these features prior to constructing the new data set.

Computation of Contiguous Subsequences. Given $k \geq 1$, a *contiguous subsequence* of length k is a subsequence formed by taking k consecutive symbols from the original sequence. For example, *gc* is a contiguous subsequence of length 2 in the sequence *gcta*, while *gt* is not.

The MSG algorithm starts with the generation of contiguous subsequences. Suppose that the original sequences have length *t*, and that the contiguous subsequences to be computed have length *k*. First, each sequence is traversed and all of its contiguous subsequences of *k* symbols are extracted from starting locations 1 through $(t - k + 1)$. Then, duplicate subsequences are removed. For example, Table 1 shows the contiguous subsequence features of length 4 for the data set in Fig. 2, together with the number of occurrences of each feature in data set sequences of a given target class.

Separation. In this step, the algorithm separates the contiguous subsequences into *n* is the number of different classes in the data set. Each subsequence is assigned to a category according to which class it is most frequent in. For example, the first four features in Table 1 are assigned to Category 1, because they are more frequent in Class 1 than in Class 0, and the other four to Category 2. If a subsequence is maximally frequent equally in more than one class, it is randomly assigned to one of those classes.

Generation of Mutated Subsequences. In this step, the MGS algorithm generates mutated subsequences. The following terminology will be used in the description of this step.

Table 1. Contiguous subsequences of length 4 of the data sequences in Fig. 2, and their frequency counts in each data set class.

Feature	Class 0	Class 1
acca	0	1
ccag	0	1
acta	0	1
ctag	0	1
cact	1	0
actg	1	0
gcta	1	0
ctaa	1	0

- *Mutated Subsequence:* A mutated subsequence is a subsequence of length k ($k \geq 1$) which contains r mutative positions, where $1 \leq r \leq k$. In each mutative position, there are two or more substitutions. In this paper, we consider mutated subsequences with only one mutative position. For example, in the subsequence $g[c|a]t$, the second position is a mutative position, with two possible substitutions c and a. Thus, a mutated subsequence has many instantiations, in our example gct and gat. We say that a mutated subsequence is contained in an original sequence when any of its instantiations is contained in the original sequence. For example, $g[c|a]t$ is contained in the sequence *gata* as well as in the sequence *gata*.
- *Candidate Pair:* a candidate pair is a pair of subsequences which are different from each other in only one position. For instance, *acca* and *acta* is a candidate pair. The candidate pair could also contain mutated subsequences, like for instance *acca* and $ac[t|g]a$.
- *Joinable Checking:* joinable checking is used to determine whether or not to join together a candidate pair of subsequences depending on their correlation with the target class. In this paper, we use the Gini index to measure this correlation. For example, suppose $sub1$ is *acca* and $sub2$ is *acta* in Table 1. Their joint sequence $sub3$ is $ac[c|t]a$. The original data set in Fig. 2 can be split into two groups by each one of the subsequences in the pair: one group consists of the original data sequences which contain the subsequence, marked as $group_1^{subi}$; the other group consists of the rest of data sequences, marked as $group_2^{subi}$. Taking $sub1$ as an example, there is only one sequence in $group_1^{sub1}$, which is in class 1 (*accag*); and there are three sequences in $group_2^{sub1}$, two of which are in class 0 (*ccctg* and *gctaa*), and the other one is in class 1 (*actag*). Thus, by using the formula in Sect. 3.3, we can calculate the Gini index for $sub1$ as followings: $impurity(g_1^{sub1}) = 1 - \left(\frac{0}{1}\right)^2 - \left(\frac{1}{1}\right)^2 = 0$; $impurity(g_2^{sub1}) = 1 - \left(\frac{0}{1}\right)^2 - \left(\frac{1}{1}\right)^2 = 0.444$; $Gini(sub1) = p(g_1^{sub1}) * impurity(g_1^{sub1}) + p(g_2^{sub1}) * impurity(g_2^{sub1}) = 0 * \frac{1}{4} + 0.44 * \frac{3}{4} = 0.333$. Similarly, we calculate the values $Gini(sub2) = 0.333$, $Gini(sub3) = 0$. Since $sub3$ has the best measure value, implying that it has the strongest association with the target class, then the candidate pair $sub1$ and $sub2$ is joinable.

```
Function Mutated_Subsequences_Generation(C):
•   Input: C: set of subsequences
•   Output: set of all features, contiguous and
    mutated subsequences, that can be generated from C

Initialization: S <- {}
for each pair of subsequences sub1 and sub2 in C do
    if sub1 and sub2 is a joinable pair then
        sub3 <- sub1 ⊕ sub2
        mark sub1
        mark sub2
        for each sequence subi in C do
            if sub3 and subi is a joinable pair then
                sub3 <- sub3 ⊕ subi
                mark subi
        S <- S ∪ {sub3}
for each sequence subj in C do
    if subj is marked then
        C <- C - subj
return C ∪ S
```

Fig. 3. The pseudo code for mutated subsequences generation. $\oplus$ is the join operator. For example, $acca \oplus acta = ac[c|t]a$ and $acca \oplus ac[t|g]a = ac[c|t|g]a$.

The MSG algorithm performs joinable checking on every candidate pair within each category. Once a candidate pair is determined to be joinable, then the two subsequences are joined together to create a new subsequence, called *subi*, and they are also marked. Then the joinable checking is performed on *subi* with every other subsequence in the category. If there are other subsequences that are joinable with, then they are also joined together with *subi* and marked. Finally, *subi* is added into the mutated subsequence set. After all candidate pairs are checked, the algorithm deletes the marked subsequences and the duplicate mutated subsequences. A simplified pseudo code of mutated subsequences generation is shown in Fig. 3.

Combination of Categories. In the previous step, some contiguous subsequences might remain intact. That is, they are not joined with other subsequences. Thus, in each category, there might be two types of features: mutated subsequences and unchanged contiguous subsequences. In this step, the algorithm combines the feature sets of all categories together into one feature set. In this set, all features have length t, as defined in Step (b).

Combination of Features of Different Lengths. After all the features of the lengths in the user-defined range are generated in Step (a) through Step (d), the MSG algorithm combines these features of different lengths together and constructs a new data set. Each data instance in the new data set corresponds to a sequence in the original data set. The instance's value for each feature is a Boolean value stating whether or not the feature is a subsequence of the instance. Table 2 shows the transformed data from the data set in Fig. 2 for the subsequence length range 3−4.

Table 2. Transformed data set obtained by applying the MSG algorithm to the data set in Fig. 2, with subsequence length range 3−4. Original data sequences are depicted as columns and extracted features as rows due to formating restrictions.

	accag	actag	ccctg	gctaa
taa	0	0	0	1
gct	0	0	0	1
ctg	0	0	1	0
cac	0	0	1	0
act	0	1	1	0
cca	1	0	0	0
acc	1	0	0	0
[t|c]ag	1	1	0	0
ctaa	0	0	0	1
gcta	0	0	0	1
actg	0	0	1	0
cact	0	0	1	0
c[t|c]ag	1	1	0	0
ac[t|c]a	1	1	0	0
Class	1	1	0	0

4.2   Feature Selection

Once that the set of candidate features has been generated as described in Sect. 4.1, we use bi-directional feature selection based on the CFS evaluation (see Sect. 4) to select the best feature set from the transformed data set. This feature set is then used to build classification models.

5   Experimental Evaluation

The performance of our MSG algorithm is compared with that of other feature generation algorithms which are commonly used for sequence classification. These algorithms are described below.

- **Position-based** [11]: Each position is regarded as a feature, the value of which is the alphabet symbol in that position.
- **k-grams** [4]: A k-gram is a sequence of length k over the alphabet of the data set (see Fig. 1(a)). The value of a k-gram induced feature for a sequence S is whether the k-gram occurs in S or not.
- **k-gapped Pair** [12]: In a k-gapped pair (xy, k), xy is an ordered pair of letters over the alphabet of the data set, k is a non-negative integer (see Fig. 1(c)). The value of a k-gapped pair induced feature for a sequence S is 1 if there is a position i in S, where $s_i = x$ and $s_{i+k+1} = y$. Otherwise, the value is 0.

Notice that a k-gapped pair can be regarded as a mutated subsequence where the k symbols between the pair can be any nucleotides. Thus, in order to perform a meaningful comparison, we use the same subsequence length values for the three generation algorithms (k-grams, MSG, and k-gapped pair) in our experiments. Moreover, we use values of $k > 5$ to obtain features, since for $k \leq 5$ most of the k-gapped pairs are contained in all sequences, making them useless for classification.

To compare the performance of these algorithms, a number of experiments are carried out on three data sets, the first two are collected from UCI Machine Learning Repository [13], and the third one was collected in our prior work [14] from WormBase [15]:

- *E.coli* **Promoter Gene Sequences Data Set:** This data set consists of 53 DNA promoter sequences and 53 DNA non-promoter sequences. Each sequence has length 57. Its alphabet is {a, c, g, t}.
- **Primate Splice-junction Gene Sequences Data Set:** This data set contains 3190 DNA sequences of length 60. Also, its alphabet is {a, c, g, t}. Each sequence is in one of three classes: exon/intron boundaries (EI), intron/exon boundaries (IE), and non-splice (N). 745 data instances are classified as EI; 751 instances as IE; and 1694 instances as N.
- *C.elegans* **Gene Expression Data Set:** This data set contains 615 gene promoter sequences of length 1000.We use here expression in EXC cells as the classification target. 311 of the genes in this data set are expressed in EXC cells, and the other 304 genes are not.

The performance comparison in the following sections focuses on the prediction level of models built on the generated features, and on differences among the models. We implemented the four feature generation methods, MSG, k-grams, position-based, and k-gapped pairs, in our own Java code. To measure the prediction level, we utilize The WEKA System version 3.7.7 [16] to build three types of prediction models: J48 Decision Trees, Support Vector Machines (SVMs), and Logistic Regression (LR). We use n-fold cross validation to test the models. We regard the models' accuracy as their prediction level. To measure the difference between two models, we perform a pair t-test on their n-fold test results, and use p-values from the paired t-test to determine whether or not the difference in model performance is statistically significant.

5.1   Results on the *E.Coli* Promoter Gene Sequences Data Set

Patterns from the Literature. As found in the biological literature [17, 18], and summarized in [19], promoter sequences share some common DNA segments. Figure 4 presents some of these segments. As can be seen in the figure, these segments can contain mutated positions. Also the segments are annotated with specific locations where the segments occur in the original sequences. This is an important characteristic distinguishing these segments from the subsequences generated by our algorithm. The data set constructed by our MSG algorithm captures only presence of the subsequences (not their positions) in the original sequences.

However, after examining the occurrences of the aforementioned segments in the Promoter Gene Sequences data set, we found that for the most part each segment occurs at most once in each sequence. Hence computational models created over this

```
@-37 "cttgac"
@-36 "ttgxca"
@-36 "ttgaca"
@-36 "ttgac"
@-14 "tataat"
@-13 "taxaxt"
@-13 "tataat"
@-12 "taxxxt"
```

Fig. 4. Patterns taken from [19]. Promoter sequences share these segments at the given locations. In these segments, "x" represents the occurrence of any nucleotide. The location is specified as an offset from the Start of Transcription (SoT). For example, "-37" refers to the location 37 base pair positions upstream from SoT.

data set that deal only with presence of these patterns are expected to achieve a prediction accuracy similar to that of computational models that take location into consideration. Therefore, since the precise location of the patterns seems to be irrelevant, we expect our MSG algorithm to perform well on this data set.

Experimental Results. Each of the four feature generation methods under consideration (MSG, k-grams, position-based, and k-gapped pair) was applied to the Promoter Gene Sequence data set separately, yielding four different data sets. Parameter values used for the feature generation methods were the following: for MSG, the range for the length of transformed subsequences was $1-5$; for k -grams, $k \leq 5$; and for k-gapped, $k \leq 10$. Then, Correlation-based Feature Selection (CFS), described in Sect. 3.2, was applied to each of these data sets to further reduce the number of features. The resulting number of features in each of the data sets was: 61 for the MSG transformed data set, 43 for k-grams, 7 for position-based, and 29 for k-gapped pair. 5-fold cross validation was used to train and test the models constructed on each of these data sets. Three different model construction techniques were used: J4.8 decision trees, Logistic Regression (LR), and SVMs. Figure 5 shows the prediction accuracy of the obtained models. Table 3 depicts the statistical significance of the performance difference between the models constructed over the MSG-transformed data set and the models constructed over data sets constructed by other feature generation methods.

From Fig. 5, we can observe that the prediction levels of the models constructed over features generated by MSG are superior to those of models constructed on other features. The t-test results in Table 3 indicate that this superiority of MSG is statistically significant at the $p < 0.05$ level in the cases highlighted in the table. As expected, the MSG algorithm generates a highly predictive collection of features for this data set. This in part due to the fact that for this data set, the presence alone, and not location, of certain subsequences (or segments) discriminates well between promoter and non-promoter sequences. Table 4 shows some of the features generated by MSG.

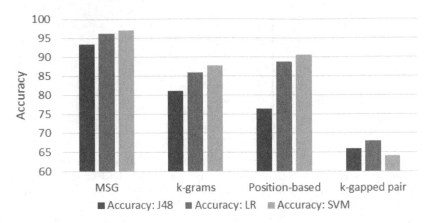

Fig. 5. Accuracy of models built on the features from four different feature generation methods (MSG, *k*-gram, Position-based, and *k*-gapped pair), using three classification algorithms on the Promoter Gene Sequences data.

Table 3. p-values obtained from t-tests comparing the prediction accuracies of models constructed over MSG-generated features and models constructed over data sets generated by the other 3 feature generation methods. t-tests were performed using 5-fold cross-validation over the Promoter Gene Sequence data set. Highlighted in the table are the cases in which the superiority of MSG is statistically significant at the $p < 0.05$ level.

Baseline: MSG	*k*-grams	Position-based	*k*-gapped pair
p-value: J48	0.08	0.02	0.03
p-value: LR	0.01	0.06	0.01
p-value: SVM	0.08	0.08	0.01

5.2 Results on the Primate Splice-Junction Gene Sequences Data Set

Patterns from the Literature. Some patterns in this data set have been identified in [20]. These patterns state that a sequence is in EI or IE if the nucleotide triplets "TAA", "TAG", and "TGA", known as stop codons, are absent in certain positions of the sequence. Conversely, if a sequence contains any stop codons in certain specified positions, the sequence is not in EI (IE). To examine the effect of position on these patterns, we generated the rules below, and calculated their confidence (i.e., prediction accuracy) on this data set.

- *Stop codons are present → not EI (74 %)*
- *Stop codons are present at specified positions → not EI (95 %)*
- *Stop codons are present → not IE (77 %)*
- *Stop codons are present at specified positions → not IE (91 %)*

As can be seen, the position information is very important in these patterns. Hence, we might expect that the MSG algorithm will not perform well on this data set, because

Table 4. Sample features constructed by the MSG algorithm over the Promoter Gene Sequences data set together with their correlation with the class feature.

MSG features	Correlation with target
t[t|a]ta	-0.61
[a|c]aaa	-0.58
ta[a|g|c|t]aa	-0.58
a[a|g|c|t]aat	-0.56
ata[t|c|a|g]t	-0.53
at[g|a|c]at	-0.51
aatt[c|a|t|g]	-0.51
aaa[g|a|t|c]t	-0.5
tta[t|a|c]a	-0.44
aaa[t|c|a|g]c	-0.44
ccc[a|g]	-0.41
ct[g|t|c|a]tt	-0.41
at[t|a]	-0.37
c[a|c|g|t]ggt	0.42
tgag[g|a]	0.43

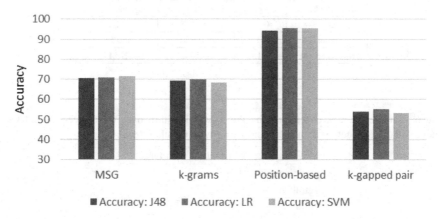

Fig. 6. Accuracy of models built on the features from four different feature generation methods (MSG, k-gram, Position-based, and k-gapped pair), using three classification algorithms on the Splice-junction Gene Sequence data.

its generated features do not contain information about the location where subsequences appear in the original sequences.

Experimental Results. Once again, each of the four feature generation methods under consideration (MSG, k-grams, position-based, and k-gapped pair) was applied to the Splice-junction Gene Sequences data set separately, yielding four different data sets.

The parameters used for the feature generation algorithms were: for MSG, the range for the length of transformed subsequences was 1–5; for k-grams, $k \le 5$; and for k-gapped, $k \le 10$. Then, Correlation-based Feature Selection (CFS), described in Sect. 3.2, was applied to each of these resulting data sets. The size of the feature set generated by the MSG algorithm was 28, by k-grams was 29, by position-based was 22, and by k-gapped pair was 49.

10-fold cross validation was used to construct and test models over these four data sets. Average accuracies of the resulting models are shown in Fig. 6, and the t-test results in Table 5. On this data set, the position-based algorithm performed the best. This is expected given that location information is relevant for the classification of this data set's sequences, as discussed above. The MSG generated features yielded prediction performance at the same level of that of k-grams; and statistically significantly higher performance (at the $p < 0.05$ significance level) than that of k-gapped pair. Some of these MSG generated features are shown in Table 6.

Table 5. p-values obtained from t-tests comparing the prediction accuracies of models constructed over MSG-generated features and models constructed over data sets generated by the other 3 feature generation methods. t-tests were performed using 10-fold cross-validation over the Splice-junction Gene Sequences data set. Highlighted in the table are the cases in which the superiority of MSG is statistically significant at the $p < 0.05$ level.

Baseline: MSG	k-grams	Position-based	k-gapped pair
p-value: J48	0.23	6.5E-18	2.12E-13
p-value: LR	0.72	6.78E-16	7.3E-13
p-value: SVM	0.15	2.3E-17	3E-14

Table 6. Sample features constructed by the MSG algorithm over the Splice-junction Gene Sequences data set together with their correlation with the class feature.

MSG features	Correlation with target
gt[a|g]ag	-0.5
ggt[a|g]a	-0.44
ggt[a|g]	-0.38
gt[a|g]a	-0.35
gtg[c|a]g	-0.35
aggt[a|g]	-0.35
gta[g|a]g	-0.34
[g|t|a]ggta	-0.32
tc[t|c]t	-0.03
[t|c]ag	-0.02
t[c|t]tc	0.01

Table 7. A PWM for PHA-4, found in [23]. It records the likelihood of each nucleotide at each position of the PHA-4 motifs.

	A	C	G	T
1	0.097	0.144	0.52	0.238
2	0.003	0.755	0.003	0.238
3	0.003	0.097	0.003	0.896
4	0.003	0.896	0.097	0.003
5	0.003	0.99	0.003	0.003
6	0.849	0.003	0.144	0.003
7	0.99	0.003	0.003	0.003
8	0.614	0.05	0.191	0.144

5.3   Results on the *C.Elegans* Gene Expression Data Set

Patters from the Literature. Motifs are short subsequences in the promoter sequences that have the ability to bind transcription factors, and thus to affect gene expression. For example, a transcription factor CEH-6 is necessary for the gene aqp-8 to be expressed in the EXC cell, by binding to a specific subsequence (ATTTGCAT) in the gene promoter region [22]. The binding sites for a transcription factor are not completely identical, as some variation is allowed. These potential binding sites are represented as a position weight matrix (PWM), see Table 7 for an example. A motif is a reasonable matching subsequence according to a specific PWM.

It has been shown that motifs at different positions in the promoter have different importance in controlling transcription [23], and that the order of multiple motifs and the distance between motifs can also affect gene expression [14].

Experimental Results. Each of the four feature generation methods under consideration (MSG, k-grams, position-based, and k-gapped pair) was applied to the *C.elegans* Gene Expression data set separately, yielding four different feature vector data sets. The parameters used for the feature generation algorithms were: for MSG, the range for the length of transformed subsequences was 1−6; for k-grams, $k \leq 6$; and for k-gapped, $k \leq 10$. After CFS, the size of the feature set generated by the MSG algorithm was 97, by k-grams was 63, by position-based was 122, and by k-gapped pair was 4. The average accuracies of the resulting models with 10-fold cross validation are shown in Fig. 7.

On this data set, MSG achieved the best performance among the methods tested. The p-values in Table 8 indicate that MSG prediction performance is significantly better than those of position-based and k-gapped pair at the $p < 0.05$ significance level. MSG performed slightly better than k-grams, but not significantly better.

5.4   Discussion

Computational Complexity Comparison of the Methods. Suppose that a data set consists of n sequences of length l over an alphabet B, where $|B| = d$ (for the three data

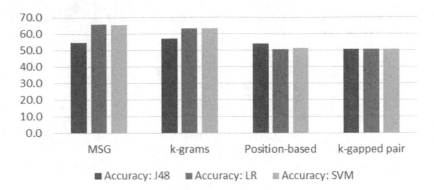

Fig. 7. Accuracy of models built on the features from four different feature generation methods (MSG, k-gram, Position-based, and k-gapped pair), using three classification algorithms on the Gene Expression data.

Table 8. p-values obtained from t-tests comparing the prediction accuracies of models constructed over MSG-generated features and models constructed over data sets generated by the other 3 feature generation methods. t-tests were performed using 10-fold cross-validation over the Gene Expression data set. Highlighted in the table are the cases in which the superiority of MSG is statistically significant at the $p < 0.05$ level.

Baseline: MSG	k-grams	Position-based	k-gapped pair
p-value: J48	0.18	0.41	0.01
p-value: LR	0.18	6.89E-6	1.45E-5
p-value: SVM	0.21	8.39E-5	4.59E-5

sets considered in this paper, $d = 4$). The position-based method has the lowest computational complexity out of the four feature generation methods employed in this paper. It takes $O(l)$ time to extract each location as a feature for each sequence, so its total complexity is $O(nl)$. The k-gapped pair method needs to compute $l - k - 1$ pairs of symbols for each sequence for a given gap size k. In our experiments, we considered pairs with gap $\leq k$, and since $k \ll l$, the time complexity for each sequence is $O(kl)$. Its total complexity is $O(nkl)$. Similarly, the k-gram method takes $O(nkl)$ time complexity to generate features of length $\leq k$.

The MSG method has the highest computational complexity among the four methods. Suppose that m subsequences are given as input to MSG (pseudo code in Fig. 3). There are two outer loops and one inner loop in this process. The first outer loop goes over all the $\binom{m}{2}$ pairs of subsequences, and its inner loop takes at most m iterations. The second outer loop traverses m subsequences to delete the marked ones. So the time complexity of this method is $C\left(m * \binom{m}{2} + m\right) = o(m^3)$ subsequences from the sequence data. Thus, its computational complexity is $O(d^{3k})$ in the worst case.

Experimental Comparison of the Methods. The experimental results on the three data sets provide evidence of the usefulness of the MSG algorithm. As we discussed above, patterns in the *E.coli* promoter gene sequences data set are position-independent, while patterns in the primate splice-junction gene sequences data set are position-dependent. Given that the MSG-generated features do not take location into consideration, MSG was expected to perform very well on the first data set but not on the second data set. Our experimental results confirm this hypothesis. In summary, MSG-generated features are most predictive in domains in which location is irrelevant or plays a minor role. Nevertheless, even in domains in which location is important, our MSG algorithm performed at the same level, or higher, than other feature generation algorithms from the literature.

In the *C.elegans* gene expression data set, patterns are much more complex than in the other two data sets considered. The MSG algorithm does not produce high classification accuracies on this data set. However, when compared to the other algorithms under consideration, MSG generates features that yield more accurate prediction models. One aspect that contributes to MSG's comparably better performance on this data set is its ability to represent mutations in the data sequences.

6 Conclusion and Future Work

In this work we present a novel feature generation method for sequence classification, Mutated Subsequence Generation (MSG). This method considers subsequences, possibly containing mutations, as potential features for the classification of the original sequences. It uses the Gini index to select the best features. We compare this method with other feature generation methods on three genetic data sets, focusing on the accuracy of the classification models built on the generated features. The experimental results show that MSG outperforms other feature generation methods in domains where presence, not specific location, of features within a sequence is relevant; and can perform at the same level or higher than other non-position-based feature generation methods in domains in which specific location, as well as presence, is important. Additionally, MSG is capable of identifying one-position mutations in the subsequence features that are highly associated with the classification target.

Future work includes further experimentation on much larger data sets; refinement of our MSG algorithm to reduce its time complexity; extension of MSG to allow for mutations in more than one subsequence position; and investigation of approaches to and the effects of incorporating location information in the MSG generated features.

References

1. Xing, Z., Pei, J., Keogh, E.: A brief survey on sequence classification. ACM SIGKDD Explor. **12**(1), 40–48 (2010)
2. Damashek, M.: Gauging similarity with n-grams: language-independent categorization of text. Science **267**(5199), 843–848 (1995)

3. Ji, X., Bailey, J., Dong, G.: Mining minimal distinguishing subsequence patterns with gap constraints. In: Proceedings of the Fifth IEEE International Conference on Data Mining (2005)
4. Chuzhanova, N.A., Jones, A.J., Margetts, S.: Feature selection for genetic sequence classification. Bioinformatics **14**(2), 139–143 (1998)
5. Huang, S.-H., Liu, R.-S., Chen, C.-Y., Chao, Y.-T., Chen, S.-Y.: Prediction of outer membrane proteins by support vector machines using combinations of gapped amino acid pair compositions. In: Proceedings of the 5th IEEE Symposium on Bioinformatics and Bioengineering (BIBE 2005) (2005)
6. Leslie, C.S., Eskin, E., Cohen, A., Weston, J., Noble, W.S.: Mismatch string kernels for discriminative protein classification. Bioinformatics **20**(4), 467–476 (2004)
7. Amaldi, E., Kann, V.: On the approximability of minimizing nonzero variables or unsatisfied relations in linear systems. Theor. Comput. Sci. **209**(1–2), 237–260 (1998)
8. Kohavi, R., Johnb, G.H.: Wrappers for feature selection. Artif. Intell. **97**(1–2), 273–324 (1997)
9. Hall, M.A., Smith, L.A.: Feature selection for machine learning: comparing a correlation-based filter approach to the wrapper. In: Proceedings of the Twelfth International FLAIRS Conference, Orlando (1999)
10. Gini, C.: Italian: Variabilità e mutabilità "(Variability and Mutability)," C. Cuppini, Bologna, p. 156. In: Pizetti, E., Salvemini, T. (eds.) Memorie di metodologica statistica. Libreria Eredi Virgilio Veschi, Rome (1912) (1955, reprinted)
11. Dong, G., Pei, J.: Sequence Data Mining, pp. 47–65. Springer, US (2009)
12. Park, K.-J., Kanehisa, M.: Prediction of protein subcellular locations by support vector machines using compositions of amino acids and amino acid pairs. Bioinformatics **19**(13), 1656–1663 (2003)
13. Bache, K., Lichman, M.: UCI Machine Learning Repository. University of California, School of Information and Computer Science, Irvine (2013). (http://archive.ics.uci.edu/ml)
14. Wan, H., Barrett, G., Ruiz, C., Ryder, E.F.: Mining association rules that incorporate transcription factor binding sites and gene expression patterns in C. elegans. In: Proceeding Fourth International Conference on Bioinformatics Models, Methods and Algorithms BIOINFORMATICS2013, pp. 81–89. SciTePress, Barcelona (2013)
15. WormBase, 1 April 2012. (http://www.wormbase.org/)
16. Hall, M., Frank, E., Holmes, G., Pfahringer, B., Reutemann, P., Witten, I.H.: The WEKA data mining software: an update. SIGKDD Explor. **11**(1), 10–18 (2009)
17. Hawley, D.K., McClure, W.R.: Compilation and analysis of Escherichia coli promoter DNA sequences. Nucleic Acids Res. **11**(8), 2237–2255 (1983)
18. Harley, C.B., Reynolds, R.P.: Analysis of E. coli promoter sequences. Nucleic Acids Res. **15**(5), 2343–2361 (1987)
19. Towell, G.G., Shavlik, J.W., Noordewier, M.O.: Refinement of approximate domain theories by knowledge-based neural networks. In: Proceedings of the Eighth National Conference on Artificial Intelligence (1990)
20. Noordewier, M.O., Towell, G.G., Shavlik, J.W.: Training knowledge-based neural networks to recognize genes in DNA sequences. Adv. Neural Inf. Process. Syst. **3**, 530–536 (1991)
21. Mah, K., Tu, D.K., Johnsen, R.C., Chu, J.S., Chen, N., Baillie, D.L.: Characterization of the octamer, a cis-regulatory element that modulates excretory cell gene-expression in Caenorhabditis elegans. BMC Mol. Biol. **11**(1), 19 (2010)
22. Reece-Hoyes, J.S., Shingles, J., Dupuy, D., Grove, C.A., Walhout, A.J., Vidal, M., Hope, I.A.: Insight into transcription factor gene duplication from Caenorhabditis elegans Promoterome-driven expression patterns. BMC Genom. **8**(1), 27 (2007)

23. Ao, W., Gaudet, J., Kent, W., Muttumu, S., Mango, S.E.: Environmentally induced foregut remodeling by PHA-4/FoxA and DAF-12/NHR. Science **305**, 1743–1746 (2004)
24. Tan, P.-N., Kumar, V., Steinbach, M.: Introduction to Data Mining. Addison-Wesley, Boston (2005)

Bio-inspired Systems and Signal Processing

Exploring Expiratory Flow Dynamics to Understand Chronic Obstructive Pulmonary Disease

Marko Topalovic[1(✉)], Vasileios Exadaktylos[2], Jean-Marie Aerts[2],
Thierry Troosters[1,3], Marc Decramer[1], Daniel Berckmans[2],
and Wim Janssens[1]

[1] Respiratory Division, Department of Clinical and Experimental Medicine,
University Hospital Leuven, KU Leuven, Leuven, Belgium
`marko.topalovic@biw.kuleuven.be`, {`thierry.troosters,`
`marc.decramer,wim.janssens`}`@med.kuleuven.be`
[2] Division Animal and Human Health Engineering, Department of Biosystems,
Faculty of Bioscience Engineering, KU Leuven, Leuven, Belgium
{`vasileios.exadaktylos,jean-marie.aerts,`
`daniel.berckmans`}`@biw.kuleuven.be`
[3] Faculty of Kinesiology and Rehabilitation Sciences,
Department of Rehabilitation Sciences, KU Leuven, Leuven, Belgium

Abstract. Chronic obstructive pulmonary disease (COPD) is one of the most common respiratory diseases and a leading cause of morbidity and mortality. It is characterized by irreversible airflow limitations. We aimed to explore whether the dynamics of expiration could serve as a descriptor of airflow limitations. Additionally, we explored the relationship between dynamic components and the presence of COPD. A data-based model was developed using data from 474 subjects. Significant difference ($p < 0.0001$) was found comparing a group of diseased patients with healthy for each dynamic component (namely the two poles, the steady state gain (SSG) and the time constant). Moreover difference was observed for each severity stage of disease. When ranking all components, SSG and pole1 are highlighted as the best COPD descriptors. We concluded that more detailed analysis of the forced expiration can be used to expand the understanding of COPD. Furthermore, the obtained parameters may improve current COPD assessment.

Keywords: Data based modeling · Transfer function · Chronic obstructive pulmonary disease · Spirometry · Forced expiration

1 Introduction

Chronic Obstructive Pulmonary Disease (COPD) is a leading cause of morbidity and mortality and therefore one of the major health challenges of the next decades [1, 2]. COPD is characterised by airflow limitation that is not fully reversible. It is usually progressive and associated with an abnormal inflammatory response of lungs to noxious particles or gases, most often from cigarette smoke [3]. Among all possible

G. Plantier et al. (Eds.): BIOSTEC 2014, CCIS 511, pp. 233–245, 2015.
DOI: 10.1007/978-3-319-26129-4_15

diseases, COPD is currently the 4th leading cause of death, while prediction of the World Health Organization is that it will become the 3rd leading cause of death in less than 2 decades from now [4–6]. According to the latest surveys, up to almost one fifth of the adults older than 40 years have present mild airflow obstruction [7]. One of the challenges in such a prevalent disease is to identify patients at risk for rapid deterioration and to develop diagnostic tools which are directly clinically important [8, 9].

Indications of COPD are production of sputum, signs of dyspnea, chronic cough or/and a history of exposure to tobacco smoke [1]. However, the current diagnosis is based on lung volumes measured by a spirometer, as neither most common signs of COPD nor patient history can accurately reflect presence of COPD. Current diagnosis is simple and inexpensive to perform, but also lately debatable due to ability of over or underdiagnosing [10]. In the past, various approaches have been developed to diagnose and characterize COPD. Attempts were made to automatize the interpretation of the forced oscillation technique with a few interesting algorithms [15], or to look into computed tomography images of chest [11, 12], or even measurements of volatile organic compounds in the exhaled air [13, 14]. Nevertheless, none of the methods entered clinical practice, due to their complexity, costly undertaking or unsatisfactory results.

Until now, mathematical data-based modelling was not employed in exploring characteristics of COPD. Starting from the base that COPD, by definition, is flow limited [3, 16] we hypothesized that modelling of the flow dynamics during exhalation may provide a more detailed description of airway obstruction and thus more accurate identification of presence of COPD.

In the present study our objective was firstly to develop a mathematical data-based model for the decline of the forced expiratory flow during expiration and secondly, to investigate the relationship between the dynamic components from the model and the presence of COPD.

2 Methods

2.1 Study Population

This study included data of 474 individuals who performed a complete pulmonary function testing (PFT) at cohort entry, including post-bronchodilator spirometry, body plethysmography and diffusing capacity. All subjects were tested between October 2007 and January 2009 at the University Hospital of Leuven (Belgium), as described earlier [17, 18]. In short, all participants were current or former heavy smokers with at least 15 pack-years and minimal age of 50 years. As COPD is a smoking disease per se, restricting our study to only smoking individuals increased chances to observe more abnormal pulmonary functions and patients with higher risk for COPD. Individuals with suspicion or diagnosis of asthma were excluded, as well as patients with exacerbations due to COPD within last 6 weeks and patients with other respiratory diseases. The study was approved by the local ethical committee of the University Hospital Leuven, (KU Leuven, Belgium). All patients included in the study provided informed consent. The study design of the LEUVEN COPD cohort can be found on www. clinicaltrials.gov (NCT00858520).

According to the international COPD GOLD guidelines [1], patients with COPD were identified when the post-bronchodilator FEV1/FVC ratio was < 0.7, furthermore they were lined over different severity stages. The population consisted of 336 patients with diagnosed COPD compared to 138 healthy controls. Stratified for disease severity from mild (GOLD I) to moderate (GOLD II), severe (GOLD III) and very severe (GOLD IV), the COPD population was comprised of 77, 101, 97 and 61, patients respectively. Table 1 describes the population characteristics within two separate groups, revealing typical characteristics for smoking and demographics of COPD patients admitted in hospitals.

Table 1. Study population characteristics; Values are median and IQR; BMI = body mass index; M = male; F = female; FEV1 = forced expiratory volume in one second; FVC = forced vital capacity; M = male; %pred. = percent predicted of normal population reference values.

	Healthy	COPD
Patients, N	138	336
Sex, M/F	110/28	260/76
Age, years	60.7(57.3–64.6)	65.1(59.5–72.1)
Smoking, pack yr.	38.0(29.3–52.0)	45.0(32.6–60.0)
BMI, kg/m^2	26.4(24.0–28.7)	25.0(22–28)
FEV$_1$, %pred.	104.0(94–112)	53.0(35–78)
FVC, %pred.	108.0(100–118)	89.0(71–106)
FEV$_1$/FVC	0.75(0.73– 0.78)	0.47(0.37– 0.62)

2.2 Pulmonary Function Tests

All pulmonary function tests were performed with standardized equipment (Masterlab, Erich Jeager, Würzburg, Germany) by experienced respiratory technicians, according to the ATS/ERS guidelines [19]. Spirometry data are post-bronchodilator measures and expressed as percent predicted of normal reference values [20].

2.3 Data Based Modelling

For the development of the data-based model, MATLAB (7.14, The MathWorks, Natick, Massachusetts) and the CAPTAIN toolbox for non-stationary time series analysis, system identification, signal processing and forecasting were used [21]. In all individuals the best expiratory curve (rule of highest sum of FEV1 and FVC [19]), within one spirometry, was exported from the Masterlab system at a sampling rate of 125 Hz. By extracting data points it was possible to reconstruct the best expiratory manoeuvre in MATLAB. To observe the dynamics of the expiration, only the declining

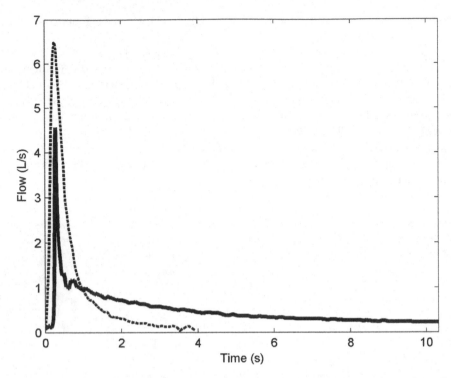

Fig. 1. Two examples of expiratory manoeuvres; The solid line shows the expiratory flow of an individual with diagnosed very severe COPD, while the dashed line shows the expiratory flow of a healthy individual. Decline is considered the section when the flow starts dropping from its maximum back to its minimum, over time.

phase of expiration was analysed. Declining is the area that starts at the peak flow and ends at the end of the expiration, far right tail (Fig. 1).

When starting with data-based modelling, the appropriate model structure is determined using objective methods of time series analysis from a generic model class. The goal is to describe the data in a parametrically efficient way, but still having simplicity in the sense of model parameters and model order. Considering our study and data, the most appropriate model was a discrete-time transfer function (TF) model for a single input single output (SISO) system. The general form of such a system is:

$$y_t = \frac{B(L)}{A(L)} u_t + \xi_t \tag{1}$$

where y_t is the output; u_t is the input; ξ_t is additive noise, assumed to be zero mean; L is the backward shift operator; A(L) and B(L) are polynomials defined by the order of the model in the following form:

$$A(L) = 1 + a_1L + \ldots + a_nL^n \qquad (2)$$

$$B(L) = b_0 + b_1L + \ldots + b_mL^m \qquad (3)$$

where n represents the order of the system: $a_1, \ldots, a_n$ and $b_0, b_1, \ldots, b_m$ are the TF denominator and numerator parameters, respectively.

Once the input-output data are available, the TF parameters (Eqs. (2, 3)) can be identified using statistical procedures. For the input data, we used an artificial unit step-down, while the output signals were the original measurements obtained from spirometry. An example of the input and output signals is given in Fig. 2. The parameters of a TF model can be estimated using various methods of identification and estimation procedures [22, 23]. In this study the Simplified Refined Instrumental Variable (SRIV) algorithm was used as a method for model identification. The advantage of SRIV lays not only in yielding consistent estimates of the parameters, but also in exhibiting close to optimum performance in the model order reduction context.

An equally important problem to the parameter estimation is the identification of the objective model order which will result in low complexity. The process of model

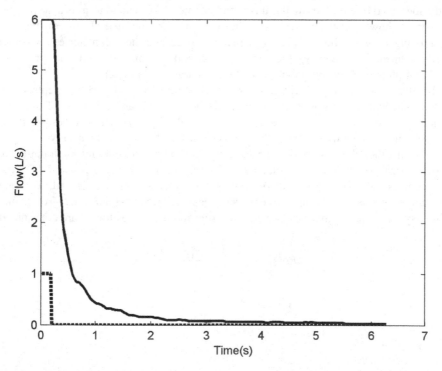

Fig. 2. The artificial unit step down (dashed line) used for each model as input signal; The solid line shows an example of declination, meaning output signal (different for each individual). Ten data points were initially inserted at the beginning of the measured signal in order to ensure stable initial conditions for the SRIV algorithm.

order identification can be performed by the use of well-chosen mathematical measures which indicate the presence of over parameterization. An often used identification procedure to select the most appropriate model structure is based on the minimisation of the Young identification criterion, (YIC) [24] (Eq. (4)).

$$YIC = \ln \frac{\sigma^2}{\hat{\sigma}_y^2} + \ln \left(\frac{1}{np} \sum_{i=1}^{np} \frac{\hat{\sigma}^2 \hat{P}_{ii}}{\hat{a}_i^2} \right) \qquad (4)$$

where $\hat{\sigma}^2$ is the sample variance of the model residuals; σ_y^2 is the sample variance of the measured system output about its mean value; np is the total number of model parameters; $\hat{a}_i^2$ is the square of the i-th element in the parameter vector $\hat{a}$; $\hat{p}_{ii}$ is the i-th diagonal element of the inverse cross product matrix P(N); $\hat{\sigma}^2 p_{ii}$ can be considered as an approximate estimate of the variance of the estimated uncertainty on the i-th parameter estimate.

YIC is a heuristic statistical criterion which consists of two terms, as shown in Eq. (4). The first term provides a normalised measure of how well the model fits the original data: the smaller the variance of the model residuals, in relation to the variance of the measured output, the smaller this term becomes. The second term is a normalised measure of how well the model parameter estimates are defined. This term tends to become bigger when the model is over-parameterised and the parameter estimates are poorly defined. Consequently, the best model should minimise the YIC and provide a good compromise between goodness of fit and parametric efficiency.

Finally, upon passing all listed steps, a derivation of dynamic components which describe the exhaled airflow was feasible. Firstly, by using an individual TF for each subject, we were able to derive poles of the model (Eq. (5)). These poles are direct representatives of the dynamics of the observed model. Secondly, the steady-state gain (SSG) of the model is also derived. SSG is the ratio of the output and the input of the model in steady state, and it is obtained by Eq. (6). Lastly, the time constant (Tc) that characterizes the response to a step input of first-order model is extracted. The time constant represents the time that system's step response needs to reach 63.2 % of its final value. By interpolating this rule to a second-order model, values for Tc are determined.

$$Pole_{1,2} = \frac{a_1 \pm \sqrt{a_1^2 - 4a_2}}{2} \qquad (5)$$

$$SSG = \frac{\Delta y}{\Delta u} = \frac{\sum_{i=1}^{m} b_i}{1 + \sum_{i=1}^{n} a_i} \qquad (6)$$

2.4   Statistical Analysis

Statistical analysis was performed using GraphPad Prism version 5.01, (GraphPad Software, La Jolla, California, USA). The Shapiro-Wilk test was used to control for normality of the datasets, while a Mann Whitney and T-test was used to evaluate differences between subjects with and without COPD within disease severity stages. To evaluate the extracted parameters we used information gain (IG) with ranking (Eq. (7)), which is known to provide the most effective results [25]. This was executed in Weka software version 3.6.10 [26].

$$IG(Ex, a) = H(Ex) - \left(\sum_{v \in m(a)} \frac{|\{x \in Ex | m(x, a) = v\}|}{|Ex|} * H(\{x \in Ex | m(x, a) = v\}) \right)$$

(7)

where Ex represents the whole dataset, a is the observed parameter, x is specific example of attribute, and m(x,a) is value of specific example. H defines the entropy.

3   Results

3.1   Expiration Model

Using the YIC, a second-order model was chosen. Looking into the complete dataset, a second-order model explains the data with YIC of -14.5 (-15.7–-13.1) and R_T^2 of 0.997 (0.994–0.998) (values are median and IQR). R_T^2 represents goodness of fit between original data and model. Confirmation of the good model order identification is presented in Fig. 3, where the original output signal is compared with the simulated

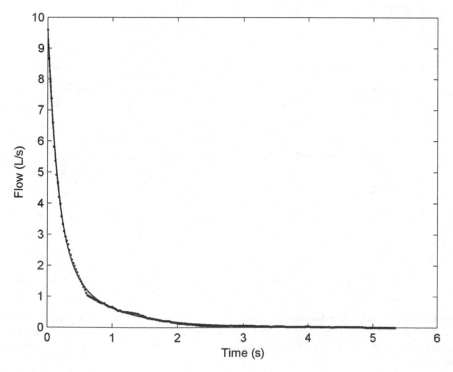

Fig. 3. High agreement between the original (dotted line) and the simulated (solid line) output signal is observed ($R_T^2 = 0.999$, YIC $= -18.081$) when using second-order model.

one using the estimated parameters from second-order model. Equation (8) shows second-order model with model parameters:

$$H(z) = \frac{b_0 + b_1 z^{-1}}{1 + a_1 z^{-1} + a_2 z^{-2}} \tag{8}$$

In total, analysis was performed employing two poles (from the second-order model) and SSG of the model from 423 individuals. From the 474 individuals included, 51 (= 10.8 %) were excluded, where 32 (= 6.8 %) due to missing data from the PFT and 19 (= 4 %) due to model instability.

3.2   Comparison of Dynamic Components

More detailed investigation of the poles of the model, that represents the dynamics of the airflow exhalation, resulted in clear difference when comparing subjects with and without COPD (Fig. 4, panels C and D). The first pole was higher when COPD was present, indicating that the system starts faster when disease occurs. Median (IQR) poles

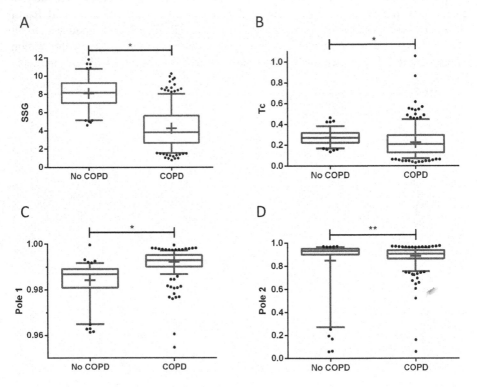

Fig. 4. Comparison of the dynamic components between groups of subjects with and without diagnosed COPD. SSG = Steady Stage Gain, Tc = Time Constant; Graphs are represented with median (IQR) with whiskers on 5–95 percentile, * = p < 0.0001, ** = p < 0.001; + indicates mean value.

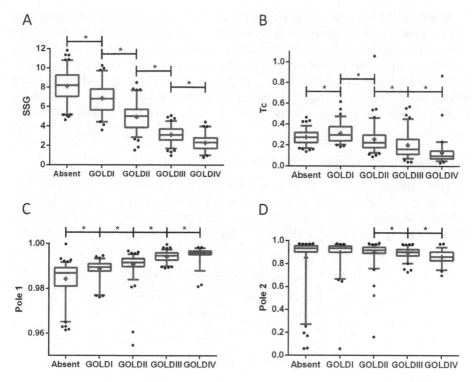

Fig. 5. Comparison of the dynamic components over each stage of disease severity. SSG = Steady Stage Gain, Tc = Time Constant; Graphs are represented with median (IQR) with whiskers on 5–95 percentile, * = p < 0.0001; + indicates mean value.

in subjects without disease were 0.9868 (0.9810–0.9892) and 0.9333 (0.9010–0.9529), respectively for first and second pole, compared to 0.9929 (0.9901–0.9952) and 0.9082 (0.8669–0.9398) in subjects with COPD (p < 0.0001 for first pole and p < 0.001 for second pole). When stratifying for disease severity, the same shift in poles with disease progression was noticed (Fig. 5, panels C and D). For the first pole, significant difference (p < 0.0001) was recorded for each GOLD stage, while the second pole showed significant difference only from the second stage on. This pointed that the dynamics of the system become faster with higher severity. Median poles (pole1 and pole2) were 0.9895 and 0.9346 for GOLD I, 0.9916 and 0.9160 for GOLD II, 0.9946 and 0.9009 for GOLD III and finally for GOLD IV 0.9959 and 0.8615.

When focusing the analysis on the SSG of the model, similar conclusions to the one with poles can be made. Namely, median (IQR) SSG in subjects with COPD was significantly lower 3.9 (2.7-5.6) compared to 8.2 (7.1–9.3) in subjects without COPD, (p < 0.0001) (Fig. 4, panel A). When stratifying over severity of COPD, SSG decreased significantly (p < 0.0001) with each GOLD stage: 6.8 (5.7–7.8), 5.0 (3.9–5.7), 3.1 (2.6–3.7) and 2.3 (1.7–2.8), respectively (Fig. 5, panel A). This is manifested due to lower flow change that occurs when the lungs are obstructive compared to healthy lungs.

Moreover, Tc was also significantly different (p < 0.0001) in disease vs. no disease (Fig. 4, panel B). Median (IQR) Tc in subjects with COPD was 0.27 (0.22–0.32) compared to 0.21 (0.13–0.30) when smoking induced disease is not recorded. Furthermore, analysis over each severity stage (Fig. 5, panel B) revealed that patients with GOLD I have highest Tc, which is followed with a significant drop over each disease severity stage. The larger a time constant is, the slower the fall of the expired airflow. Therefore, indicating that the more severe COPD is, expired flow will decline faster.

3.3   Ranking of Dynamic Components

Using the information gain, we evaluated the amount of useful information that each component provides for prediction of a disease by knowing its presence or absence. SSG was highlighted as the best descriptor of COPD presence with IG of 0.3788, which is closely followed by a Pole 1 with IG of 0.3041 (Table 2).

Table 2.   Ranking by information gain method.

Rank	Information gain	Dynamic component
1.	0.3788	Steady state gain
2.	0.3041	Pole 1
3.	0.1154	Pole 2
4.	0.0945	Time constant

4   Discussion and Conclusions

Our study demonstrates that the dynamics from a data-based model of forced expiration can be used as descriptor of chronic obstructive pulmonary disease. We show that the dynamic components relate highly with the presence and severity of COPD. In addition, our method may bring diagnosis for COPD back to its definition by quantifying flow limitations.

As far as we know, our on-going project is the first to explore the concept of modelling airflow dynamics and associating it with COPD presence in a larger group of individuals comprising COPD patients of all severity stages, as well as smoking controls. In our population, we found that poles, steady state gain and time constant match well with severity of COPD. Moreover, we see a clear trend of the data (statistically significant) for each of the components, over every incrementing stage of the disease. Statistical ranking of the developed components, based on measure of decrease in entropy (COPD vs. No COPD), revealed SSG together with pole 1 as the best dynamic descriptors of COPD presence.

The concept we are introducing opens new opportunities for research in the field of respiratory mechanism and respiratory diseases. In general, by using airflow dynamics, we see significant drop of steady state gain with worsening of the disease. It confirms that obstructive lungs are having more difficulties to exhale high flows with subsequent lower speeds. We see that components anticipating faster dynamics of the system (pole 1) are associated with the presence of COPD. Further, the system dynamics will also increase with the increase of COPD severity stage. One can conclude that bigger obstruction of lungs will result in very fast emptying of the certain amount of air (mainly from trachea) and then very slow emptying the rest of the air (seen from pole 2). The evolution of the time constant also supports such a conclusion, as more obstruction means less time to reach 63.2 % of the total exhaled air. Various reasons influence such an occurrence. Firstly it is common to observe that airway narrowing or airway collapse cause suddenly diminished airflow [27]. Additionally, in COPD, the greatest reduction in air flow occurs during expiration, as the pressure in the chest compress the airways [28]. One would assume that loss of lung tissue elasticity, typical for emphysematous type of COPD, plays an additional role in accelerating exhalation dynamics, as it might be the case that lungs reach faster their limits while exhaling [29].

When comparing with other alternative approaches, the advantage of our method is that parameters obtained from model-based method can have physiological validity. Further, when used with routine spirometry during patient examination, this method is simplest, fastest and cheapest to perform.

An additional strength of this study is the fact that observing dynamics of the flow decay represents the same approach that many researchers had performed in the past, but based only on a visual basis of typical patterns [30, 31]. Today routinely, clinicians are capable to presume presence of Chronic Obstructive Pulmonary Disease, on the basis of visual assessment of flow decay, whereas with this study we offer a quantitative and automated way of inspection. We believe that the concept which we are introducing, is easy to understand and linked to physiological behaviour of the lungs.

Finally, our method failed to provide valid measurements in 4 % of the cases. This occurrence is inevitable, as we tried to the automatize process where data selection and estimation algorithm are not always the optimal ones. Certainly, this could be avoided in most of the cases, if ensuring that exhalation ends with a plateau (having stable ending).

To conclude, our data provide strong evidence that more detailed analysis of forced expiration can be used to expand understanding of the chronic obstructive pulmonary disease. Moreover, if characterized as in our model, flow decline may improve current COPD assessment by spirometry.

Acknowledgements. The authors would like to thank Geert Celis and co-workers (Respiratory Division, University Hospital Leuven, Belgium) for helping in collection of patient data and their technical support in extracting data from the Masterlab.

References

1. Rabe, K.F., Hurd, S., Anzueto, A., Barnes, P.J., Buist, S.A., Calverley, P., Fukuchi, Y., Jenkins, C., Rodriguez-Roisin, R., van Weel, C., Zielinski, J.: Global strategy for the diagnosis, management, and prevention of chronic obstructive pulmonary disease: GOLD executive summary. Am. J. Respir. Crit. Care Med. **176**(6), 532–555 (2007)
2. Rennard, S.I., Vestbo, J.: Natural histories of chronic obstructive pulmonary disease. Proc. Am. Thorac. Soc. **5**(9), 878–883 (2008)

3. Decramer, M., Janssens, W., Miravitlles, M.: Chronic obstructive pulmonary disease. Lancet **379**(9823), 1341–1351 (2012)
4. Mathers, C.D., Loncar, D.: Projections of global mortality and burden of disease from 2002 to 2030. PLoS Med. **3**(11), e442 (2006)
5. Murray, C.J., Lopez, A.D.: Alternative projections of mortality and disability by cause 1990-2020: global burden of disease study. Lancet **349**(9064), 1498–1504 (1997)
6. WHO, World Health Statistic, EN_WHS08_Full.pdf. 2012 (2008). http://www.who.int/whosis/whostat/
7. Mannino, D.M., Buist, A.S.: Global burden of COPD: risk factors, prevalence, and future trends. Lancet **370**(9589), 765–773 (2007)
8. Agusti, A., Calverley, P.M., Celli, B., Coxson, H.O., Edwards, L.D., Lomas, D.A., MacNee, W., Miller, B.E., Rennard, S., Silverman, E.K., Tal-Singer, R., Wouters, E., Yates, J.C., Vestbo, J.: Characterisation of COPD heterogeneity in the ECLIPSE cohort. Respir. Res. **11**, 122 (2010)
9. Miravitlles, M., Soler-Cataluna, J.J., Calle, M., Soriano, J.B.: Treatment of COPD by clinical phenotypes: putting old evidence into clinical practice. Eur. Respir. J. **41**(6), 1252–1256 (2013)
10. Garcia-Rio, F., Soriano, J.B., Miravitlles, M., Munoz, L., Duran-Tauleria, E., Sanchez, G., Sobradillo, V., Ancochea, J.: Overdiagnosing subjects with COPD using the 0.7 fixed ratio: correlation with a poor health-related quality of life. Chest **139**(5), 1072–1080 (2011)
11. Bodduluri, S., Newell, Jr., J.D., Hoffman, E.A., Reinhardt, J.M.: Registration-based lung mechanical analysis of chronic obstructive pulmonary disease (COPD) using a supervised machine learning framework. Acad. Radiol. 20(5), 527–536 (2013)
12. Sorensen, L., Nielsen, M., Lo, P., Ashraf, H., Pedersen, J.H., de Bruijne, M.: Texture-based analysis of COPD: a data-driven approach. IEEE Trans. Med. Imaging **31**(1), 70–78 (2012)
13. Fens, N., Zwinderman, A.H., van der Schee, M.P., de Nijs, S.B., Dijkers, E., Roldaan, A.C., Cheung, D., Bel, E.H., Sterk, P.J.: Exhaled breath profiling enables discrimination of chronic obstructive pulmonary disease and asthma. Am. J. Respir. Crit. Care Med. **180**(11), 1076–1082 (2009)
14. Phillips, C.O., Syed, Y., Parthalain, N.M., Zwiggelaar, R., Claypole, T.C., Lewis, K.E.: Machine learning methods on exhaled volatile organic compounds for distinguishing COPD patients from healthy controls. J. Breath. Res. **6**(3), 036003 (2012)
15. Amaral, J.L., Lopes, A.J., Jansen, J.M., Faria, A.C., Melo, P.L.: Machine learning algorithms and forced oscillation measurements applied to the automatic identification of chronic obstructive pulmonary disease. Comput. Methods Programs Biomed. **105**(3), 183–193 (2012)
16. Dellaca, R.L., Santus, P., Aliverti, A., Stevenson, N., Centanni, S., Macklem, P.T., Pedotti, A., Calverley, P.M.: Detection of expiratory flow limitation in COPD using the forced oscillation technique. Eur. Respir. J. **23**(2), 232–240 (2004)
17. Lambrechts, D., Buysschaert, I., Zanen, P., Coolen, J., Lays, N., Cuppens, H., Groen, H.J., Dewever, W., van Klaveren, R.J., Verschakelen, J., Wijmenga, C., Postma, D.S., Decramer, M., Janssens, W.: The 15q24/25 susceptibility variant for lung cancer and chronic obstructive pulmonary disease is associated with emphysema. Am. J. Respir. Crit. Care Med. **181**(5), 486–493 (2010)
18. Topalovic, M., Exadaktylos, V., Peeters, A., Coolen, J., Dewever, W., Hemeryck, M., Slagmolen, P., Janssens, K., Berckmans, D., Decramer, M., Janssens, W.: Computer quantification of airway collapse on forced expiration to predict the presence of emphysema. Respir. Res. **14**, 131 (2013)

19. Miller, M.R., Hankinson, J., Brusasco, V., Burgos, F., Casaburi, R., Coates, A., Crapo, R., Enright, P., van der Grinten, C.P., Gustafsson, P., Jensen, R., Johnson, D.C., MacIntyre, N., McKay, R., Navajas, D., Pedersen, O.F., Pellegrino, R., Viegi, G., Wanger, J.: Standardisation of spirometry. Eur. Respir. J. **26**(2), 319–338 (2005)
20. Quanjer, P.H., Tammeling, G.J., Cotes, J.E., Pedersen, O.F., Peslin, R., Yernault, J.C.: Lung volumes and forced ventilatory flows. work group on standardization of respiratory function tests. european community for coal and steel. official position of the european respiratory society. Rev. Mal. Respir. **11**(Suppl 3), 5–40 (1994)
21. Taylor, C.J., Pedregal, D.J., Young, P.C., Tych, W.: Environmental time series analysis and forecasting with the Captain toolbox. Environ. Model Softw. **22**(6), 797–814 (2007)
22. Ljung, L.: System Identification: Theory for the User. Prentice-Hall, Englewood Cliffs (1987)
23. Young, P.C.: Recursive Estimation and Time-Series Analysis: An Introduction. Springer, Heidelberg (1984)
24. Young, P.: Parameter-estimation for continuous-time models - a survey. Automatica **17**(1), 23–39 (1981)
25. Yang, Y., Pedersen, J.P.: A comparative study on feature selection in text categorization. In: Proceedings of the International Conference on Machine Learning (ICML 1997), pp. 412–420 (1997)
26. Hall, M., Frank, E., Holmes, G., Pfahringer, B., Reutemann, P., Witten, I.H.: The WEKA data mining software: an update. SIGKDD Explor. **11**(1), 10–18 (2009)
27. Healy, F., Wilson, A.F., Fairshter, R.D.: Physiologic correlates of airway collapse in chronic airflow obstruction. Chest **85**(4), 476–481 (1984)
28. Koulouris, N.G., Hardavella, G.: Physiological techniques for detecting expiratory flow limitation during tidal breathing. Eur. Respir. Rev. **20**(121), 147–155 (2011)
29. Papandrinopoulou, D., Tzouda, V., Tsoukalas, G.: Lung compliance and chronic obstructive pulmonary disease. Pulm. Med. **2012**, 542769 (2012)
30. Bass, H.: The flow volume loop: normal standards and abnormalities in chronic obstructive pulmonary disease. Chest **63**(2), 171–176 (1973)
31. Jayamanne, D.S., Epstein, H., Goldring, R.M.: Flow-volume curve contour in COPD: correlation with pulmonary mechanics. Chest **77**(6), 749–757 (1980)

Velar Movement Assessment for Speech Interfaces: An Exploratory Study Using Surface Electromyography

João Freitas[1,2]($\boxtimes$), António Teixeira[2], Samuel Silva[2],
Catarina Oliveira[3], and Miguel Sales Dias[1,4]

[1] Microsoft Language Development Center, Lisbon, Portugal
{t-joaof,miguel.dias}@microsoft.com
[2] Department Electronics Telecommunications and Informatics/IEETA,
University of Aveiro, Aveiro, Portugal
{ajst,sss}@ua.pt
[3] Health School/IEETA, University of Aveiro, Aveiro, Portugal
coliveira@ua.pt
[4] ISCTE-University Institute of Lisbon (ISCTE-IUL), Lisbon, Portugal

Abstract. In the literature several silent speech interfaces based on Surface Electromyography (EMG) can be found. However, it is yet unclear if we are able to sense muscles activity related to nasal port opening/closing. Detecting the nasality phenomena, would increase the performance of languages with strong nasal characteristics such as European Portuguese. In this paper we explore the use of surface EMG electrodes, a non-invasive method, positioned in the face and neck regions to explore the existence of useful information about the velum movement. For an accurate interpretation and validation of the proposed method, we use velum movement information extracted from Real-Time Magnetic Resonance Imaging (RT-MRI) data. Overall, results of this study show that differences can be found in the EMG signals for the case of nasal vowels, by sensors positioned below the ear between the mastoid process and the mandible in the upper neck region.

Keywords: Velum movement detection · Surface electromyography · Silent speech interfaces

1 Introduction

Automatic Speech Recognition (ASR) has suffered a significant evolution in the past decades. However, issues like environmental noise and irregular speech patterns such as the ones found in elderly speech remain a challenge for the speech community [1]. One of the reasons behind these problems is due to the fact that conventional speech interfaces strongly rely on the acoustic signal. Hence, ASR becomes inappropriate when used in the presence of environmental noise, such as in office settings, or when used in situations where privacy or confidentiality is required. For the same reason, speech-impaired persons such as those who were subjected to a laryngectomy are unable to use this type of interface. Hence, robust speech recognition and improved

© Springer International Publishing Switzerland 2015
G. Plantier et al. (Eds.): BIOSTEC 2014, CCIS 511, pp. 246–260, 2015.
DOI: 10.1007/978-3-319-26129-4_16

user experience, with this type of interfaces, are still an attractive research topic [2–4]. A Silent Speech Interface (SSI) can be viewed as an alternative or complementary solution since it allows for communication to occur in the absence of an acoustic signal and, although they are still in an early stage of development, recent results show that this type of interface can be used to tackle these issues. For an overview about SSIs the reader is forwarded to Denby et al. [5].

Surface Electromyography (EMG) is one of the approaches reported in literature that is suitable for implementing an SSI, having achieved promising results [6, 7]. A known challenge in SSIs, including those based on surface EMG, is the detection of the nasality phenomena in speech production and it remains unclear if information on nasality is present [5]. Nasality detection is a known challenge in the field of SSI and an important one for languages with nasal characteristics, such as European Portuguese (EP) [8], which is the selected language for the experiments here reported. Additionally, no SSI exists for Portuguese and, as previously discussed by Freitas et al. [9], nasality can cause severe accuracy degradation for this language. Given the particular relevance of nasality for EP, we have conducted an experiment that seeks to expand knowledge in this field, determining the possibility of detecting nasality in EMG-based speech interfaces, consequently improving interaction with these systems. The rationale behind this experiment consists in crossing two types of data containing information about the velum movement: (1) images collected using Real Time Magnetic Resonance Imaging (RT-MRI) and (2) the myoelectric signal collected using Surface EMG. By combining these two sources, ensuring compatible scenario conditions and proper time alignment, we are able to accurately estimate when the velum moves, under a nasality phenomenon, and establish the differences between nasal and oral vowels using surface EMG.

The remainder of this article is organized as follows: Sect. 2 provides a brief contextualization regarding nasality and the muscles associated with velar activity; Sect. 3 presents an overview regarding the use of EMG in the context of SSIs; Sect. 4 describes the main features of our experimental setting; Sect. 5 presents and discusses notable results; finally, Sects. 6 and 7 provide the conclusions and ideas that should guide further developments.

2 Background

For a vowel to be perceived as nasal, the velopharyngeal port needs to open and the velum to be lowered, creating an additional air passage and making the air go through the oral and nasal cavities. The air passage for the nasal cavity is essentially controlled by the velum that when lowered, enables resonance in the nasal cavity. The production of oral sounds occurs when the velum is raised and the access to the nasal cavity is closed [8]. The process of moving the velum involves several muscles [10–12]. The muscles responsible for elevating the velum are the following: *Levator veli palatini*; *Musculus uvulae* [13]; *Superior pharyngeal constrictor*; and the *Tensor veli palatini*. Along with gravity, relaxation of the above-mentioned muscles, the *Palatoglossus* and the *Palatopharyngeous* are responsible for the lowering of the velum.

2.1 Nasality in European Portuguese

Nasality is an important characteristic not only of European Portuguese but also of a vast number of languages with around 20 % of the world languages having nasal vowels [14]. In EP there are five nasal vowels ([ĩ], [ẽ], [ɐ̃], [õ], and [ũ]); three nasal consonants ([m], [n], and [ɲ]); and several nasal diphthongs [wɐ̃] (e.g. quando), [wẽ] (e.g. aguentar), [jɐ̃] (e.g. fiando), [wĩ] (e.g. ruim) and triphthongs [wɐ̃w] (e.g. enxaguam). Nasal vowels in EP diverge from other languages, such as French, in its wider variation in the initial segment and stronger nasality at the end [15, 16]. Doubts still remain regarding tongue positions and other articulators during nasals production in EP, namely, nasal vowels [17]. Differences at the pharyngeal cavity level and velum port opening quotient were also detected by Martins et al. [18] when comparing EP and French nasal vowels articulation. In EP, nasality can distinguish consonants (e.g. the bilabial stop consonant [p] becomes [m]), creating minimal pairs such as [patu]/[matu] and vowels, in minimal pairs such as [titu]/[tĩtu].

3 Related Work

In previous studies, the application of EMG to measure the level of activity of the muscles involved in the velum movement has been performed by means of intramuscular electrodes [12, 19] using surface electrodes positioned directly on the oral surface of the soft palate [20, 21]. One of the distinctive features of our work is the use of surface electrodes placed in the face and neck regions, a significantly less invasive approach and quite more realistic and representative of the SSIs case scenarios. Also, although intramuscular electrodes may offer more reliable myoelectric signals, they also require considerable medical skills. For both reasons, intramuscular electrodes were not considered for this study.

No literature exists in terms of detecting the muscles involved in the velopharyngeal function with surface EMG electrodes placed on the face and neck regions. However, previous studies in the lumbar spine region have shown that if proper electrode positioning is considered a representation of deeper muscles can be acquired [22] thus raising a question that is currently unanswered: is surface EMG able to detect activity of the muscles related to nasal port opening/closing and consequently detect the velar movement? Another related question that can be raised is how we can show, with some confidence, that the signal we are seeing is in fact the myoelectric signal generated by the velum movement and not spurious movements caused by neighboring muscles unrelated to the velopharyngeal function.

3.1 EMG-Based Speech Interfaces

Our method relies on Surface EMG sensors to detect nasality. This technique has also been applied to audible speech and silent speech recognition (e.g. Schultz and Wand [6]). Relevant results in this area were first reported in 2001 by Chan et al. [23] where surface EMG sensors were used to recognize ten English digits, achieving accuracy rates as high as 93 %. In 2003, Jorgensen et al. [7] achieved an average accuracy rate of 92 % for a vocabulary with six distinct English words, using a single pair of electrodes for non-audible speech. In 2007, Jou et al. [24] reported an average accuracy of 70.1 % for a 101-word vocabulary in a speaker dependent scenario. In 2010, Schultz and Wand [6] reported similar average accuracies using phonetic feature bundling for modelling coarticulation on the same vocabulary and an accuracy of 90 % for the best-recognized speaker. Latest research in this area has been focused on the differences between audible and silent speech and how to decrease the impact of different speaking modes [25]; the importance of acoustic feedback [26]; EMG-based phone classification [27]; session-independent training methods [28]; adapting to new languages [9]; and EMG recording systems based on multi-channel electrode arrays [29].

Our technique can, in theory, be used as a complement to a surface EMG-based speech interface by adding a new sensor, or be combined with other silent speech recognition techniques such as Ultrasonic Doppler Sensing [30], Video [31], etc. by using a multimodal approach.

4 Methodology

To determine the possibility of detecting nasality using surface EMG we need to know when the velum is moving, avoiding signals from other muscles, artifacts and noise, to be misjudged as signal coming from the target muscles. To overcome this problem we take advantage of a previous data collection based on RT-MRI [32], which provides an excellent method to interpret EMG data and estimate when the velum is moving. Recent advances in MRI technology allow real-time visualization of the vocal tract with an acceptable spatial and temporal resolution. This sensing technology enables us to have access to real time images with relevant articulatory information for our study, including velum raising and lowering. In order to make the correlation between the two signals, audio recordings were performed in both data collections by the same speakers. Notice that EMG and RT-MRI data cannot be collected together, so the best option is to collect the same corpus for the same set of speakers, at different times, reading the same prompts in EMG and RT-MRI.

4.1 Corpora

The two corpora collected (RT-MRI and EMG) share a subset of the same prompts. This set of prompts, shown in Table 1, is characterized by several non-sense words that contain five EP nasal vowels ([ɐ̃, ẽ, ĩ, õ, ũ]) in isolated, word-initial, word-internal and word-final context (e.g. ampa [ɐ̃pɐ], pampa [pɐ̃pɐ], pam [pɐ̃]). The nasal vowels were flanked by the bilabial stop or the labiodental fricative. For comparison purposes the set of prompts also includes isolated oral vowels and in context. In the EMG data collection a total of 90 utterances per speaker were recorded. A detailed description of the RT-MRI corpus can be found in [32].

Three female EP speakers aged 22, 22 and 33 participated in this study. No history of hearing or speech disorders is known for all of them. One of the speakers is a professor in the area of Phonetics and the remaining speakers are students in the area of Speech Therapy.

Table 1. Prompt subset used in EMG and RT-MRI data collection. IPA phonetic notation was used.

Vowel	Context	Prompt
ɐ̃	p	ampa, pampa, pã [ɐ̃pɐ, pɐ̃pɐ, pɐ̃]
ẽ	p	empa, pempa, pem [ẽpɐ, pẽpɐ, pẽ]
ĩ	p	impa, pimpa, pim [ĩpɐ, pĩpɐ, pĩ]
õ	p	ompa, pompa, pom [õpɐ, põpɐ, põ]
ũ	p	umpa, pumpa, pum [ũpɐ, pũpɐ, pũ]
ɐ̃	f	anfa, fanfa, fan [ɐ̃fɐ, fɐ̃fɐ, fɐ̃]
ĩ	f	infa, finfa, fin [ĩfɐ, fĩfɐ, fĩ]
ũ	f	unfa, funfa, fun [ũfɐ, fũfɐ, fũ]
a	p	pápa [papɐ]
ɐ	p	pâpa [pɐpɐ]
e	p	pépa [pepɐ]
ɛ	p	pêpa [pɛpɐ]
i	p	pipa [pipɐ]
ɔ	p	pópa [pɔpɐ]
o	p	pôpa [popɐ]
u	p	pupa [pupɐ]
[ɐ̃, ẽ, ĩ, õ, ũ]	Isolated	ã, em, im, om, um [ɐ̃, ẽ, ĩ, õ, ũ]
[a, ɛ, i, ɔ, u]	Isolated	a é i ó u [a, ɛ, i, ɔ, u]
[ɐ, e, o]	Isolated	â ê ô [ɐ, e, o]
ɐ	m	ama, ana, anha [ɐmɐ, ɐnɐ, ɐɲɐ]
i	m	imi, ini, inhi [imi, ini, iɲi]
u	m	umu, unu, unhu [umu, unu, uɲu]

4.2   RT-MRI Data

The RT-MRI data collection was previously conducted for the three speakers at IBILI/Coimbra in the context of ongoing research focusing on nasality in European Portuguese. Images were acquired in the midsagittal and coronal oblique (encompassing the oral and nasal cavities) planes of the vocal tract at a frame rate of 14 frames/second. For additional information concerning the image acquisition protocol the reader is forwarded to Silva et al. [33]. The audio was recorded simultaneously with the real-time images, inside the scanner, at a sampling rate of 16000 Hz, using a fiber optic microphone. For synchronization purposes a TTL pulse was generated from the RT-MRI scanner [32].

Extraction of Information on Nasal Port from RT MRI Data. For the mid-sagittal RT-MRI sequences of the vocal tract, since the main interest was to interpret velum position/movement from the sagittal RT-MRI sequences, instead of measuring distances, we opted for a method based on the area variation between the velum and pharynx, closely related to velum position. These images allowed deriving a signal over time that describes the velum movement (shown in Fig. 1 and depicted as dashed line in Fig. 4). As can be observed, minima correspond to a closed velopharingeal port (oral sound) and maxima to an open port (nasal sound). Coronal oblique real-time images were also processed to extract information of the nasal port (refer to Fig. 2 for notable examples). After segmentation of the nasal cavity the corresponding area was computed and a variation curve was derived, similar to the one obtained from the sagittal sequences: local maxima corresponding to an open velar port and minima to a closed port. Additional details concerning the segmentation of the oblique real-time images for velum movement extraction can be found in Silva et al. [33] and resulted in similar variation curves.

4.3 Surface EMG Data Collection

To take advantage of the RT-MRI information, the same speakers that have performed the RT-MRI recordings were considered. The EMG recordings, for each speaker, took place in a single session in order to ensure that the sensors position were maintained throughout the recordings. While uttering the prompts no other movement, besides the one associated with speech production, was made, including any kind of neck movement. The recordings took place in an isolated quiet room. An assistant was responsible for pushing the record button and also stopping the recording after the prompt in order to avoid unwanted muscle activity. The prompts were presented to the speaker in a random order and were selected based on the already existent RT-MRI corpus [32]. In this data collection two signals were acquired synchronously: myoelectric and audio.

The used acquisition system, from Plux [34], consisted of 5 pairs of EMG surface electrodes where the result is the amplified difference between each pair of electrodes. These electrodes measure the myoelectric activity using bipolar and monopolar electrode configuration, always using a reference electrode located in a place with low or negligible muscle activity. In the monopolar configuration, instead of placing the electrode pair along the muscle fiber, only one of the electrodes is placed on the articulatory muscles while the other electrode is placed in an area without muscle activity. The sensors were attached to the skin using single-use 2.5 cm diameter clear plastic self-adhesive surfaces and approximately considering 2 cm spacing between the electrodes center. No specific background literature in speech science exists to support surface EMG sensor position placement in order to detect the muscles referred in Sect. 2. Hence, we determined a set of positions that should cover, as best as possible, the most probable positions for detecting the targeted muscles based on the anatomy and physiology literature (e.g., Hardcastle [11]) along with preliminary trials. As depicted on Fig. 3, the 5 sensor pairs were positioned in a way that covers the upper neck area, the area above the mandibular notch and the area below the ear between the mastoid process and the mandible. The reference electrodes were placed in the mastoid portion of the temporal bone and in the cervical vertebrae. Even though the goal is to detect signals from the muscles involved in the velopharyngeal function it is also expected to acquire unwanted myoelectric signals due to the superposition of muscles in these areas, such as the jaw muscles. However, in spite of the muscles of the velum being remote from this peripheral region, we expect to be able to select a sensor location that enables us to identify and classify the targeted muscle signal with success.

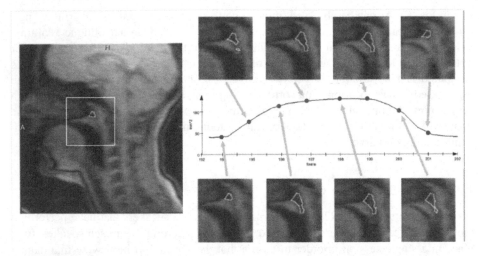

Fig. 1. Mid-sagittal RT-MRI images of the vocal tract for several velum positions, over time, showing evolution from a raised velum, to a lowered velum and back to initial conditions. The presented curve, used for analysis, was derived from the images.

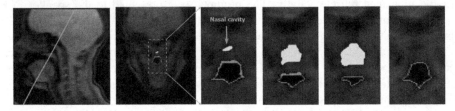

Fig. 2. Coronal oblique RT-MRI images. From left to right, orientation of the oblique plane depicted over a mid-sagittal image, coronal oblique image depicting the location of the region of interest and a set of four detail images showing different velar port apertures. In the rightmost image, the velar port is completely closed.

The technical specifications of the acquisition system [34] include snaps with a diameter of 14.6 mm and 6.2 mm of height, a voltage range that goes from 0.0 V to 5.0 V and a voltage gain of 1000. The recording signal was sampled at 600 Hz and 12 bit samples were used.

The audio recordings were performed using a laptop integrated dual-microphone array using a sample rate of 8000 Hz, 16 bits per sample and a single audio channel. Since the audio quality was not a requirement in this collection we opted for this solution instead of a headset microphone which could cause interference with the EMG signal.

4.4 Signal Synchronization

In order to address the nasality detection problem we need to synchronize the EMG and RT-MRI signals. We start by aligning both EMG and the information extracted from

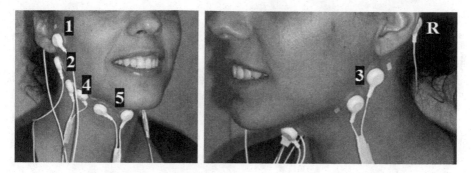

Fig. 3. EMG electrodes positioning and the respective channels (1 to 5) plus the reference electrode (R). EMG channels 1 and 2 use a monopolar configuration and channels 3, 4 and 5 use bipolar configurations.

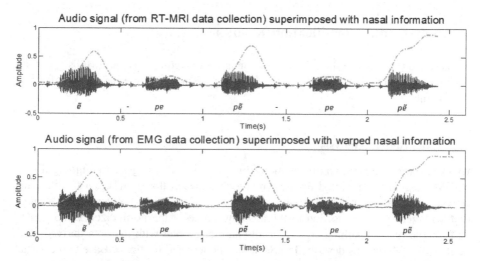

Fig. 4. Notable example that depicts the warped signal containing the nasal information extracted from RT-MRI (dashed line) superimposed on the speech recorded during the corresponding RT-MRI and EMG acquisition, for the sentence [ɐ̃pɐ, pɐ̃pɐ, pɐ̃].

the RT-MRI with the corresponding audio recordings. Next, we resample the audio recordings to 12000 Hz and apply Dynamic Time Warping (DTW) to the signals, finding the optimal match between the two sequences. Based on the DTW result we map the information extracted from RT-MRI to the EMG time axis, establishing the needed correspondence between the EMG and the RT-MRI information, as depicted in Fig. 4.

Based on the information extracted from the RT-MRI signal and after signal alignment, we are able to segment the EMG signal into nasal and non-nasal, as depicted in Fig. 5. Considering the normalized RT-MRI signal x, based on an empirical analysis of the signals of all users, we determine that $x(n) \geq \bar{x} + \left(\frac{\sigma}{2}\right)$ indicates a nasal event and

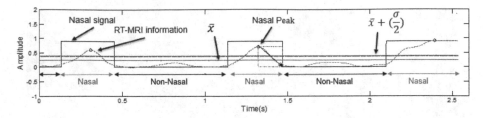

Fig. 5. Nasal and non-nasal zones definition based on the RT-MRI velum information.

$x(n) < \bar{x} + \left(\frac{\sigma}{2}\right)$ indicates a non-nasal event. Then by we use the angle between the nearest peak and the points where the $x(n) = \bar{x}$ to calculate the nasal zone boundaries, making the transitional part of the signal (i.e., lowering and raising of the velum) to be included in the nasal zones.

5   Analysis and Classification Results

In this section we present the results of the analysis that combines the EMG signal with the information extracted from the RT-MRI signal and a frame-based classification experiment.

5.1   EMG Signal Analysis

After extracting the required information from the RT-MRI images and aligning it with the EMG signal we explored possible relations between the signals. To facilitate the analysis the EMG signal was pre-processed by applying a 12-point moving average filter with zero-phase distortion to the absolute value of the normalized EMG signal. An example of the resulting signal for all channels, along with the data derived from the RT-MRI, aligned as described in Sect. 4.4, is depicted in Fig. 6. Based on a visual analysis, it is worth noticing that several peaks anticipate the nasal sound, especially in channels 2, 3 and 4. These peaks are most accentuated for the middle and final word position.

By using surface electrodes the risk of acquiring myoelectric signal superposition is relatively high, particularly for muscles related with the movement of the lower jaw and the tongue considering the electrodes position. However, if we analyze an example of a close vowel such as [ĩ], where the movement of the jaw is less prominent, the peaks found in the signal still anticipate the RT-MRI velar information for channels 3 and 4. Channel 5 also exhibits notable activity in this case which might be caused by its position near the tongue muscles and the tongue movement associated with the artic-ulation of the [ĩ] vowel. If the same analysis is considered for isolated nasal vowels ([ẽ, ẽ, ĩ, õ, ũ]) of the same speaker, EMG Channel 1 signal exhibits a more clearer signal apparently with less muscle crosstalk and peaks can be noticed before the nasal vowels.

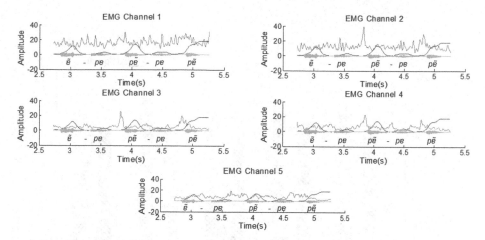

Fig. 6. Filtered EMG signal for the several channels (pink), the aligned RT-MRI information (blue) and the corresponding audio signal for the sentence [ẽpɐ, pẽpɐ, pẽ] from speaker 1.

For the remaining channels there is not a clear relation with all the vowels, although signal amplitude variations can be noticed in the last three vowels for EMG channel 3.

The fact that all seemed to point for the presence of differences between the two classes (nasal and non-nasal) motivated an exploratory classification experiment based on Support Vector Machines (SVMs), which have presented an acceptable performance in other applications, even when trained with small data sets.

5.2   Frame-Based Nasality Classification

In a real use situation the information about the nasal and non-nasal zones extracted from the RT-MRI signal is not available. As such, in order to complement our study and because we want to have a nasality feature detector, we have conducted an experiment where we split the EMG signal into frames and classify them as one of two classes: nasal or non-nasal. For estimating classifier performance we have applied 10-fold cross-validation technique to a set of frames from the 3 speakers, which includes the utterances with the nasal vowels flanked by a bilabial stop. A total of 1572 frames (801 nasal and 771 non-nasal) were considered. From each frame we extract 9 first order temporal features similar to the ones used by Hudgins et al. [35]. Our feature vector is then composed by mean, absolute mean, standard deviation, maximum, minimum, kurtosis, energy, zero-crossing rate and mean absolute slope. We have considered 100 ms frames and a frame shift of 20 ms. Both feature set and frame sizes were determined after several experiments. For classification we have used SVMs with a Gaussian Radial Basis Function.

In our classification experiments we start by using the data from all speakers. Results of 4 relevant metrics are depicted in Fig. 7. Besides the mean value of the 10-fold, 95 % confidence intervals are also included. Results indicate a best result for EMG Channel 3 with 32.5 % mean error rate, an F-score of 66.3 %, a mean sensitivity

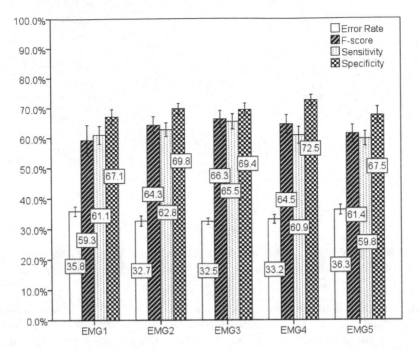

Fig. 7. Classification results (mean value of the 10-fold for error rate, F-score, sensitivity and specificity) for all channels and speakers. Error bars show a 95 % confidence interval.

of 65.5 % and a mean specificity of 69.4 %. Channels 4 and 2 achieved second and third best error rates with mean error rates of 32.7 % and 33.2 % and an F-score of 64.5 % and 64.3 %.

We have also run the same experiment for each individual speaker. EMG channel 3 obtained the best overall result with 24.3 % mean error rate and 62.3 % mean F-score. The best results for each individual speaker were found for Speaker 3 with 23.4 % and 23.6 % mean error rate and F-score values of 72.1 % and 70.3 % in EMG channels 4 and 3, respectively. For Speaker 1 and 2, EMG channel 3 presents the best results with 25.7 % and 23.7 % mean error rate and F-score values of 76.1 % and 51.9 %. However, if we look into the data of Speaker 2 a higher amount of nasal frames is found, explained by common breathings between words, which imply an open velum.

On a different perspective, if we subtract the global mean error rate from all channels then, as seen in Fig. 8, EMG channel 3 exhibits a mean error rate 4.1 % below this mean, followed by EMG channel 4 with 1.0 % below the global mean.

To assess if any advantage could be extracted from using channel combination to improve classification we have also experienced classification with multiple EMG channels. However, no improvements were verified when comparing with the previously obtained results, a fact that might suggest information overlap among channels.

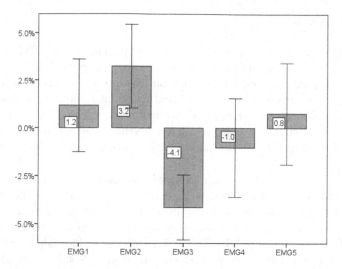

Fig. 8. Difference between the mean error rate of all channels and the corresponding result of each channel for all speakers. Error bars show a 95 % confidence interval.

6 Conclusions

To address the challenge of nasality detection in silent speech interaction, we propose a solution based on surface EMG that uses RT-MRI, a distinct source of information, as a ground truth modality. The information extracted from the RT-MRI images allows us to know when to expect nasal information. Thus, by synchronizing both signals, based on simultaneously recorded audio signals from the same speaker, we are able to explore the existence of useful information in the EMG signal about the velum movement. The global results of this study, although preliminary in the sense that further validation is required, point to the fact that the selected approach can be used to reduce the error rate caused by nasality in languages where this characteristic is particularly relevant such as Portuguese, also providing background for future studies in terms of sensor positioning. The results of this study show that, in a real use situation, error rates as low as 23.4 % can be achieved for sensors positioned below the ear between the mastoid process and the mandible in the upper neck region, and that careful articulation, positioning of the sensors or even anatomy of the speaker may influence nasality detection results. Also, although the methodology used in this study partially relies on RT-MRI information for scientific substantiation, a technology which requires a complex and expensive setup, the proposed solution to detect nasality is solely based on a single sensor of surface EMG. Thus, the development of an SSI based on EMG for EP, with language adapted sensor positioning, seems to be now a possibility.

7 Future Work

For future work our aim is to integrate this solution into a multimodal SSI for European Portuguese. We will also analyze other non-invasive modalities such as Ultrasonic Doppler [30] or general electromagnetic motion sensor (GEMS) [36], which combined with this approach, may improve the achieved results and allow for a more accurate detection. It would also be interesting to explore an event-based technique that upon muscle activation would be able to trigger a velum state change. However, the superimposition of muscles in the studied areas make it a complex achievement to attain.

The work presented in this article explicitly addresses nasality. Nevertheless, the work being carried out encompasses a wider range of studies. The multimodal acquisition setting used for the presented experiments is an instantiation of a generic framework designed to support exploratory studies concerning the applicability of surface EMG in the context of SSIs. This framework can include additional modalities to provide reference data targeting different articulators. Notable examples of synchronously acquired modalities already being used in our different studies include 3D video using Kinect (lips) and ultrasound (tongue movements). The data collected from these different settings, testing different EMG sensor placements, should help us gain further insight into the value of surface EMG, paving the way towards the best possible setting (i.e. covering the most aspects of speech production possible, with surface EMG, using a minimal number of sensors).

Acknowledgements. This work was partially funded by Marie Curie IAPP Golem (ref.251415, FP7-PEOPLE-2009-IAPP), Marie Curie IAPP IRIS (ref. 610986, FP7-PEOPLE-2013-IAPP) and by FEDER through the Operational Program Competitiveness factors - COMPETE under the scope of QREN 5329 FalaGlobal, by National Funds through FCT (Foundation for Science and Technology) in the context of the Project HERON II (PTDC/EEA-PLP/098298/2008) and by project Cloud Thinking (funded by the QREN Mais Centro program: CENTRO-07-ST24-FEDER-002031).

References

1. Huang, X., Acero, A., Hon, H.: Spoken Language Processing. Prentice Hall PTR, Upper Saddle River (2001)
2. Flynn, R., Jones, E.: Combined speech enhancement and auditory modelling for robust distributed speech recognition. Speech Commun. **50**(10), 797–809 (2008)
3. Stark, A., Paliwal, K.: MMSE estimation of log-filterbank energies for robust speech recognition. Speech Commun. **53**(3), 403–416 (2011)
4. Yang, C., Brown, G., Lu, L., Yamagishi, J., King, S.: Noise-robust whispered speech recognition using a non-audible-murmur microphone with VTS compensation. In: 2012 8th International Symposium on Chinese Spoken Language Processing (ISCSLP), pp. 220–223 (2012)
5. Denby, B., Schultz, T., Honda, K., Hueber, T., Gilbert, J.M., Brumberg, J.S.: Silent speech interfaces. Speech Commun. **52**(4), 270–287 (2009)

6. Schultz, T., Wand, M.: Modeling coarticulation in large vocabulary EMG-based speech recognition. Speech Commun. **52**(4), 341–353 (2010)
7. Jorgensen, C., Lee, D., Agabon, S.: Sub auditory speech recognition based on EMG signals. In: Proceedings of the International Joint Conference on Neural Networks (IJCNN), pp. 3128–3133 (2003)
8. Teixeira, J.S.: Síntese Articulatória das Vogais Nasais do Português Europeu [Articulatory Synthesis of Nasal Vowels for European Portuguese]. Ph.D. Thesis, Universidade de Aveiro (2000)
9. Freitas, J., Teixeira, A. Dias, M.S.: Towards a silent speech interface for portuguese: surface electromyography and the nasality challenge. In: International Conference on Bio-Inspired Systems and Signal Processing, Vilamoura, Algarve, Portugal (2012)
10. Seikel, J.A., King, D.W., Drumright, D.G.: Anatomy and Physiology for Speech, Language, and Hearing. Delmar Learning, Clifton Park (2010)
11. Hardcastle, W.J.: Physiology of Speech Production: An Introduction for Speech Scientists. Academic Press, London (1976)
12. Fritzell, B.: The velopharyngeal muscles in speech: an electromyographic and cineradiographic study. Acta Otolaryngolica **250**, 1–81 (1969)
13. Kuehn, D.P., Folkins, J.W., Linville, R.N.: An electromyographic study of the musculus uvulae. Cleft Palate J. **25**(4), 348–355 (1988)
14. Rossato, S., Teixeira, A., Ferreira, L.: Les Nasales du Portugais et du Français: une étude comparative sur les données EMMA. In: XXVI Journées d'Études de la Parole, Dinard, France (2006)
15. Lacerda, A., Head, B.F.: Análise de sons nasais e sons nasalizados do Português. Revista do Laboratório de Fonética Experimental (de Coimbra), vol. 6, pp. 5–70 (1996)
16. Trigo, R.L.: The inherent structure of nasal segments. In: Huffman, M.K., Krakow, R.A. (eds.) Nasals, Nasalization, and the Velum, Phonetics and Phonology, vol. 5, pp. 369–400. Academic Press Inc., London (1993)
17. Teixeira, A., Moutinho, L.C., Coimbra, R.L.: Production, acoustic and perceptual studies on European portuguese nasal vowels height. In: Internat. Congress Phonetic Sciences (ICPhS), pp. 3033–3036 (2003)
18. Martins, P., Carbone, I.C., Pinto, A., Silva, A., Teixeira, A.: European Portuguese MRI based speech production studies. Speech Commun. **50**(11/12), 925–952 (2008). ISSN 0167-6393
19. Bell-Berti, F.: An electromyographic study of velopharyngeal function. Speech J. Speech Hearing Res. **19**, 225–240 (1976)
20. Kuehn, D.P., Folkins, J.W., Cutting, C.B.: Relationships between muscle activity and velar position. Cleft Palate J. **19**(1), 25–35 (1982)
21. Lubker, J.F.: An electromyographic-cinefluorographic investigation of velar function during normal speech production. Cleft Palate J. **5**(1), 17 (1968)
22. McGill, S., Juker, D., Kropf, P.: Appropriately placed surface EMG electrodes reflect deep muscle activity (psoas, quadratus lumborum, abdominal wall) in the lumbar spine. J. Biomech. **29**(11), 1503–1507 (1996)
23. Chan, A.D.C., Englehart, K., Hudgins, B., Lovely, D.F.: Hidden Markov model classification of myoelectric signals in speech. In: Proceedings of the 23rd Annual International Conference of the IEEE Engineering in Medicine and Biology Society, vol. 2, pp. 1727–1730 (2001)
24. Jou, S., Schultz, T., Waibel, A.: Continuous electromyographic speech recognition with a multi-stream decoding architecture. In: Proceedings of the IEEE International Conference on Acoustics, Speech, and Signal Processing, ICASSP 2007, Honolulu, Hawaii, US (2007)

25. Wand, M., Schultz, T.: Investigations on speaking mode discrepancies in emg-based speech recognition. In: Interspeech 2011, Florence, Italy (2011)
26. Herff, C., Janke, M., Wand, M., Schultz, T.: Impact of different feedback mechanisms in EMG-based speech recognition. In: Interspeech 2011, Florence, Italy (2011)
27. Wand, M., Schultz, T.: Analysis of phone confusion in EMG-based speech recognition. In: IEEE International Conference on Acoustics, Speech and Signal Processing, ICASSP 2011, Prague, Czech Republic (2011)
28. Wand, M., Schultz, T.: Session-independent EMG-based speech recognition. In: International Conference on Bio-Inspired Systems and Signal Processing 2011, Biosignals 2011, Rome, Italy (2011)
29. Wand, M., Schultz, C., Janke, M., Schultz, T.: Array-based electromyographic silent speech interface. In: 6th International Conference on Bio-Inspired Systems and Signal Processing, Biosignals 2013, Barcelona, Spain (2013)
30. Freitas, J., Teixeira, A., Vaz, F., Dias, M.S.: Automatic speech recognition based on ultrasonic doppler sensing for European portuguese. In: Torre Toledano, D., Ortega Giménez, A., Teixeira, A., González Rodr\'ıguez, J., Hernández Gómez, L., San Segundo Hernández, R., Ramos Castro, D. (eds.) IberSPEECH 2012. CCIS, vol. 328, pp. 227–236. Springer, Heidelberg (2012)
31. Galatas, G., Potamianos, G., Makedon, F.: Audio-visual speech recognition incorporating facial depth information captured by the Kinect. In: Proceedings of the 20th European Signal Processing Conference (EUSIPCO), pp. 2714–2717, 27–31 August 2012 (2012)
32. Teixeira, A., Martins, P., Oliveira, C., Ferreira, C., Silva, A., Shosted, R.: Real-Time MRI for portuguese. In: Caseli, H., Villavicencio, A., Teixeira, A., Perdigão, F. (eds.) PROPOR 2012. LNCS, vol. 7243, pp. 306–317. Springer, Heidelberg (2012)
33. Silva, S., Martins, P., Oliveira, C., Silva, A., Teixeira, A.: Segmentation and analysis of the oral and nasal cavities from MR time sequences. In: Campilho, A., Kamel, M. (eds.) ICIAR 2012, Part II. LNCS, vol. 7325, pp. 214–221. Springer, Heidelberg (2012)
34. Plux Wireless Biosignals, Portugal. http://www.plux.info/
35. Hudgins, B., Parker, P., Scott, R.: A new strategy for multifunction myoelectric control. IEEE Trans. Biomed. Eng. 40(1), 82–94 (1993)
36. Quatieri, T.F., Brady, K., Messing, D., Campbell, J.P., Campbell, W.M., Brandstein, M.S., Weinstein, C., Tardelli, J., Gatewood, P.D.: Exploiting nonacoustic sensors for speech encoding. IEEE Trans. Audio, Speech, Lang. Process. 14(2), 533–544 (2006)

Observer Design for a Nonlinear Minimal Model of Glucose Disappearance and Insulin Kinetics

Driss Boutat[1]($\boxtimes$), Mohamed Darouach[2], and Holger Voos[3]

[1] INSA Centre Val de Loire, University of Orléans, PRISME EA 4229,
88 Boulevard Lahitolle, 18020 Bourges, Cedex, France
driss.boutat@insa-cvl.fr
[2] IUT de Longwy, CRAN-CNRS, UHP Nancy I, 186, Rue de Lorraine,
54400 Cosnes-et-romain, France
Mohamed.Darouach@univ-lorraine.fr
[3] Faculté des Sciences, de la Technologie Et de la Communication,
Université du Luxembourg, 6, Rue Richard Coudenhove-Kalergi,
1359 Luxembourg City, Luxembourg
holger.voos@uni.lu

Abstract. This work deals with an observer design for a nonlinear minimal dynamic model of glucose disappearance and insulin kinetics (GD-IK). At first, the model is transformed into a nonlinear observer normal form. Then, using the knowledge of the plasma blood glucose level, we estimate the state variables that are not directly available from the system, i.e. the remote compartment insulin utilization, the plasma insulin deviation and the infusion rate. In addition, we estimate the amount of absorbed glucose by means of the inverse dynamics.

Keywords: Nonlinear dynamical systems · Observer design · Insulin kinetics

1 Introduction

Diabetes is a serious disease during which the body's production and use of insulin is impaired, which causes deviations of the glucose concentration level in the bloodstream from normal values. Significant and prolonged deviations from the normal level may give rise to numerous pathologies with serious clinical impact [10]. Today, diabetes represents a major threat to public health with alarmingly rising trends of incidence and severity in recent years. In general, diabetes type 1 and type 2 can be distinguished. Diabetes type 1 is characterized by insulin deficiency due to an autoimmune destruction of the pancreatic insulin-secreting β-cells. Diabetes type 2 patients suffer from either a reduced (but present) insulin secretion or abnormal increased peripheral insulin resistance, or both. The major difference between diabetes type 1 and type 2 is that diabetes type 1 patients cannot survive without exogenous insulin, and therefore the primary focus of this paper is on this type of diabetes.

© Springer International Publishing Switzerland 2015
G. Plantier et al. (Eds.): BIOSTEC 2014, CCIS 511, pp. 261–273, 2015.
DOI: 10.1007/978-3-319-26129-4_17

The most common treatment of diabetes type 1 is the measurement of the glucose level using suitable sensing devices and to regulate this level with an injection of insulin. Currently, insulin doses are adjusted by the patient himself according to blood glucose levels intermittently four to six times per day. This control principle is neither efficient nor optimal, it usually does not restore the stability of the blood glucose level [11]. More advanced solutions are applying continuous subcutaneous glucose monitoring devices and continuous subcutaneous infusion of insulin using an insulin pump. While subcutaneous measurement and injection is currently the most comfortable solution for the patient, the insulin pump and hence the injection rates are still under manual control of the human patient. To increase the effectiveness of the pump therapy, the implementation of a closed-loop control of the insulin infusion has been proposed, also known as artificial pancreas [3,11], and thus automatically control the glucose level in the blood. The system includes a controller that sends a signal to the insulin pump in function of the measured blood glucose level and the reference, see e.g. [3,10,11] for a comprehensive overview of the technological aspects.

Unfortunately, all currently available artificial pancreas solutions are far from being optimal. The subcutaneous glucose sensing suffers from large deviations between glucose concentration in the interstitium and in blood compartment, especially during rapidly changing conditions (such as after a meal), and varying delays (4–10 min) between blood and interstitium glucose level, see [11]. Regarding the insulin pump solution, the subcutaneous injection also causes delays and deviations comparing to a direct injection in the bloodstream. From a control engineering perspective, the system under control is a very complex and nonlinear dynamic system and both the measurements as well as actuations are suffering from the mentioned large deviations and delays. Many control approaches have been applied, ranging from PID control to LQR and model predictive control, see also [3,11] for a comprehensive overview. One main problem of most control solutions is the fact that not all important variables are known or measurable and hence observers play a very important role in this control task.

However, the development of suitable observers as well as control algorithms requires the derivation of a dynamic mathematical model of the system under control, i.e. the complex dynamics of glucose disappearance and insulin kinetics(GD-IK). During the last decades, considerable research has been devoted to the derivation and improvement of such models, and many of them have already been described in the literature ranging from simple expressions to very complex nonlinear mathematical models [10]. One model which is commonly used in the literature is the so called minimal model which is a single input-output nonlinear dynamic system with three states. Most research which is so far interested in control or observer synthesis using the minimal GD-IK model is based on linearisation of this model, see [4,18,19,24,26–28] and references therein. More recent work on the same theme can be found in [16,31].

In this paper however, we will design an observer for the nonlinear minimal GD-IK model without any simplification. Indeed, we will transform this model into a nonlinear observer normal form. The considered model also contains an unknown amount of glucose absorption from the gut which will be estimated by the dynamic inversion method. Furthermore, we distinguish two situations in this work: the case where the amount of glucose absorption from the gut is known and the case where this is considered as an unknown disturbance rate. On the one hand, our contribution is based on the works of [6,8,9,22,29] in order to derive a change of coordinates that transforms the minimal GD-IK model into an observer nonlinear normal from. This enables us to design a robust observer. On the other hand, it is based on works of [5,15,21,23,32] to build an observer based on unknown disturbances.

This paper is organized as follows. The next section presents the nonlinear minimal GD-IK model and states the problem to be solved. The third section deals with the change of coordinates and describes the observer nonlinear normal form. The fourth section is devoted to the design of two types of observers, a full order observer by assuming that the glucose absorption from the gut is known and a reduced observer in the case where this absorption is unknown.

2   Dynamic Model of the GD-IK

Most of the controller design methods for automatic glucose regulation as well as observer design require the derivation of a mathematical model of the complex dynamic insulin-glucose interaction. Many models have already been described in the literature, ranging from simple expressions that relate glucose and insulin to very complex mathematical models [10]. The three general groups of mathematical models are (1) linear models, (2) nonlinear models, and (3) comprehensive models. Linear models apply sets of linear time-invariant differential equations, which are adequate when the intrinsic dynamics of the metabolic system are essentially linear [10]. There are many linear models proposed in the literature to-date, see e.g. [10,11] for an overview. One first possibility to derive a linear model is simple compartmental analysis, which describes the mass balance equations for each compartments and the relations describing the rate of material transfer between compartment. For example, the model of Ackerman [1], one of the most cited linear models, consists of a system of equations in which the parameters have been lumped into two dependent variables: glucose concentration and blood hormone concentration (including insulin). Linear models allow the application of linear control but have the disadvantage of a gross oversimplification of the underlying insulin-glucose interaction in the actual human body [10].

Nonlinear models range from less complex ones to comprehensive ones [10]. Comprehensive models attempt to transfer the knowledge of metabolic regulations into a generally large, nonlinear model of high order, with a large number of model parameters. In general however, comprehensive models cannot be easily identified. The nonlinear model which is generally used in the literature is the one developed by Bergman [4], also called the minimal model for its simplicity. It describes the insulin-glucose interactions by a system of three nonlinear differential equations and will be described in more detail in the following.

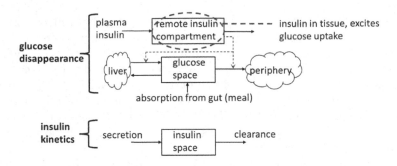

Fig. 1. Structure of the Bergman nonlinear minimal model of GD-IK.

Extensions of the minimal model are proposed, see e.g. [2,14,25], which are extending the model to include changes of the patient dynamics and the time delays between injection and absorption. Other extensions include effects based on the meal composition, see [13,30]. In addition to the Bergman model, other nonlinear more general but also much more complex models can be found in the literature, see e.g. [12,17,20]. Those models are rarely used in control approaches because of its complexity and large order.

The model of the dynamic insulin-glucose interaction which is used in this paper is based on the Bergman nonlinear minimal model [4] and is a combination of models extracted from papers of [19,28]. The minimal model consists of two parts. The first part describes the minimal model of the glucose disappearance (GD) and considers the blood plasma glucose level $g(t)$ and a variable $v(t)$ which is proportional to the insulin in a compartment remote from plasma and which enhances glucose disappearance. The second part describes the minimal model of the insulin kinetics (IK) and considers the plasma insulin concentration level $i(t)$. The interaction between these two parts which together form the model of glucose disappearance and insulin kinetics (GD-IK) is depicted in Fig. 1.

The considered nonlinear minimal GD-IK model can be derived as (see also [19,28]):

$$\begin{cases} \dot{g} = -P_1 g(t) - g(t)v(t) + P_1 g_b + g_M(t) \\ \dot{v} = -P_2 v(t) + P_3 i(t) - P_3 i_b \\ \dot{i} = -ni(t) + \gamma(g(t) - h)t \\ y = g(t) \end{cases} \tag{1}$$

Herein, g_b is the basal blood glucose level, g_M is the rate of glucose absorption from meal (glucose absorption from the gut) and i_b is the basal insulin level. Parameter P_1 represents glucose effectiveness, P_2 denotes the decreasing level of insulin, P_3 is the rate at which insulin action is increased as the level of insulin deviates from the corresponding baseline, γ is the rate at which insulin is produced, n denotes the fractional insulin clearance and h denotes the pancreatic target glycemia level. As in [19], we add to the above model the pump dynamics:

$$\dot{w} = \frac{1}{a}(-w(t) + u(t)) \tag{2}$$

where $w(t)$ represents the infusion rate, $u(t)$ the control input and a denotes the time constant of the pump. From now on, this model is rewritten in a general state variable format with four state variables $x_1(t) = g(t), x_2(t) = v(t), x_3(t) = i(t), x_4(t) = w(t)$:

$$\begin{cases} \dot{x}_1 = -P_1 x_1 - x_1 x_2 + P_1 g_b + g_M(t) \\ \dot{x}_2 = -P_2 x_2 + P_3 x_3 - P_3 i_b \\ \dot{x}_3 = -n x_3 + x_4 + \gamma(x_1 - h)t \\ \dot{x}_4 = -\frac{1}{a} x_4 + \frac{1}{a} u \\ y = x_1 \end{cases} \qquad (3)$$

3 Nonlinear Observer Normal Form of GD-IK

3.1 Reformulation of the GD-IK Dynamic Model

This dynamic system can be further expressed in the following compact form:

$$\begin{cases} \dot{x} = f(x) + B_1 u + \nu(t, y) + g_M(t) B_2 \\ y = h(x) \end{cases} \qquad (4)$$

where

- $x = (x_1, x_2, x_3, x_4)^T$ is the vector of state variables and $h(x) = x_1$ is the output,
- $f(x) = \left(-x_1 x_2, -P_2 x_2 + P_3 x_3, -n x_3 + x_4, -\frac{1}{a} x_4 \right)^T$ is the drift vector field
- $B_1 = \left(0, 0, 0, \frac{1}{a} \right)^T$ is the control direction,
- $B_2 = (1, 0, 0, 0)^T$ is the unknown direction,
- $\nu(t, y) = \left(P_1 g_b - P_1 y, -P_3 i_b, \gamma(y - h)t, 0 \right)^T$ is a direction depending on the output y and time t.

In this work, we consider the following problem: How can we find a change of coordinates $z = \phi(x)$ in order to transform (3) into a nonlinear observer normal form, i.e.

$$\begin{cases} \dot{z} = A_O z + \beta(y, t) + \overline{B}_1 u + \alpha(y) g_M(t) \\ \overline{y} = C_O z = z_4 \end{cases} \qquad (5)$$

where $A_O = \begin{pmatrix} 0\,0\,0\,0 \\ 1\,0\,0\,0 \\ 0\,1\,0\,0 \\ 0\,0\,1\,0 \end{pmatrix}$, $C_O = (0\,0\,0\,1)$, and the new output $\overline{y} = \varphi(y)$ is a diffeomorphism of the output y. In addition, this nonlinear observer normal form enables us to deal with the following problems: (i) Design an observer-based feedback if $g_M(t)$ is known or if $g_M(t) = 0$ and (ii) Design an observer-based feedback by the concept of inversion dynamics if $g_M(t)$ is unknown.

3.2   Transformation Algorithm

There are several sophisticated geometrical algorithms that enable us to transform the dynamic system (4) into a nonlinear observer normal form (5), see [6,7,9,22,29]. In this paper, thanks to the special form of the proposed system, we can establish an algorithm based on matrix calculus. At the same time, we provide an algorithm to compute change of coordinates. For this purpose, let us consider a single input-output dynamic system with the following form:

$$\begin{cases} \dot{x} = Ax + \mu(y, t, u, s(t)) \\ y = Cx \end{cases} \tag{6}$$

with the vector of state variables $x \in \mathbb{R}^n$, the output $y \in \mathbb{R}$ and the function $\mu(y, t, u, s(t))$ which does not depend on the unmeasured state. We assume that the pair (C, A) is observable. Thus, the matrix

$$O = \begin{pmatrix} C \\ CA \\ \cdots \\ CA^{n-1} \end{pmatrix}$$

is of full rank n. Let $p(s) = s^n + a_{n-1}s^{n-1} + a_{n-2}s^{n-2} + \cdots + a_1 s + a_0$ be the characteristic polynomial of the matrix A. We recall that the Cayley-Hamilton theorem states that $p(A) = 0$. Then the following result holds.

Theorem 1. *The following linear change of coordinates*

$$\begin{aligned} z_n &= Cx \\ z_{n-i} &= CA^i x + \sum_{k=1}^{i} a_{n-k} CA^{i-k} x \quad for \;\; i = 1 : n - 1 \end{aligned} \tag{7}$$

transforms the dynamic system (6) into the following observer normal form:

$$\begin{cases} \dot{z} = A_O z + \overline{\mu}(y, t, u, v(t)) \\ y = C_O z = z_n \end{cases} \tag{8}$$

where the pair (C_O, A_O) is in Brunovsky canonical form and $\overline{\mu}$ is defined by its components as follows

$$\begin{aligned} \overline{\mu}_n &= C\mu - a_{n-1}y \\ \overline{\mu}_{n-i} &= CA^i \mu + \sum_{k=1}^{i} a_{n-k} CA^{i-k} \mu - a_{n-i-1}y \\ & for \;\; i = 1 : n - 1 \end{aligned} \tag{9}$$

Proof. We proceed by successive derivation of the change of coordinates given in (7). Then, we obtain:

$$\dot{z}_n = CAx + C\mu = z_{n-1} - a_{n-1}y + C\mu$$
$$\dot{z}_{n-i} = z_{n-i-1} - a_{n-i-1}y + CA^i\mu +$$
$$+ \sum_{k=1}^{i} a_{n-k}CA^{i-k}\mu \tag{10}$$
$$\text{for} \quad i = 1 : n - 2 \tag{11}$$
$$\dot{z}_1 = -a_0 y + CA^{n-1}\mu + \sum_{k=1}^{n-1} a_{n-k}CA^{n-1-k}\mu$$

where the last equation is obtained by using the Cayley-Hamilton theorem.

3.3   Application to the GD-IK

In this subsection, we will apply the results obtained in the previous section to the GD-IK model. Let us consider the nonlinear dynamic system (3). We start by transforming it first into the form (6). For this we use the concept of diffeomorphism on the output (see [6–9,29]). In our case we define the new output $\bar{y} = -\ln(y)$. Hence, if we consider the new variable $\xi = -\ln(x_1)$, then the dynamic system (3) is rewritten as follows:

$$\begin{cases} \dot{\xi} = x_2 + P_1 - P_1 e^{\bar{y}} g_b - e^{\bar{y}} g_M \\ \dot{x}_2 = -P_2 x_2 + P_3 x_3 - P_3 i_b \\ \dot{x}_3 = -n x_3 + x_4 + \gamma(e^{-\bar{y}} - h)t \\ \dot{x}_4 = -\frac{1}{a} x_4 + \frac{1}{a} u \\ \bar{y} = \xi = -\ln y \end{cases} \tag{12}$$

With the definition of the matrix

$$A = \begin{pmatrix} 0 & 1 & 0 & 0 \\ 0 & -P_2 & P_3 & 0 \\ 0 & 0 & -n & 1 \\ 0 & 0 & 0 & -\frac{1}{a} \end{pmatrix}$$

and the vector $C = (1\,0\,0\,0)$, the dynamic system given in (12) can be written in the desired form given in (6):

$$\dot{X} = AX + B_1 u + B_2(y)g_M + \beta(y, t)$$
$$\bar{y} = CX = \xi$$

with $X = (\xi, x_2, x_3, x_4)^T$ and $\mu = B_1 u + B_2(\bar{y})g_M + \beta(\bar{y}, t)$.

As the pair (C, A) is observable, we can use Theorem 1. The characteristic polynomial of A is given by $s^4 + \left(n + P_2 + \frac{1}{a}\right) s^3 + \left(\frac{1}{a}(n + P_2) + nP_2\right) s^2 + \frac{1}{a}nP_2 s$, then the change of coordinates can be given by the following expression:

$$z_1 = \tfrac{1}{a}P_3 x_3 + \tfrac{n}{a}x_2 + P_3 x_4 - \tfrac{1}{a}nP_2 \ln x_1$$
$$z_2 = P_3 x_3 + \left(n + \tfrac{1}{a}\right) x_2 - \left(\tfrac{1}{a}\left(n + P_2\right) + nP_2\right) \ln x_1$$
$$z_3 = x_2 + \left(n + P_2 + \tfrac{1}{a}\right) \ln x_1$$
$$z_4 = -\ln x_1 = \xi$$

Therefore, we obtain the nonlinear observer normal form (6) for the nonlinear dynamic system (3) as follows:

$$\begin{cases} \dot{z} = A_O z + \beta(y,t) + \overline{B}_1 u + \alpha(y) g_M(t) \\ \overline{y} = C_O z = z_4 \end{cases} \tag{13}$$

where

- $\beta(\overline{y}, t) = \left(\beta_1, \beta_2, \beta_3, \beta_4\right)^T$ with $\beta_1 = \left(\tfrac{1}{a}\left(e^{-\overline{y}} - h\right) t\gamma P_3 - \tfrac{1}{a}nP_3 i_b\right) + \tfrac{1}{a}nP_2 P_1 + \tfrac{P_1}{\overline{y}}\tfrac{1}{a}nP_2 g_b$, $\beta_2 = \tfrac{1}{a}nP_2 \ln x_1 + P_3\gamma(e^{-\overline{y}} - h)t - \left(n + \tfrac{1}{a}\right)P_3 i_b + \left(\tfrac{1}{a}\left(n + P_2\right) + nP_2\right)P_1 + \tfrac{P_1}{\overline{y}}\left(\tfrac{1}{a}\left(n + P_2\right) + nP_2\right)g_b$, $\beta_3 = -\left(\tfrac{1}{a}\left(n + P_2\right) + nP_2\right)\overline{y} - P_3 i_b + \left(n + P_2 + \tfrac{1}{a}\right)P_1 + \tfrac{P_1}{\overline{y}}\left(n + P_2 + \tfrac{1}{a}\right)g_b$, $\beta_4 = -\left(n + P_2 + \tfrac{1}{a}\right)\overline{y} + P_1 + \tfrac{P_1}{\overline{y}}g_b$
- $\overline{B}_1 = \left(\tfrac{P_3}{a}, 0, 0, 0\right)^T$
- $\alpha(\overline{y}) = \tfrac{1}{\overline{y}}\left(\tfrac{1}{a}nP_2, \tfrac{1}{a}\left(n + P_2\right) + nP_2, n + P_2 + \tfrac{1}{a}, 1\right)^T$

4   Observer Design

In this section we will present two types of observers. The first one assumes that g_M is known and the second one assumes that g_M is unknown. In the last case we will design an observer to estimate both the state and g_M. First, it should be noted that (13) is controllable.

4.1   Full Order Observer

In the first case, we consider (13) and we define the following observer-based feedback:

$$\dot{\hat{z}} = A_O \hat{z} + \beta(y,t) + \overline{B}_1 u + \alpha(y) g_M + K(\hat{\overline{y}} - \overline{y}). \tag{14}$$

If we set the observation error $e = \hat{z} - z$, we can obtain that its dynamics is linear and given by $\dot{e} = (A_O + KC_O)e$. As the pair (C_O, A_O) is observable we can find a gain K such that $A_O + KC_O$ is asymptotically stable.

We provide also an observer-based feedback with $u = K\hat{z}$ such that the output $g(t)$, the glucose level, reaches the glucose basal level ($99\,mg/dl$), see also Fig. 1. The estimations of the states as well as the actual values obtained in the simulation are given in Figs. 2, 3 and 4, respectively. The parameters and initial states used in the simulations are: $P_1 = 0, P_2 = 0.81/100, P_3 = 4.01/1e6, n = 0.23, a = 2, gb = 99, ib = 8, \gamma = 2.4/1000, h = 93, x_1(0) = 337, x_2(0) = 0, x_3(0) = 192, x_4(0) = 2$. These parameters and initial states are the same as in [19].

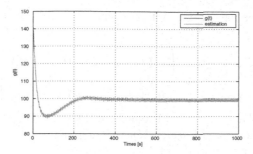

Fig. 2. Evolution of $g(t)$.

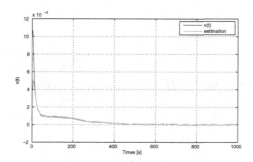

Fig. 3. Evolution of $v(t)$.

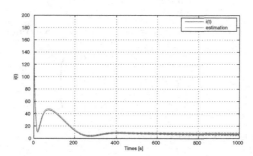

Fig. 4. Evolution of $i(t)$.

4.2   Observer for Unknown Input

In the second case we assume that g_M is an unknown input and we will design an observer to estimate both the state and g_M. If we consider g_M as an unknown input, we can follow [5, 15, 21, 23, 32] which leads to a decomposition of the state of the observer normal form (13) into two parts, namely the unmeasurable and the measurable part: $z = (I - MC)z + MCz = q + My$, where

$$M = \frac{1}{C_O \alpha} \alpha = \left(\tfrac{1}{a} n P_2, \tfrac{1}{a} (n + P_2) + n P_2, n + P_2 + \tfrac{1}{a}, 1 \right)^T$$

is a constant matrix even if α is not constant. Therefore we have the following projector:

$$\tilde{\Pi} = I - MC = \begin{pmatrix} 1 & 0 & 0 & -\frac{1}{a}nP_2 \\ 0 & 1 & 0 & -\left(\frac{1}{a}(n+P_2)+nP_2\right) \\ 0 & 0 & 1 & -\left(n+P_2+\frac{1}{a}\right) \\ 0 & 0 & 0 & 0 \end{pmatrix}.$$

Now, we consider the dynamics of the unknown part q. Thanks to the fact that $\tilde{\Pi}\alpha = 0$, we obtain $\dot{q} = \tilde{\Pi}\left(A_O q - My + \overline{B}_1 u + \beta(y,t)\right)$. An observer for this last dynamic system is derived as follows:

$$\dot{\hat{q}} = \tilde{\Pi}\left(A_O\hat{q} - My + \overline{B}_1 u + \beta(y,t)\right)$$
$$\qquad - \tilde{\Pi}\left(LC_O\left(\hat{q}-q\right)\right) \tag{15}$$
$$\hat{z} = \hat{q} + My \tag{16}$$

Therefore, the dynamics of the error $e_q = \hat{q}-q$ is given by $\dot{e}_q = \tilde{\Pi}\left(A_O - LC_O\right)e$. In order to write the projector $\tilde{\Pi}$ in the canonical form, we proceed as in the algorithms described in [5,15,21,23,32], and we consider the change of coordinates given by the following matrix:

$$Q = \begin{pmatrix} 1 & 0 & 0 & \frac{1}{a}nP_2 \\ 0 & 1 & 0 & nP_2+\frac{1}{a}(n+P_2) \\ 0 & 0 & 1 & n+P_2+\frac{1}{a} \\ 0 & 0 & 0 & 1 \end{pmatrix}$$

In these new coordinates the projector $\tilde{\Pi} = I - MC$ becomes:

$$\Pi = Q^{-1}\tilde{\Pi}Q = \begin{pmatrix} 1 & 0 & 0 & 0 \\ 0 & 1 & 0 & 0 \\ 0 & 0 & 1 & 0 \\ 0 & 0 & 0 & 0 \end{pmatrix}$$

and the matrix A_O is decomposed into four blocs:

$$\tilde{A}_O = Q^{-1}A_O Q = \begin{pmatrix} \tilde{A}_{1,1} & \tilde{A}_{1,2} \\ \tilde{A}_{2,1} & \tilde{A}_{2,2} \end{pmatrix}$$

where $\tilde{A}_{1,1} = \begin{pmatrix} 0 & 0 & -\frac{1}{a}nP_2 \\ 1 & 0 & -\frac{1}{a}(n+P_2)-nP_2 \\ 0 & 1 & -P_2-\frac{1}{a}-n \end{pmatrix}$ $\tilde{A}_{2,1} = \begin{pmatrix} 0 & 0 & 0 \end{pmatrix}$ $\tilde{A}_{1,2} =$

$$\begin{pmatrix} -\frac{1}{a}nP_2\left(n+P_2+\frac{1}{a}\right) \\ \frac{1}{a}nP_2+\frac{1}{a}\left(n+P_2+\frac{1}{a}\right)\left(-n-P_2-anP_2\right) \\ nP_2+\frac{1}{a}(n+P_2)+\frac{1}{a}\left(n+P_2+\frac{1}{a}\right)\left(-an-aP_2-1\right) \end{pmatrix} \tilde{A}_{2,2} = 0, \tilde{C} = CQ =$$

$\left(\tilde{C}_1,\tilde{C}_2\right)$ with $\tilde{C}_1 = \begin{pmatrix} 0 & 0 & 0 \end{pmatrix}$ and $\tilde{C}_2 = 1$. The following result is widely established in [5,15,21,23,32]:

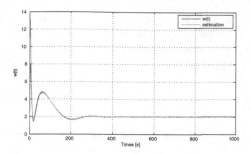

Fig. 5. Evolution of $w(t)$.

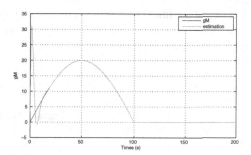

Fig. 6. Estimation by inverse dynamics of g_M.

Theorem 2. *As* $rank(C_O\alpha) = rank(\alpha)$ *and the pair* $(\widetilde{A}_{1,1}, \widetilde{C}_1)$ *is detectable (because* $\widetilde{A}_{1,1}$ *is asymptotically stable for all initial condition* $q(0) = Pz(0)$*), (15) is an asymptotic observer.*

Remark 1. The observer normal for (14) becomes under the change of coordinates $\tilde{z} = Qz$ as follows:

$$\dot{\tilde{z}} = \tilde{A}_O \tilde{z} + \tilde{\beta}(y,t) + \overline{B}_1 u + \tilde{\alpha}(y)G_M \qquad (17)$$

where $\overline{B}_1$ has not changed, $\tilde{\alpha} = Q^{-1}\alpha = (0,0,0,\frac{1}{y})^T$, and $\tilde{\beta} = Q^{-1}\beta = \beta + \beta_4 Q^{-1}(0,0,0,1)^T - (0,0,0,\beta_4)^T$.

Now, we are ready to compute the inverse dynamics of the observer normal form (13). For this, let us denote $z_r = (\tilde{z}_1, \tilde{z}_2, \tilde{z}_3)^T$, $\tilde{\beta}_r = (\tilde{\beta}_1, \tilde{\beta}_2, \tilde{\beta}_3)^T$, and $\overline{B}_{1,r} = (\overline{B}_{1,1}, 0, 0)^T$, then the inverse dynamics is as follows:

$$\begin{cases} \dot{z}_r = \tilde{A}_{1,1} + \overline{B}_{1,r}u + \tilde{\beta}_r(y,t) \\ g_M = e^{-\overline{y}}(\dot{\overline{y}} - \beta_4) \end{cases} \qquad (18)$$

Using the same parameters and initial states given in the previous subsection, an estimation of the unknown g_M is performed. The results of this simulation are depicted in Fig. 5.

Remark 2. The existing papers dealing with the observer of the GD-IK model given by the nonlinear dynamic system (3), only estimated the glucose level $g(t)$. However, in this work we estimate also $i(t)$ and $\nu(t)$. Moreover, we estimate by inverse dynamics g_M which has not been addressed anywhere yet.

5   Conclusions

To the best of our knowledge this paper is the first one which has dealt with observer an design for the minimal model GD-IK using the nonlinear observer form concept. Moreover, it has applied the inverse dynamics of the GD-IK model in the case where the amount of glucose absorption is unknown or considered as a meal disturbance input. First simulation results have underlined the correctness and applicability of this novel approach. Furthermore, this observer can be used to design a controller to regulate the glucose level.

References

1. Ackerman, E., Rosevear, J., McGuckin, W.: A mathematical model of the glucose tolerance test. Phys. Med. Biol. **9**(2), 203–213 (1964)
2. Benett, D., Gourley, S.: Asymptotic properties of a delay differential equation model for the interaction of glucose with plasma and interstitial insulin. Appl. Math. Comput. **151**(1), 189–207 (2003)
3. Bequette, B.: Challenges and recent progress in the development of a closed-loop artificial pancreas. Annu. Rev. Control **36**, 255–266 (2012)
4. Bergman, R., Ider, Y., Bowden, C., Cobelli, C.: Quantitative estimation of insulin sensitivity. Am. J. Physiol.-Endocrinol. Metab. **236**(6), E667 (1979)
5. Bhattacharyya, S.: Observer design for linear systems with unknown inputs. IEEE Trans. Autom. Control **23**(3), 483–484 (1978)
6. Boutat, D.: Geometrical conditions for observer error linearization via $\int 0, 1, ..., (N-2)$. In: 7th IFAC Symposium on Nonlinear Control Systems Nolcos (2007)
7. Boutat, D.: Extended nonlinear observer normal forms for a class of nonlinear dynamical systems. Int. J. Robust Nonlinear Control (2013). http://dx.doi.org/10.1002/rnc.3102
8. Boutat, D., Benali, A., Hammouri, H., Busawon, K.: New algorithm for observer error linearization with a diffeomorphism on the outputs. Automatica **45**(10), 2187–2193 (2009)
9. Boutat, D., Busawon, K.: On the transformation of nonlinear dynamical systems into the extended nonlinear observable canonical form. Int. J. Control **84**(1), 94–106 (2011)
10. Chee, F., Fernando, T.: Closed-Loop Control of Blood Glucose. Springer, Berlin (2007)
11. Cobelli, C., Bernard, E., Kovatcher, B.: Artificial pancreas, past, present, future. Diabetes **60**, 2672–2682 (2011)
12. Cobelli, C., et al.: An integrated mathematical model of the dynamics of blood glucose and its hormonal control. Math. Biosci. **58**, 27–60 (1982)
13. Dalla Man, C., et al.: A model of glucose production during a meal. In: 28th IEEE EMBS Annual International Conference, pp. 5647–5650. New York (2006)

14. Dalla Man, C., Caumo, A., Cobelli, C.: The oral glucose minimal model: estimation of insulin sensitivity from a meal test. IEEE Trans. Biomed. Eng. **49**(5), 419–429 (2002)
15. Darouach, M., Zasadzinski, M., Xu, S.: Full-order observers for linear systems with unknown inputs. IEEE Trans. Autom. Control **39**(3), 606–609 (1994)
16. Eberle, C., Ament, C.: Identifiability and online estimation of diagnostic parameters with in the glucose insulin homeostasis. Biosystems **107**(3), 135–141 (2012). http://www.sciencedirect.com/science/article/pii/S0303264711001857
17. Fabietti, P., et al.: Control oriented model of insulin and glucose dynamics in type 1 diabetes. Med. Biol. Eng. Comput. **44**, 69–78 (2006)
18. González, P., Femat, R.: Control of glucose concentration in type 1 diabetes mellitus with discrete-delayed measurements. In: 18th IFAC World Congress Milano (Italy) (2011)
19. Hariri, A., Wang, Y.: Observer-based state feedback for enhanced insulin control of type idiabetic patients. Open Biomed. Eng. J. **5**, 98 (2011)
20. Hovorka, R., et al.: Partitioning glucose distribution, transport, disposal and endogenous production during IVGTT. Am. J. Physiol. Endocrinol. Metab. **282**, 992–1007 (2002)
21. Hui, S., Zak, S.: Observer design for systems with unknown inputs. Int. J. Appl. Math. Comput. Sci. **15**(4), 431 (2005)
22. Krener, A., Isidori, A.: Linearization by output injection and nonlinear observers. Syst. Control Lett. **3**(1), 47–52 (1983)
23. Kudva, P., Viswanadham, N., Ramakrishna, A.: Observers for linear systems with unknown inputs. IEEE Trans. Autom. Control **25**, 113–115 (1980)
24. Kovcs, L., Palncz, B., Benyo, Z.: Design of luenberger observer for glucose-insulin control via mathematica. In: 29th Annual International Conference of the IEEE Engineering in Medicine and Biology Society (2007)
25. Lin, J., et al.: Adaptive bolus-based set-point regulation of hyperglycemia in critical care. In: 26th IEEE EMBS Annual International Conference, pp. 3463–3466 (2004)
26. Magni, L., Raimondo, D., Bossi, L., Dalla Man, C., De Nicolao, G., Kovatchev, B., Cobelli, C.: Artificial pancreas: closed-loop control of glucose variability in diabetes: model predictive control of type 1 diabetes: an in silico trial. J. Diabetes Sci. Technol. **1**(6), 804 (2007)
27. Parker, R., Doyle III, F., Peppas, N.: A model-based algorithm for blood glucose control in type I diabetic patients. IEEE Trans. Biomed. Eng. **46**(2), 148–157 (1999)
28. Percival, M., Zisser, H., Jovanovič, L., Doyle III, F.: Closed-loop control and advisory mode evaluation of an artificial pancreatic β cell: use of proportional-integral-derivative equivalent model-based controllers. J. Diabetes Sci. Technol. **2**(4), 636 (2008)
29. Respondek, W., Pogromsky, A., Nijmeijer, H.: Time scaling for observer design with linearizable error dynamics. Automatica **40**(2), 277–285 (2004)
30. Roy, A., Parker, R.: Dynamic modeling of free fatty acid, glucose and insulin: an extended minimal model. Diabetes Technol. Ther. **8**, 617–626 (2006)
31. Villafaña-Rojas, J., González-Reynoso, O., Alcaraz-González, V., González-García, Y., González-Álvarez, V., Solís-Pacheco, J.R., Aguilar-Uscanga, B., Gómez-Hermosillo, C.: Asymptotic observers a tool to estimate metabolite concentrations under transient state conditions in biological systems: determination of intermediate metabolites in the pentose phosphate pathway of saccharomyces cerevisiae. Chem. Eng. Sci. **104**, 73–81 (2013). http://www.sciencedirect.com/science/article/pii/S0009250913006301
32. Yang, F., Wilde, R.: Observers for linear systems with unknown inputs. IEEE Trans. Autom. Control **33**(7), 677–681 (1988)

Comparison of Angle Measurements Between Integral-Based and Quaternion-Based Methods Using Inertial Sensors for Gait Evaluation

Takashi Watanabe[1(✉)], Yuta Teruyama[1], and Kento Ohashi[2]

[1] Graduate School of Biomedical Engineering, Tohoku University, Sendai, Japan
{nabet,yuta.teruyama}@bme.tohoku.ac.jp
[2] Graduate School of Engineering, Tohoku University, Sendai, Japan
kento.ohashi@bme.tohoku.ac.jp

Abstract. The measurement method of lower limb angles, which were based on the integral of angular velocity with Kalman filter, was validated previously in measurement of angles in a plane with a rigid body model of double pendulum and shown to be practical in measurement of angles in the sagittal plane during gait of healthy subjects. In this paper, in order to realize practical measurements of 3 dimensional (3D) movements with inertial sensors, the previously developed integral-based angle calculation method with variable Kalman gain and a quaternion-based one with fixed gain were tested. First, the quaternion-based method showed higher measurement accuracies for angles in the sagittal and the frontal planes during 3D movements of a rigid body model of lower limb than that of the integral-based one. Second, the integral-based method was shown to be effective in measurement of angles in the sagittal plane of healthy subjects during treadmill walking compared to the quaternion-based one, which suggested effectiveness of the variable Kalman gain method. Third, the integral-based method was shown to cause inappropriate results in some cases of calculation of lower limb angles of hemiplegic subjects during gait, while some of such cases could be measured appropriately with the quaternion-based one. The results of this paper suggested that the quaternion-based method would be more effective with a variable Kalman-gain for measurement of 3D movements. The integral-based method was also suggested to be useful for measurement of angles in the sagittal plane of healthy subjects during gait.

Keywords: Angle · Inertial sensor · Kalman filter · Integral · Quaternion · Gait · Rehabilitation

1   Introduction

Lower limb motor functions are important to prevent bedridden and to make independence in daily living and social participation. Therefore, motor disabled persons or elderly people with decreased motor function need rehabilitation training of their lower limbs. In the rehabilitation, it is important to evaluate a level of motor function of subjects in order to make rehabilitation program and to instruct it.

© Springer International Publishing Switzerland 2015
G. Plantier et al. (Eds.): BIOSTEC 2014, CCIS 511, pp. 274–288, 2015.
DOI: 10.1007/978-3-319-26129-4_18

Generally, therapists perform the evaluation of motor function in rehabilitation by simple manual methods such as watching movements, measurement of the range of motion (ROM) with a manual goniometer, or measurement of time and counting the number of steps in 10 m walking test. Although these simple, manual evaluation methods are effective in limited space and time for rehabilitation training, those evaluation results depend on therapists. On the other hand, for quantitative and objective evaluation of movements, motion measurement system such as a camera-based system or electric goniometer has been used. Rehabilitation program based on the quantitative and objective evaluations with motion measurement system is expected to increase rehabilitation effect and to decrease rehabilitation term. However, those motion measurement systems are mainly used in research works in laboratories, because the systems require large space for setting the system and time-consuming setup process, and are expensive.

Recently, use of inertial sensors (accelerometers and gyroscopes) has been studied in measurement and analysis of movements focusing on its shrinking in size, low cost and easiness for settings. In evaluation of motor functions, segment inclination angles and joint angles have important information for therapists and patients. Therefore, many studies have been performed on measurement of joint angles or segment tilt angles with inertial sensors [1–6].

A motion measurement system using inertial sensors has to give joint or segment inclination angles calculating from angular velocities and/or acceleration signals. In addition, measurement of total lower limb movements such as simultaneous measurement of hip, knee and ankle joint angles is required for clinical evaluation in rehabilitation support. In our previous study, a joint angle calculation method based on the integral of angular velocity using Kalman filter was applied to all the joint angles of the lower limbs. Measurement of gait with healthy subjects suggested that the method can be used practically in measurement of those angles in the sagittal plane [7–9]. The integral-based method was modified to have variable Kalman gain and was found to have high measurement accuracies in measurements of angles during 2-dimensional movements of a double pendulum rigid body model and angles in the sagittal plane during treadmill walking with healthy subjects [10].

Angle measurement of 3-dimensional (3D) movements has been required for evaluation of motor function. For measurement of 3D angles with inertial sensors, a method of using attitude angle representation by quaternion was proposed [5]. However, measurement of Euler angle was tested in that study. On the other hand, the integral of angular velocity can be expanded to measure 3D movements, and it is possible to provide simply angles both in the sagittal plane and in the frontal plane.

The question focused in this paper was whether there are any differences in angle calculation between the integral-based method and the quaternion based one or not. Therefore, this paper aimed to evaluate angle measurement accuracies of the different calculation methods. For this purpose, the previously developed integral-based method with variable Kalman gain [10] was expanded to measure angles in the sagittal and the frontal planes. Then, an angle calculation method using quaternion with a fixed gain Kalman filter was developed. The quaternion-based and the extended integral-based methods were evaluated in measurements of angles during 3D movements in the sagittal and the frontal planes using the rigid body model that represented the lower limb.

In addition, the two methods were compared in calculation of lower limb angles measured during treadmill walking with healthy subjects and during gait with hemiplegic subjects.

2 Angle Calculation Methods

In this paper, two calculation methods shown in Fig. 1 were tested in angle measurements. Both methods are described below.

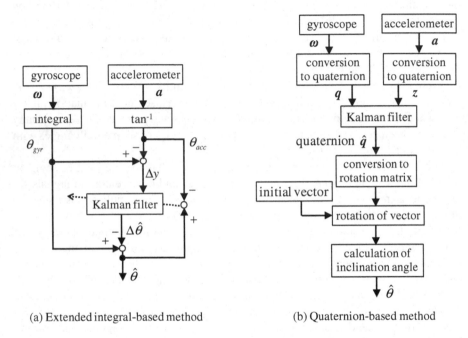

(a) Extended integral-based method   (b) Quaternion-based method

Fig. 1. Outline of tested angle calculation methods.

2.1 Extended Integral-Based Method

Figure 1(a) shows outline of the integral-based method of calculating segment inclination angle. In this paper, the previously developed integral-based method with variable Kalman gain [10] was extended to calculate angles in the sagittal and the frontal planes. Basically, a segment inclination angle is calculated by the integral of angular velocity (an output of a gyroscope). That is, segment inclination angle $\theta_{inc}(t)$ is calculated by

$$\theta_{inc}(t) = \int_0^t \omega(\tau)d\tau + \theta_{inc}(0) \tag{1}$$

where $\omega(t)$ shows angular velocity measured with a gyroscope. $\theta_{inc}(0)$ is the initial joint angle calculated from acceleration data. For instance, the angle in the sagittal plane is calculated from acceleration signal, a_x and a_z, by the following equation.

$$\theta_{inc}(0) = \tan^{-1} \frac{a_z(0)}{a_x(0)} \tag{2}$$

Joint angle $\theta_{joint}(t)$ are calculated from 2 inclination angles of the adjacent segments. For example,

$$\theta_{joint}(t) = \theta_{inc1}(t) - \theta_{inc2}(t) \tag{3}$$

Here, the calculated angle is corrected by Kalman filter using angle measured with an accelerometer in order to remove accumulated calculation error in the integral [7–9]. Kalman filter estimates error in the angle calculated from the output of a gyroscope ($\Delta\hat{\theta}$) by using the difference between angles obtained by a gyroscope and by an accelerometer (Δy). Then, angle ($\hat{\theta}$) is calculated. The difference of angles in the sagittal plane is calculated by

$$\Delta y(t) = \theta_{gyro}(t) - \theta_{acc}(t) = \theta_{inc}(t) - \tan^{-1} \frac{a_z(t)}{a_x(t)} \tag{4}$$

The state equation and the observation equation are shown by following equations using the error of the angle measured with gyroscopes ($\Delta\theta$) and bias offset (Δb):

$$\begin{bmatrix} \Delta\theta_{k+1} \\ \Delta b_{k+1} \end{bmatrix} = \begin{bmatrix} 1 & \Delta t \\ 0 & 1 \end{bmatrix} \begin{bmatrix} \Delta\theta_k \\ \Delta b_k \end{bmatrix} + \begin{bmatrix} \Delta t \\ 1 \end{bmatrix} w \tag{5}$$

$$\Delta y_k = \begin{bmatrix} 1 & 0 \end{bmatrix} \begin{bmatrix} \Delta\theta_k \\ \Delta b_k \end{bmatrix} + v \tag{6}$$

where w and v are errors in measurement with the gyroscope and with the accelerometer, respectively.

Kalman filter repeats correction by Eq. (7) and prediction by Eq. (8):

$$\begin{bmatrix} \Delta\hat{\theta}_k \\ \Delta\hat{b}_k \end{bmatrix} = \begin{bmatrix} \Delta\hat{\theta}_k^- \\ \Delta\hat{b}_k^- \end{bmatrix} + \begin{bmatrix} K_1 \\ K_2 \end{bmatrix} (\Delta y_k - \Delta\hat{\theta}_k^-) \tag{7}$$

$$\begin{bmatrix} \Delta\hat{\theta}_{k+1}^- \\ \Delta\hat{b}_{k+1}^- \end{bmatrix} = \begin{bmatrix} 1 & \Delta t \\ 0 & 1 \end{bmatrix} \begin{bmatrix} \Delta\hat{\theta}_k \\ \Delta\hat{b}_k \end{bmatrix} \tag{8}$$

where K_1 and K_2 are Kalman gain for $\Delta\theta$ and Δb, respectively. The hat upon a character and the superscript minus represent estimated value and predicted value, respectively. For the initial state, $\Delta\theta_0$ was set at zero and Δb_0 was set at the value at the last measurement.

Value of Kalman gain was determined by the noise ratio that was adjusted based on the difference between the angle estimated by the Kalman filter and the angle calculated from acceleration signals. Here, the angle difference was used approximately as the magnitude of influence of impact and motion accelerations, because large angle difference is considered to involve large error caused by impact and movement accelerations in the angle calculated from accelerations. That is, the value of the noise ratio n was adjusted by the following equation:

$$n = n_0 e^a \cdot \left| \hat{\theta} - \theta_{acc} \right| \tag{9}$$

where, $\left| \hat{\theta} - \theta_{acc} \right|$ represent the angle difference between the angle estimated by the Kalman filter $\hat{\theta}$ and the angle calculated from acceleration signals θ_{acc}. n_0 and a are parameters whose values were determined by trial and error method, respectively. In this case, value of Kalman gain decreases as the noise ration increases.

2.2 Quaternion-Based Method

Quaternion can be used to represent the attitude of each segment of a rigid body model and human body. As shown in Fig. 1(b), two quaternions are calculated from acceleration signals and from angular velocity signals measured with an inertial sensor. First, attitude angle representation by quaternion was obtained from the angular velocity. Then, Kalman filter was applied to correct the error using attitude angle representation by quaternion obtained from the gravitational acceleration.

Using the triaxial angular velocity $\boldsymbol{\omega} = (\omega_x, \omega_y, \omega_z)$, quaternion $\boldsymbol{q}$ is propagated according to the differential equation [11]:

$$\dot{q} = \frac{1}{2} \begin{bmatrix} 0 & -\omega_x & -\omega_y & -\omega_z \\ \omega_x & 0 & \omega_z & -\omega_y \\ \omega_y & -\omega_z & 0 & \omega_x \\ \omega_z & \omega_y & -\omega_x & 0 \end{bmatrix} q \tag{10}$$

The state equation shown by Eq. (11) is the time integration of Eq. (10), where $\boldsymbol{w}$ is the process noise in measurement with a gyroscope.

$$q_{k+1} = \frac{1}{2} \begin{bmatrix} 2 & -\Delta t \omega_{xk} & -\Delta t \omega_{yk} & -\Delta t \omega_{zk} \\ \Delta t \omega_{xk} & 2 & \Delta t \omega_{zk} & -\Delta t \omega_{yk} \\ \Delta t \omega_{yk} & -\Delta t \omega_{zk} & 2 & \Delta t \omega_{xk} \\ \Delta t \omega_{zk} & \Delta t \omega_{yk} & -\Delta t \omega_{xk} & 2 \end{bmatrix} q_k + w \tag{11}$$

The observation equation is given by the following equation considering the observation noise $\boldsymbol{v}$ in measurement with an accelerometer.

$$z_k = \begin{bmatrix} 1 & 0 & 0 & 0 \\ 0 & 1 & 0 & 0 \\ 0 & 0 & 1 & 0 \\ 0 & 0 & 0 & 1 \end{bmatrix} q_k + v = I q_k + v \tag{12}$$

where the observation vector is the quaternion-based attitude representation z that is obtained from the gravitational acceleration. Then, correction and prediction are represented by

$$\hat{q}_k = \hat{q}_k^- + K(z_k - \hat{q}_k^-) \tag{13}$$

$$\hat{q}_{k+1}^- = \frac{1}{2} \begin{bmatrix} 2 & -\Delta t \omega_{xk} & -\Delta t \omega_{yk} & -\Delta t \omega_{zk} \\ \Delta t \omega_{xk} & 2 & \Delta t \omega_{zk} & -\Delta t \omega_{yk} \\ \Delta t \omega_{yk} & -\Delta t \omega_{zk} & 2 & \Delta t \omega_{xk} \\ \Delta t \omega_{zk} & \Delta t \omega_{yk} & -\Delta t \omega_{xk} & 2 \end{bmatrix} \hat{q}_k \tag{14}$$

The attitude angle representation by quaternion z can be obtained by the followings [12].

$$z_k = \left[\cos\left(\frac{\theta_k}{2} \right), \ \sin\left(\frac{\theta_k}{2} \right) \times \left[\frac{A_k}{\|A_k\|} \right] \right] \tag{15}$$

where the angle θ_k and axis of rotation A_k are obtained from the inner and the cross products of a measured acceleration vector a_k and the acceleration vector defined as the initial attitude of the sensor a_0. That is,

$$\theta_k = \cos^{-1} \left(a_k \cdot a_0 \right) \tag{16}$$

$$A_k = a_k \times a_0 \tag{17}$$

Using a rotation matrix calculated from the corrected quaternion $\hat{q}_k$, longitudinal vector of each body segment is rotated. Then, the rotated vector is projected onto the sagittal and the frontal planes of the global coordinate system. Inclination angles are obtained from the inner product of the projected vector and an unit vector in each plane.

3   Evaluation of Angle Calculation Methods

3.1   Measurement of 3D Movements with Rigid Body Model

Experimental Method. A rigid body model consisted of steel prop body and tubular aluminium prismatic bar with a ball joint as shown in Fig. 2, which represented the thigh with the hip joint. A wireless inertial sensor (WAA-010, Wireless Technologies) was attached to the rigid body model with double-sided adhesive tapes. The inertial sensor includes a 3-axis gyroscope (IDG-3200, InvenSense) and a 3-axis accelerometer (ADXL345, Analog Devices). The inertial sensor communicates with a personal computer using Bluetooth (Ver 2.0 + EDR, Class 2). Markers for the optical motion measurement system (OPTOTRAK, Northern Digital Inc.) were also attached on the prismatic bar with double-sided adhesive tapes in order to measure reference angles for evaluation of measurement accuracy.

The sensor signals and the marker positions were measured simultaneously with a personal computer at a sampling frequency of 100 Hz. Measured acceleration signals were filtered with Butterworth low-pass filter with the cut-off frequency of 10 Hz in order to remove high frequency noise. Then, inclination angles were calculated by the 2 methods shown in Fig. 1.

280   T. Watanabe et al.

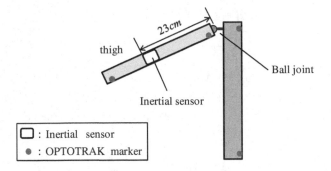

Fig. 2. Rigid body model used in measurement of angles during 3D movements.

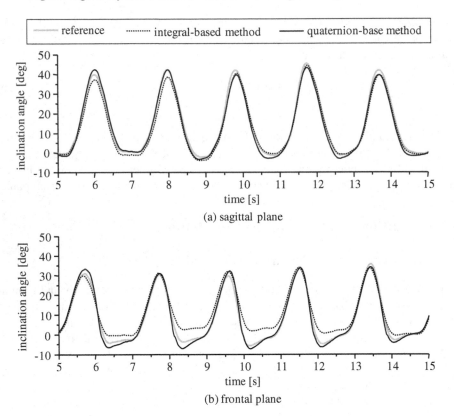

Fig. 3. An example of measured inclination angles during 3D movements of the rigid body model (2 s of cycle period).

The initial position of the thigh part was in the direction of the gravity. In the measurements, the thigh part was moved repeatedly simulating the circumduction gait. That is, the thigh part was moved to flexed position of about 45° in the sagittal plane through adducted position of about 45° in the frontal plane from the initial position, and then the thigh part was moved to the initial position by extension movement in the sagittal plane. This movement was performed manually with a cycle period of 2 s, 4 s and 8 s, respectively. The sensor was faced almost the frontal plane during the movement. The movement was performed repeatedly during a measurement trial of 30 s. Since prolonged measurements did not increase measurement error in our previous tests [8], the number of measurement trial was increased (10 trials) with 30 s of measurement time for each trial in this paper. In the evaluation, the differences between the sensor and marker settings were measured before measurement trial and removed.

Results. Figure 3 shows an example of measured inclination angles during the 3D movements. Inclination angles in the sagittal plane calculated by the 2 methods showed similar waveforms as that of reference angle waveform. Angles in the frontal plane calculated by the integral-based method, however, showed larger difference in angle waveform with the reference angle waveform than that by the quaternion-based one.

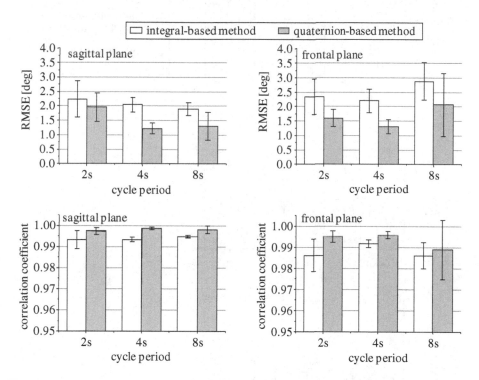

Fig. 4. Evaluation of angle calculation methods in measurement of 3D movements of the rigid body model.

Figure 4 shows evaluation results of the calculation methods for measurements of 3D movements of the rigid body model. The RMSE (root mean square error) and the CC (correlation coefficient) values showed that the quaternion-based method increased measurement accuracy both for angles in the sagittal and the frontal planes. For fast movements (2 s of cycle period), variations of RMSE and CC values also decreased with the quaternion-based method. However, for slow movements (8 s of cycle period), the variation of RMSE and CC values increased with the quaternion-based method although measurement accuracy increased with the quaternion-based one.

3.2   Measurement of Movement During Treadmill Walking with Healthy Subjects

Method.   As the first test of evaluation of the two calculation methods in measurement of human movements, previously measured treadmill walking [10] were analyzed. The

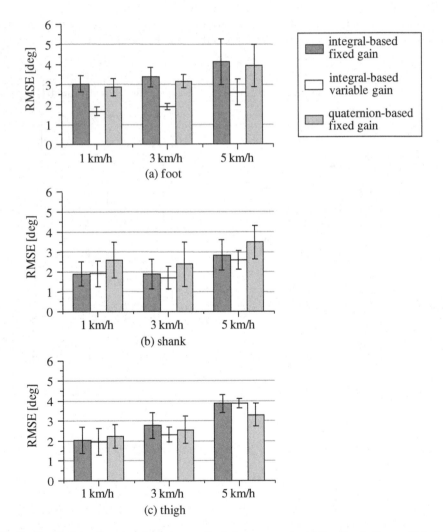

Fig. 5. Evaluation of angle calculation methods in measurement of angles in the sagittal plane during treadmill walking of healthy subjects.

walking were measured with 3 healthy subjects (male, 22-23 y.o.) during 90 s under 3 different walking speed conditions (1 km/h, 3 km/h, and 5 km/h) with wireless inertial sensors (WAA-010, Wireless Technologies) and 3D motion analysis system (OPTOTRAK, Northern Digital Inc.) simultaneously. Five trials were performed for each walking speed.

Result. Figure 5 shows average RMSE values for angles in the sagittal plane calculated by different 3 methods: integral-based method with fixed Kalman gain, integral-based method with variable Kalman gain and quaternion-based method. The RMSE values of the quaternion-based method were larger than the integral-based method with variable Kalman gain except for the thigh angle under the fast speed condition (5 km/h). The quaternion-based method showed almost similar RMSE values as those by the integral-based method with fixed Kalman gain, although shank angles were measured with better accuracy using the integral-based methods.

3.3   Preliminary Tests in Measurement of Gait with Hemiplegic Subjects

Method. Gait movements of 2 hemiplegic subjects were measured with wireless iner-tial sensors (WAA-010, Wireless Technologies). In this measurement, each subject walked about 10 m on level floor. The measured foot inclination angle was compared to the angle calculated from acceleration signals only. The angle calculated from accel-eration only involves measurement error caused by motion acceleration. However, angles under the quiet condition can be measured from gravitational acceleration with good measurement accuracy.

Result. Average values of measured motion acceleration of the foot during quiet standing and at around the foot flat condition were shown in Table 1. Here, the foot flat condition was detected by the angular velocity measured with the sensor attached to the foot [13]. This result means that the foot inclination angle calculated from acceleration only can provide appropriate angle values at the foot flat condition because the motion acceleration was small as same as that under the quiet standing condition.

Table 1. Average magnitude of motion acceleration of the foot during quiet standing condition (before walking) and during foot flat condition.

	Quiet standing [G]	Foot flat [G]
Subject 1	0.058	0.065
Subject 2	0.056	0.088

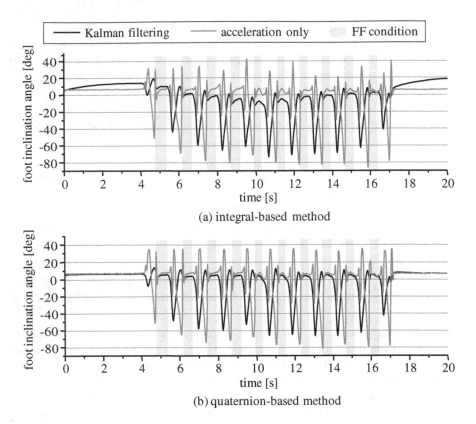

Fig. 6. Examples of measured inclination angle of the paralyzed foot during gait with a hemiplegic subject (subject 1).

Figures 6 and 7 show examples of measured inclination angles of the paralyzed foot in the sagittal plane. As seen in Fig. 6, the angle calculated by the integral-based method were different from the angle calculated from acceleration before and after the walking, and also at around the foot flat shown by shaded area. The quaternion-based method seems to measure appropriately the angle at around the foot flat in addition to those before and after the walking as shown in Fig. 6(b). For the other hemiplegic subject, however, the quaternion-based method could not remove measurement error at around the foot flat and after the walking although it decreased the errors compared to the integral-based method as shown in Fig. 7.

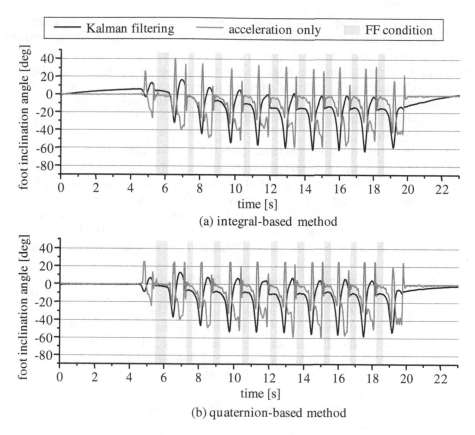

Fig. 7. Examples of measured inclination angle of the paralyzed foot during gait with a hemiplegic subject (subject 2).

4   Discussions

As shown in Fig. 4, the quaternion-based angle calculation method showed higher measurement accuracy in measurement of 3D movements of the rigid body model than that of the integral-based method with variable Kalman gain. Average RMSE values of the integral-based and the quaternion-based methods were less than 2.5° and 2.0°, respectively, and average CC values were larger than 0.993 and 0.997, respectively. For angles in the frontal plane during 3D movements, average RMSE values were less than 3.0° and 2.1° and average CC values were larger than 0.986 and 0.988 for the integral-based and the quaternion-based methods, respectively. In our previous study, the integral-based method with variable Kalman gain showed that average values of RMSE and correlation coefficient in measurement of 2D movements of a double pendulum rigid body model were less than 1.5° and larger than 0.998, respectively [10]. The quaternion-based method would be more effective by improving measurement accuracies for 3D movements up to those for 2D movements.

The quaternion-based angle calculation method is considered to be effective compared to the integral-based one. As shown in Fig. 3, the integral-based method caused difference in waveform of angle in the frontal plane. Measurement accuracy of the quaternion-based method was higher than the integral-based one as shown in Fig. 4. In addition, as seen in Figs. 6(a) and 7(a), the integral-based method calculated angle inappropriately in some cases of measurement of gait of hemiplegic subjects. The quaternion-based method could measure angles appropriately in some of such cases as shown in Fig. 6(b).

Angle measurement accuracy obtained from the analysis of treadmill walking measured with healthy subjects showed that the integral-based method with variable Kalman gain would be useful for measurement of angles in the sagittal plane during gait of healthy subjects. It can improve manual measurement with goniometer that is used with resolution larger than about 5°. The sensor system also makes possible to measure angles during movements in addition to measurement of range of motion (ROM). These suggest that the angle calculation methods tested in this paper can be practical in measurement of movements.

In the measurement of angles during treadmill walking, the quaternion-based method showed almost similar RMSE values as those of the integral-based method with fixed Kalman gain, which were less than those of the integral-based method with variable Kalman gain. For measurement of lower limb angles during gait, the variable Kalman gain method is considered to be effective because it could remove error in Kalman filtering caused by impact and motion acceleration and also increase Kalman filtering effect under low motion acceleration conditions. Since the variable Kalman gain method used in the integral-based method can not be applied directly to the quaternion-based one, it is necessary to develop and test variable Kalman gain method applicable to the quaternion-based method.

Figure 4 showed that measurement accuracy of angles in the sagittal plane was high for slow movements, especially with the quaternion-based method. For fast movements (2 s of cycle period), it is considered that the integral-based method could decrease the error by applying the variable Kalman gain method. Since the quaternion-based method with constant Kalman gain showed higher measurement accuracy than that of the integral-based method, it is expected that the quaternion-based method further improves measurement accuracy for fast movements by applying variable gain method. However, for the angles in the frontal plane, the measurement accuracy for slow movements (8 s of cycle period) decreased. Although the cause of the error increase in measurement of angles in the frontal plane is not clear, there is a possibility that there were some differences between fast and slow movements.

The integral-based method showed inappropriate angle calculation results in some cases of measurement of gait movement of hemiplegic subjects. The quaternion-based method was shown to have a possibility of improving the measurement accuracy. Since such inappropriate angle calculation was mainly seen in angles of paralyzed leg, it is considered that rotation movement of lower limb as seen in circumduction gait caused the inappropriate results. Evaluation tests of angle measurement accuracies for various gait movements are necessary to validate a measurement method of 3D movements.

5   Conclusions

In this paper, the integral-based angle calculation method with variable Kalman gain and the quaternion-based method with fixed Kalman gain were compared in angle measurements during 3D movements with the rigid body model and in measurements of lower limb angles during treadmill walking measured with healthy subjects and during gait measured with hemiplegic subjects. The quaternion-based method showed higher measurement accuracies for angles in the sagittal and the frontal planes during 3D movements of the rigid body model than those of the integral-based one. Although the quaternion-based method showed less measurement accuracy for angles during tread-mill walking of healthy subjects than that of the integral-based one, the quaternion-based one was suggested to be effective in angle measurement during gait of hemiplegic subjects. Since the results of measurement angles with the rigid body model and during treadmill walking suggested effectiveness of variable Kalman gain method, it is expected that the quaternion-based method improves the measurement accuracy for treadmill walking by applying the variable Kalman gain method. The quaternion-based angle calculation method would be more effective for measurement of angles in the sagittal and the frontal planes during gait by improving measurement accuracies for 3D movements up to those of the 2D movements.

Acknowledgements.   This work was supported in part by the Ministry of Education, Culture, Sports, Science and Technology of Japan under a Grant-in-Aid for challenging Exploratory Research.

References

1. Tong, K., Granat, M.H.: A practical gait analysis system using gyroscopes. Med. Eng. Phys. **21**, 87–94 (1999)
2. Dejnabadi, H., Jolles, B.M., Aminian, K.: A new approach to accurate measurement of uniaxial joint angles based on a combination of accelerometers and gyroscopes. IEEE Trans. Biomed. Eng. **52**, 1478–1484 (2005)
3. Findlow, A., Goulermas, J.Y., Nester, C., Howard, D., Kenney, L.P.: Predicting lower limb joint kinematics using wearable motion sensors. Gait Posture **28**, 120–126 (2008)
4. Cooper, G., Sheret, I., McMillian, L., Siliverdis, K., Sha, N., Hodgins, D., Kenney, L., Howard, D.: Inertial sensor-based knee flexion/extension angle estimation. J. Biomech. **42**, 2678–2685 (2009)
5. Sabatini, A.M.: Quaternion-based extended Kalman filter for determining orientation by inertial and magnetic sensing. IEEE Trans. Biomed. Eng. **53**(7), 1346–1356 (2006)
6. Mazzà, C., Donati, M., McCamley, J., Picerno, P., Cappozzo, A.: An optimized Kalman filter for the estimate of trunk orientation from inertial sensors data during treadmill walking. Gait Posture **35**(1), 138–142 (2012)
7. Saito, H., Watanabe, T.: Kalman-filtering-based joint angle measurement with wireless wearable sensor system for simplified gait analysis. IEICE Trans. Inf. Syst. **E94-D**, 1716–1720 (2011)

8. Watanabe, T., Saito, H., Koike, E., Nitta, K.: A preliminary test of measurement of joint angles and stride length with wireless inertial sensors for wearable gait evaluation system. Comput. Intel. Neurosci. **2011**, Article ID: 975193 (2011). doi:10.1155/2011/975193
9. Watanabe, T., Saito, H.: Tests of wireless wearable sensor system in joint angle measurement of lower limbs. In: 33rd IEEE Engineering in Medicine and Biology Society, pp. 5469–5472 (2011)
10. Teruyama, Y., Watanabe, T.: Effectiveness of variable-gain Kalman filter based on angle error calculated from acceleration signals in lower limb angle measurement with inertial sensors. Comput. Math. Methods Med. **2013**, Article ID 398042 (2013). http://dx.doi.org/10.1155/2013/398042
11. Chou, J.C.K.: Quaternion kinematic and dynamic differential equations. IEEE Trans. Robot. Automat. **8**(1), 53–64 (1992)
12. Favre, J.: Quaternion-based fusion of gyroscopes and accelerometers to improve 3d angle measurement. Electron. Lett. **42**(11), 612–614 (2006)
13. Watanabe, T., Endo, S., Murakami, K., Kumagai, Y., Kuge, N.: Movement change induced by voluntary effort with low stimulation intensity fes-assisted dorsiflexion: a case study with a hemiplegic subject. In: 6th International IEEE EMBS Conference on Neural Engineering, pp. 327–330 (2013)

Using Illumination Changes to Synchronize Eye Tracking in Visual Paradigms

Daniel Siboska[✉] and Henrik Karstoft

Department of Engineering, Aarhus University, Aarhus, Denmark
{dasi,hka}@eng.au.dk

Abstract. This paper presents a novel method for synchronizing the recording of a subject's gaze from an eye tracker (ET) to the display of visual stimuli. The method consists of embedding a signal used as a common time base in a small area of the visual stimuli, measuring this signal with an optical detector attached to the presentation screen, and modulating the global illumination used by the eye tracker synchronously to this measured signal. The timing signal generated with this method can be used to synchronize other data sources, such as electroencephalography (EEG) to the presentation of the visual stimuli as well. The prototype system where this method was implemented achieved a single sample of jitter for both the EEG and ET data.

Keywords: Electroencephalography · Eye tracking · Gaze estimation · Synchronization

1 Introduction

Recording of brain activity through electroencephalography (EEG) combined with measurements of eye movements gives researchers a powerful tool for analyzing the human visual system (HVS) [1]. Such tools have been used for developing methods for enhancing the everyday life of severely disabled people, who have no other means of communication than modulating their eye movement and brain wave patterns [2, 3]. Applications for ordinary users, such as image searching and classification, are emerging as well [4, 5], and the combination of eye tracking (ET) and EEG recording holds promise as one of the fundamental technologies in developing augmented memory applications [6, 7] in the near future.

Multiple researchers [8, 9] have set up EEG/ET-systems where the visual stimuli is presented on a computer display, the eye movements are measured with a video-based eye tracker, and the EEG is recorded with a digital recording device connected to a computer. The synchronization of these signals (stimuli, EEG, and eye movement) is of profound importance if any causality between the stimuli and the response is to be analyzed, and is a major challenge.

This challenge is addressed by the authors in [9], who use accurate control of the timing of the appearance of each frame of stimuli on the monitor and of the recording of each sample from the eye tracker and EEG to achieve synchronization of the stimuli,

© Springer International Publishing Switzerland 2015
G. Plantier et al. (Eds.): BIOSTEC 2014, CCIS 511, pp. 289–298, 2015.
DOI: 10.1007/978-3-319-26129-4_19

EEG and eye movement. This strategy requires low level control of the graphics hardware, as well as a special purpose video recording device for the eye tracker.

In commercial state of the art systems such as the RED500 with EEG headset from SensoMotoric Instruments GmbH (SMI) and Emotiv, or the Smart Eye Pro from Electrical Geodesics Inc. (EGI) and Smart Eye AB, a similar low level control of the sampling time of both EEG and ET is used.

In this article we propose an alternative strategy for solving the synchronization issue in an EEG/ET-system. We embed a synchronization signal in the visual stimuli and record this signal with both the eye tracker and EEG recording device. This offers greater freedom in the choice of stimulus display and video recording equipment for eye tracking, and enables a more flexible generation of stimuli, without the strict need for low level hardware programming. Hereby already available hardware can be used when a researcher wants access to a combined EEG/ET system, instead of having to acquire new hardware.

The following section describes the hardware and software algorithms used in the combined EEG/ET setup. Subsequently we evaluate the performance of this setup regarding spatial accuracy of the eye tracker as well as timing jitter between the stimulation presentation and EEG as well as eye movement recordings.

2 Methods and Materials

The developed system is outlined in Fig. 1, and comprises the following subsystems:

- **Eye tracker** - consisting of two infrared (IR) light emitting diodes (LED) directed towards the face of the subject, as well as a camera with infrared recording capabilities used to record the eye movement of the subject.
- **EEG recorder** - including an EEG cap with active electrodes and a 16 channel EEG amplifier with additional trigger input.
- **Visual stimulus presentation** - consisting of a large computer display with a dedicated area for embedding the synchronization signal.
- **Synchronization system** - consisting of an optical sensor measuring the signal in the dedicated synchronization area on the stimulus display and regulating the intensity of the IR LEDs based on the signal from the optical sensor. The optical sensor output is connected to the EEG amplifier as well.

    All of the subsystems are controlled by a central computer.

2.1 Eye Tracking Hardware

Eye movements were recorded with a remote video eye tracking system using the pupil-center corneal-reflection (PCCR) technique [10]. The PCCR technique requires two infrared illumination sources to produce two distinct reflections on each eyeball and to provide general illumination of the subjects face.

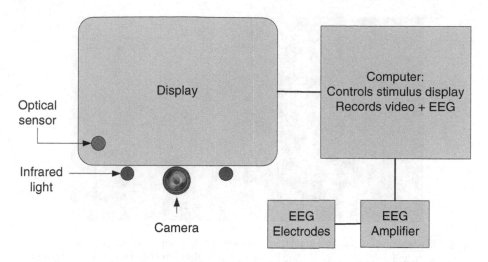

Fig. 1. Overview of the hardware used in the combined EEG/Eye tracking system.

The camera used to record the eyes was a Basler ACA640-100gc GigE camera with a resolution of 658×492 pixels and a maximum frame rate of 100 Hz. The camera used a fixed focus lens with a focal length of 16 mm, which resulted in a field of view of 10×15 cm at the operating distance of 60 cm. Attached to the lens was a Schneider Kreuznach 093 IR pass filter which helped control the exposure of the camera sensor, since the IR LEDs were the dominating source of infrared illumination in the setup.

Each of the illumination sources in our system comprised a cluster of four OSRAM SFH485 infrared (IR) light emitting diodes (LED). Since the subjects were exposed to the infrared radiation for extended periods of time, the current through each LED was limited to avoid exceeding the long term exposure limit for the retina [11]. The infrared LEDs were also used as part of the synchronization system by modulating the global illumination of the camera scene based on the signal from the optical sensor. This is described further in a later section.

2.2 Eye Tracking Algorithm

The PCCR technique uses the two reflections of the LEDs on each cornea of the eyes as well as the location of pupil center in the video stream to determine the direction of the subjects gaze.

The reflections were extracted from each video frame by calculating a difference of Gaussians (DoG) and thresholding the result, which resulted in a number of candidate clusters. The corneal reflections for each eye were found as the best fit of the distance and orientation of each pair of candidate clusters to an experimentally established mean distance and mean orientation, which was obtained by manually measuring on a video frame from 12 test subjects.

The positions of the corneal reflections were refined on a sub-pixel level by fitting a constrained 2D Gaussian to the corneal reflection using least squares.

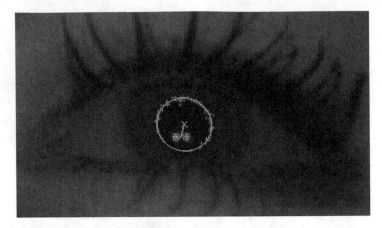

Fig. 2. Eye tracking features extracted from each eye. Blue: Corneal reflections. Yellow: PCCR vector. Green and red: Edge points found by the starburst algorithm (Red points were rejected by the RANSAC step) (Color figure online).

The pupil centers were found using a modified Starburst algorithm [12]. The first step in the Starburst algorithm is to threshold the vicinity of the corneal reflections to find a rough estimate of the pupil's location. This first estimate is used as a starting point for a search for gradients above an experimentally established threshold along rays extending from the center of the blob. If such gradients are found, the positions are added to a list of potential edge points. This search is repeated for each of the located potential edge points, with rays directed back towards the center of the blob. When the geometric center of the edge points converges, the list of points is used in a RANSAC [13] based search for an ellipse representing the edge of the pupil. The last step in the algorithm is to refine the position of the ellipse using an optimization step. The optimization searches for the strongest gradient along the edge of the pupil, with experimentally established constraints on the size and eccentricity of the ellipse. The output of the algorithm is the PCCR vector from the center of the two corneal reflections to the center of the pupil for each eye. The extracted features are shown in Fig. 2.

To map the relative locations of the corneal reflections and pupil (PCCR vector) to a point on the stimulus display, a mapping function was calculated individually for each eye of each subject. The mapping function from PCCR vector to the image coordinates is a second order polynomial in two variables of the form:

$$x_{screen} = a_1 x^2 + a_2 y^2 + a_3 xy + a_4 x + a_5 y + a_6$$
$$y_{screen} = b_1 x^2 + b_2 y^2 + b_3 xy + b_4 x + b_5 y + b_6 \tag{1}$$

where x_{screen} and y_{screen} are the coordinates in the image, x and y are the coordinates of the PCCR vector, and $a_{1...6}$ and $b_{1...6}$ are subject specific constants.

To determine the constants in the mapping function, each subject was presented with a standard calibration screen on the stimulus display with nine fixation targets in a three-by-three grid. The subjects were asked to fixate on each of the nine patterns in turn, while 100 frames (1 s) of video were recorded for each position. Since the position on the stimulus display of each pattern was known, the best fitting mapping constants could be found by calculating the least squares fit on all PCCR vectors extracted from the 900 frames of video.

2.3   EEG Recording Hardware

EEG data were recorded using a 16 channel G.tec g.USBamp with the g.GAMMAsys active electrode system. The data were recorded at a sampling rate of 1200 Hz and post processed with a band-pass filter between 0.1 Hz and 200 Hz. During recording all electrodes were referenced to the Cz electrode position and the cheek was connected to ground.

2.4   Visual Stimulus Presentation and Synchronization

The presentation of visual stimulus as well as the embedded synchronization signal was shown on a 26″ LG DT-3003X display with a resolution of 1280 × 768 pixels and a refresh rate of 60 Hz. The face of the subjects were placed 60 cm from the monitor slightly below the center.

A dedicated area of 20 × 20 pixels in the lower left corner of the stimulus display was used for the embedded synchronization signal. The synchronization signal can be viewed as a one bit wide serial data link between the stimulus display and the ET and EEG recorder. A '0' is coded by turning the dedicated area black, while a '1' is represented by a white area.

The detection of the synchronization signal on the stimulus display is achieved by the use of an optical sensor. The detected signal is sent to the EEG amplifier and is used by the synchronization hardware to modulate the illumination used in the eye tracker. The amount of light from the high and low level of this modulation is chosen to allow robust detection of the synchronization signal from the video stream, without compromising the fidelity of the video through under- or overexposure of the camera sensor.

When this is accomplished, the histogram of the video frames with low illumination can be transformed to match the frames with high illumination before the ET algorithm.

The synchronization signal was extracted from the video stream by calculating the number of pixels whose intensities changed in a positive direction and subtracting the number of pixels whose intensities changed in a negative direction between each pair of frames. The resulting signal showed a strong resemblance with the time derivative of the original synchronization signal and had strong positive and negative peaks when the global illumination of the scene changed rapidly. An example of this kind of signal is shown in Fig. 3 along with the original synchronization signal as well as the time derivative.

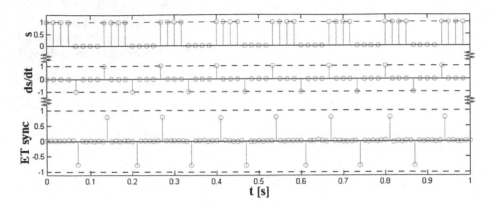

Fig. 3. Original (s) and extracted (ET sync) sync signal. The extracted signal is calculated as the normalized difference between the number of pixel intensities changed in a positive and negative direction.

3   Results

3.1   Spatial Precision of the Eye Tracker

The spatial precision of the eye tracker was evaluated by having a group of 12 subjects go through the procedure of calibrating the eye tracking system followed by a validation procedure. During the validation procedure a grid of 9 fixation patterns was shown, one after the other, and one second of eye movement was recorded while the subject fixated on each pattern in turn. The accuracy and precision of each of the 9 positions are shown in visual angle in Table 1.

Table 1. Bias and standard deviation of eye position for the 9 areas of the display measured in degrees of visual angle.

x: $-0.18 \pm 0.43°$ y: $0.05 \pm 0.31°$	x: $0.13 \pm 0.38°$ y: $0.30 \pm 0.38°$	x: $0.01 \pm 0.33°$ y: $-0.20 \pm 0.35°$
x: $0.22 \pm 0.35°$ y: $-0.17 \pm 0.32°$	x: $0.18 \pm 0.46°$ y: $-0.33 \pm 0.60°$	x: $-0.40 \pm 0.47°$ y: $0.54 \pm 0.63°$
x: $0.01 \pm 0.41°$ y: $0.18 \pm 0.31°$	x: $-0.31 \pm 0.34°$ y: $-0.13 \pm 0.31°$	x: $0.34 \pm 0.41°$ y: $-0.25 \pm 0.44°$

3.2   Measurement of Synchronization Jitter

The synchronization of the stimulus display and the eye tracker was subject to timing jitter. The primary cause of this jitter is illustrated in Fig. 4 where it can be seen that with the eye tracker's frame rate of 100 Hz, there can be a delay of up to 10 ms before a change in the synchronization signal presented on the stimulus display is recorded by the eye tracker.

To measure if any other sources of timing jitter between the stimulus display and the eye tracker were present, a simple stimulus, which is used in visual evoked potentials (VEP) experiments, was shown to one subject. The stimulus showed a checkerboard with each square alternating between black and white. One of the alternating squares resided in the dedicated synchronization area on the display.

The frequency of the alternating black and white squares must divide the display refresh rate into an integer, so a frequency of 7.5 Hz was chosen, resulting in 4 consecutive frames of the same color being displayed before changing color. The synchronization signal was extracted from the video stream (as shown in Fig. 3) and the variation in time between each period was calculated. The period of the original signal was 133.3 ms, and the maximum as well as minimum period extracted from the eye tracking signal was 130.0 ms (13 frames) and 140.0 ms (14 frames) respectively. In other words, only a single frame of jitter was present in the extracted synchronization signal. The signal-to-noise ratio of the extracted synchronization signal was 37.6 dB.

The amount of jitter in the EEG recordings was evaluated with the same experimental setup as described above, and the maximum as well as minimum period extracted from the EEG data was 132.5 ms (159 samples) and 134.1 ms (161 samples) respectively.

3.3 Real-Time Performance

To determine if it would be possible to make the stimulus display dependant on the current fixation target in a closed loop experiment, the runtime performance of the eye tracking algorithm was evaluated. The processing frame rate on a 3.4 GHz Intel Core i5 computer was 10 Hz.

4 Discussion

The focus of the combined eye tracking and EEG recording setup presented in this article is on high flexibility and acceptable performance for HVS experiments. Compared to state of the art commercial remote eye tracking and EEG solutions, such as the RED500 with EEG headset from SensoMotoric Instruments GmbH (SMI) and Emotiv, or the Smart Eye Pro from Electrical Geodesics Inc. (EGI) and Smart Eye AB, the presented system has some limitations due to the choice of hardware used in the implementation.

The methods developed are not limited to be used with the chosen hardware, therefore, it is expected that the performance of the system will improve if the hardware is improved.

The camera used in the system was able to achieve a temporal resolution of 10 ms with no more than one frame of jitter, which makes it possible to distinguish between saccades and fixations in the eye movement data. If the dynamic behavior of the saccades needs to be analyzed, a camera with a higher temporal resolution must be used, such as the 500 Hz camera used in the RED500 eye tracker.

In general the jitter will be uniformly distributed with a minimum and maximum of −10 and +10 ms respectively. This amount of jitter makes it difficult to use the eye tracking data directly for eliminating EOG artifacts in the EEG data; however it gives a rough estimate of the time of such artifacts, which can then be refined further.

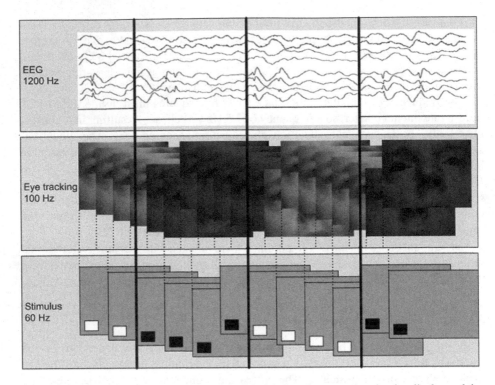

Fig. 4. Diagram showing the primary cause of timing jitter between the stimulus display and the eye tracker and EEG recording respectively.

The amount of jitter in the EEG recordings was measured to be within a single sample, and can therefore be expected to be uniformly distributed with a minimum and maximum of −0.8 and +0.8 ms respectively. Since the peaks from the event-related potentials (ERP) recorded by the EEG device are generally separated by tens or hundreds of milliseconds [14], this amount of jitter does not affect the ability to resolve individual ERPs.

The spatial accuracy of the system was measured to be 1.26° in the worst case, which is comparable to the performance of 1.01° achieved by [10] and slightly worse than the advertized accuracy of 0.4° for the RED500.

The obtained spatial resolution is sufficient to use the system in a brain computer interface (BCI), if the distance between fixation targets is kept above this lower limit. For reliable control of a general purpose computer interface designed for use by a computer mouse, a higher spatial resolution is required.

The human eye is able to attend to any target within a 2 degree cone of the fixation point without eye movement [15]. Even with higher accuracy, there will still be a high degree of ambiguity as to which target the subject is directing the attention towards within this 2 degree cone, so for many HVS studies, the obtained resolution is sufficient. The runtime performance of the software algorithms suggests the possibility of using the developed system in closed-loop experiments, where the visual stimuli depend on the fixation point, if the lower temporal resolution of 10 Hz is sufficient. A complete evaluation of the possibility of real-time performance with an optimized implementation of the algorithms resulting in higher frame rates is beyond the scope of this article.

In the experiments presented in the previous section, the synchronization signal was used as a simple clock signal, however since arbitrary data can be encoded in the synchronization signal, it is possible to use this communication channel to embed information about the stimuli directly in the ET video. One suggested use of this feature would be in the visual oddball paradigm [16], where different types of objects or characters (target, non-target and novelty) are presented to the subject sequentially. Using the synchronization signal as a data channel, the type of object currently presented to the subject could be embedded directly in the eye tracking data.

The results presented in this paper will be used as the foundation for future work involving measurements of reaction time in neurological paradigms involving visual stimulation.

5   Conclusions

In this paper we have described the general implementation of a combined EEG and eye tracking system, as well as a new flexible way to solve the problem of synchronizing the different data sources in such a system. The proposed synchronizing method forgoes the need for low level control of the sampling time of eye tracker and EEG as well as presentation time of the visual stimuli. This in turn eases the development of different types of stimuli and enables the use of a wider range of eye tracking camera and EEG recording equipment. The results obtained with the reference implementation of this method are comparable to similar systems with respect to the spatial and temporal resolution, and the amount of jitter in the system is primarily dependent on the temporal resolution of the eye tracking and EEG recording equipment.

Acknowledgements. This work was partly supported by the Sino-Danish Center for Education and Research (SDC).

References

1. Sereno, S.: Measuring word recognition in reading: eye movements and event-related potentials. Trends Cogn. Sci. **7**(11), 489–493 (2003)
2. Wang, Y., Gao, X., Hong, B., Jia, C., Gao, S.: Brain-computer interfaces based on visual evoked potentials. IEEE Eng. Med. Biol. Mag. **27**(5), 64–71 (2008)
3. Agustin, J.S.: Low-Cost Gaze Interaction: Ready to Deliver the Promises, pp. 4453–4458 (2009)
4. Wang, J., Pohlmeyer, E., Hanna, B., Jiang, Y.-G., Sajda, P., Chang, S.-F.: Brain state decoding for rapid image retrieval. In: Proceedings of the Seventeen ACM International Conference on Multimedia - MM 2009, p. 945 (2009)
5. Pohlmeyer, E.A., Wang, J., Jangraw, D.C., Lou, B., Chang, S.-F., Sajda, P.: Closing the loop in cortically-coupled computer vision: a brain-computer interface for searching image databases. J. Neural Eng. **8**(3), 036025 (2011)

6. Davies, S.: Still building the memex. Commun. ACM **54**(2), 80 (2011)
7. Bell, G., Gemmell, J.: A digital life. Sci. Am. **296**(3), 58–65 (2007)
8. Plöchl, M., Ossandón, J.P., König, P.: Combining EEG and eye tracking: identification, characterization, and correction of eye movement artifacts in electroencephalographic data. Front. Hum. Neurosci. **6**, 278 (2012)
9. Görgen, K., Walter, S.: Combining eyetracking and EEG. Publ. Inst. Cogn. Sci. **15** (2010)
10. Villanueva, A., Daunys, G., Hansen, D.W., Böhme, M., Cabeza, R., Meyer, A., Barth, E.: A geometric approach to remote eye tracking. Univers. Access Inf. Soc. **8**(4), 241–257 (2009)
11. Jäger, C.: Eye Safety of IREDs used in Lamp Applications Application Note (2010)
12. Winfield, D., Parkhurst, D.J.: Starburst: a hybrid algorithm for video-based eye tracking combining feature-based and model-based approaches. In: 2005 IEEE Computer Society Conference on Computer Vision and Pattern Recognition (CVPR 2005) - Workshops, vol. 3, pp. 79–79 (2005)
13. Fischler, M.A., Bolles, R.C.: Random sample consensus: a paradigm for model fitting with applications to image analysis and automated cartography. Commun. ACM **24**(6), 381–395 (1981)
14. Luck, S.J.: An Introduction to the Event-Related Potential Technique, vol. 25, no. 1, p. 388. A Bradford Book, Cambridge (2005)
15. Fairchild, M.: Color Appearance Models, p. 7. Addison, Wesley, & Longman, Reading (1998)
16. Courchesne, E., Hillyard, S.A., Galambos, R.: Stimulus novelty, task relevance and the visual evoked potential in man. Electroencephalogr. Clin. Neurophysiol. **39**(2), 131–143 (1975)

Diagnostics of Coronary Stenosis: Analysis of Arterial Blood Pressure and Mathematical Modeling

Natalya Kizilova[1,2(✉)]

[1] Interdisciplinary Centre for Mathematical and Computational Modeling,
Warsaw University, Ul. Prosta, 69, Warsaw, Poland
n.kizilova@gmail.com
[2] Kharkov National University, Svobody Sq., 4, Kharkiv, Ukraine

Abstract. Severity of the coronary stenoses and necessity of the percutaneous coronary intervention is usually estimated basing on analysis of the pressure and flow signals measured in vivo by a pressure gauge at certain distances before and after the stenosis. In the paper the differences in the pressure gradients at different stenosis severity are shown and discussed. A method of decomposition of the measured biosignals into the mean and oscillatory components is proposed. A mathematical model of the steady and pulsatile flow through the viscoelastic blood vessel in the presence of the rigid guiding wire is developed for biomechanical interpretation of the measured coronary blood pressure and flow signals. A novel approach for estimation the stenotic severity basing on the measured and computed data is proposed.

Keywords: Arterial blood pressure · Pulse wave · Coronary arteries · Stenosis · Mathematical modeling · Signal processing

1 Introduction

Coronary artery disease, which is also known as atherosclerotic or ischemic heart disease, has become one of the most severe diseases causing a large number of deaths each year over the world. The partial occlusion of the stenosed artery and abnormal blood flow through it to the heart cells lead to insufficient oxygen delivery, especially when the possibilities of the perfusion regulation by the resistive coronary vessels are spent [1]. The causes to the formation of atherosclerotic lesions and arterial stenosis are still unknown but it is well established that the fluid dynamics, particularly the wall shear stress (WSS) and local pressure oscillations play an important role in the genesis of the disease [2].

In the absence of stenosis, the driving pressure gradient is constant over the coronary vessels. With progressing of the stenosis severity, the pressure gradient required to impel the blood through the narrowed path increases that results in a higher blood pressure at the inlet of the stenosed artery. The heart must work harder to increase the produced pressure, and when the blood supply to the working heart is insufficient the angina and even heart attack may occur. In-time diagnostics of the

© Springer International Publishing Switzerland 2015
G. Plantier et al. (Eds.): BIOSTEC 2014, CCIS 511, pp. 299–312, 2015.
DOI: 10.1007/978-3-319-26129-4_20

stenosed coronary arteries is crucial for timely therapy or/and surgery of the coronary lesions and mathematical modeling is an important tool for that.

Coronary angiography (AG), intravascular ultrasound (IVUS) and coronary computed tomography angiography (CCTA) are commonly used for estimation of the stenosis severity by computations of the minimal lumen area (MLA) that is determined as the ratio of the minimal A_{min} to normal A_0 lumen areas: MLA = A_{min}/A_0 (%). The results of the AG, CCTA and IVUS-based MLA computations correspond well to each other [3], but not in the case of the calcified wall [4]. MLA gives geometric approximation of the stenosis and in many cases the stenoses with MLA <50 % remain insignificant and do not need stenting or bypass surgery, because sufficient perfusion is provided by autoregulation of the resistive vessels and collateral blood supply.

The functional severity of the stenosis can be estimated by the fractional flow reserve (FFR) defined as the ratio of the mean distal Pd and proximal (anterior) Pa coronary pressures measured via the pressure wire at certain distances before and after the stenosis during maximal hyperemia produced by intravenous adenosine administration that leads to relaxation of the myocardial vessels. The normal FFR = 0.94–1.0, whereas the FFR <0.75 highly correlates with insufficient perfusion and myocardial ischemia. The patients from the grey zone 0.75 < FFR < 0.8 may have had a risk of ischemia [5, 6]. FFR reveals the dangerous ischemia-producing lesions [7], and it is recognized as gold standard for assessing the hemodynamic significance of coronary stenoses [8]. The similar approach based on the flow velocities at rest and the hyperaemic state has also been developed.

Computational fluid dynamics (CFD) is widely used in advanced studies on the blood flows in rigid and compliant boundaries. The corresponding finite element and finite volume models and the computational schema have been used for the blood flow modeling in the vessels of different size up to the cellular level [9]. CFD study of the flow past symmetric and asymmetric stenoses in the straight, curved, helical and bifurcating tubes allow computations of the FFR values for every single stenosis as well as for the tandem, overlapping and bifurcational lesions. CFD approach allows virtual planning and estimation the outcomes of the surgery (stents, grafts, bypass) [10], and the model-based FFR computation is a challenge that is widely discussed in recent publications [11–13]. In the present paper some novel aspects of the FFR assessment and analysis of the measured pressure signals are proposed and discussed basing on the measurement data and the mathematical model of the blood flow in different rigid and compliant boundaries.

2 Blood Pressure Signals

2.1 The Measurement Procedure

CFD computations are based on the 3D models of the viscous incompressible blood flow in the rigid patient-specific geometry of the larger epicardial coronary vessels recognized in AG and CCTA images, while the invasive FFR calculations are based on the in vivo measurements of the blood pressures before and after the stenosis at the presence of the guiding catheter and wire.

At local anesthesia, a guide catheter (Fig. 1) is inserted into the orifice of the coronary artery through the femoral or radial artery. The pressure and flow signals in the coronary arteries can be measured by the pressure and Doppler guide wire. The diameters of the catheters can be chosen between d = 1.5–2.3 mm, while the manufactured guidewires have the diameters d = 0.35–0.89 mm.

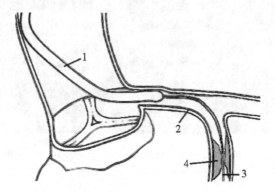

Fig. 1. Schematic representation of the intravascular coronary examination: the guide catheter (1) in the coronary artery (2) and the guidewire with pressure and flow gauges (3) in the coronary stenosis (4).

According to the measurement data [14], the main coronary arteries of adult humans have the following diameters: d = 4.5 ± 0.6 mm for the left main artery; d = 3.7 ± 0.4 mm and d = 1.9 ± 0.4 mm for the proximal and distal parts of the left anterior descending artery; d = 3.4 ± 0.5 mm for the left circumflex artery; d = 3.9 ± 0.6 mm and d = 2.8 ± 0.5 mm for the right coronary artery. The comparison of the diameters shows that both the catheter and wire can produce disturbances in the natural coronary blood flow and wave propagation.

In this study 45 data samples recorded in the epicardial coronary arteries of 32 patients with different stenosis severity diagnosed by the pressure gauge administrated via the guiding catheter have been analyzed. An example of the recorded rata digitized from the CathLab software is presented in Fig. 2. The red and green time-varying curves correspond to the pressure signals $P_a(t)$ and $P_d(t)$ accordingly, while the relatively smooth red and green lines correspond to their mean values. The measurements have been carried out during the adenosine administration which dynamics can be followed by the shift between the both oscillating and mean value curves. The FFR value indicated with yellow color has been computed automatically by the CathLab software.

2.2   Smooth and Oscillatory Signals

Depending on the presence and severity of the stenosis, the pressure gradients in the signals measured before $P_a(t)$ and behind $P_d(t)$ the stenosis have significant differences. As the stenosis severity is progressing, the pressure behind the lesion drops first in

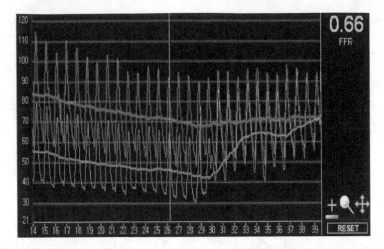

Fig. 2. An example of the pressure signals recorded in the coronary artery by the pressure gauge and analyzed by the CathLab software.

diastole, while the pressure decrease after the peak systole is the same as in the pressure signal $P_a(t)$ (Fig. 3a). Then the pressure drop in diastole becomes more significant (Fig. 3b) and the differences in the pressure gradients appear also in the systole (Fig. 3c).

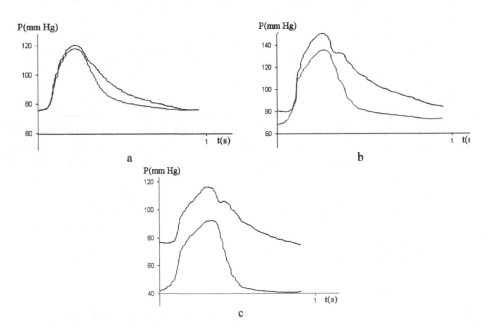

Fig. 3. Blood pressure signals $P_a(t)$ (upper lines) and $P_d(t)$ (lower lines) measured in the epicardial coronary arteries with progressing stenosis severity (a, b, c).

The contour analysis of the $P_a(t)$ and $P_d(t)$ signals characterises their relative differences in slopes and values, while some novel information important for diagnostics can be driven from the $P_a(P_d)$, pressure-flow $P(U)$, and phase curves $P'(P)$ and $U'(U)$ computed from the measured signals where the stroke sign denotes the time derivative [15]. For instance, the $P_a(P_d)$ curves computed from the $P_a(t)$ and $P_d(t)$ signals by elimination of time are presented by loops (Fig. 4) slightly varying according to the heart rate, blood pressure and flow variability [16, 17]. In spite of the heart rate and blood perfusion variability, the characteristic shape of the loop is preserved from beat to beat. When the myocardial perfusion is normal, the $P_a(P_d)$ loop is elongated and tends to the straight line (Fig. 4a).

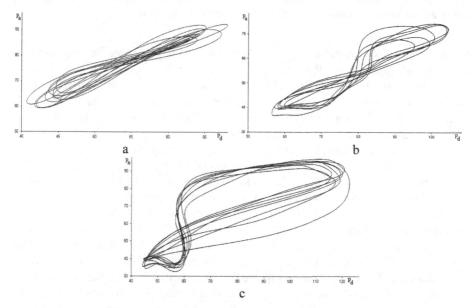

Fig. 4. $P_a(P_d)$ loops for the stenotic flows at FFR = 0.86 (a); FFR = 0.7 (b); FR = 0.53 (c).

When the stenotic flow is critical in the term of the FFR values, the loop is shaped as digit '8' and the self-intersection point is located in the middle of the loop (Fig. 4b). When the perfusion is insufficient, the FFR value is low and the urgent surgery is necessary, the $P_a(P_d)$ becomes 'thicker' and is looking as asymmetric '8' because of the asymmetric location of the self-interaction point (Fig. 4c). Similar changes in the shapes of the dependencies $(P_a–P_d)$ on P_d and $(P_a–P_d)$ on P_a with progressing stenotic severity (functional, not geometrical!) have been observed in this study.

Representation of the measured blood pressure signals as cycles allows computation of different integral parameters like the area located inside the loop and its two subparts produced by the intersection point, variability of its location and slope. The measured blood pressure signals $P_a(t)$ sometimes exhibit oscillating behaviour (Fig. 5a), while in many cases they remain relatively smooth (Fig. 3). Note that the $P_d(t)$ curves do not demonstrate such oscillating behaviour, because the stenosis serves

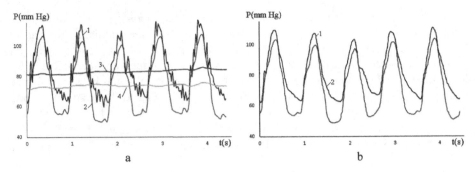

Fig. 5. In vivo measured pressure curves before (a) and after (b) the smoothing procedure. The curves are marked by numbers: $P_a(t)$ (1), $P_d(t)$ (2), $<P_a(t)>$ (3), $<P_d(t)>$ (4).

as the wave absorber producing reflected waves that propagate in the upstream direction and appear in the $P_a(t)$ signal.

Similar regularity has been found in [18]. Numerical simulations on the 1D model exhibit high-frequency, short wave-length reflected waves superimposed over the main wave front, and the computed high frequency oscillations were not a consequence of the numerical solver.

Applying the 3–5 point smoothing filters or eliminating the high harmonics from the Fourier expansion, the $P_a(t)$ signals may be transformed in the smooth curves, but the computed FFR values will be always lower for the initial $P_a(t)$ (oscillatory) signals than for the smoother ones, because the smoothing procedure cuts the high oscillations and decreases the mean values of the signals. In that way the FFR computed on the oscillating curves can overestimate the stenosis severity. The smoothing leads to elimination of information that can be complementary to the FFR value and useful for more detailed diagnostics of the stenosis rigidity or presence of the atheroma, thrombus and fibrous cap. The $P_a(P_d)$ loops computed from the oscillating (Fig. 6a) and smoothed (Fig. 6b) pressure signals (see Fig. 5a and 5b correspondingly) demonstrate the intensity of the high-frequency oscillations produced by additional wave reflection. The smoothed curves (Fig. 5b) still can be classified and explained in correspondence to the examples presented in Fig. 4, while the oscillating ones (Fig. 5a) needs elaboration of new indexes and their biomechanical interpretation.

The pulsatile component of the measured pressure signals is not taken into account in the FFR computations, so decomposition of the signal $P(t)$ into the mean $<P(t)>$ and oscillatory $P'(t)$ terms and examination of the oscillatory component may be interested for the diagnostic purposes, as well as for deeper understanding the blood flow and pressure wave propagation through the stenosis. For instance, the FFR values could be computed separately for the mean and oscillatory components as

$$FFR = <P_d(t)> / <P_a(t)> \quad \text{and} \quad FFR_{osc} = P'_d(t)/P'_a(t).$$

A comparative analysis of the FFR, FFR_{osc} and MLA values on a large representative group of the measurements in the stenosed arteries will be done in the next studies.

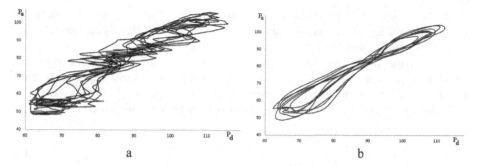

Fig. 6. $P_a(P_d)$ loops for the oscillatory (a) and smoothed (b) pressure signals presented in Fig. 5a and Fig. 5b accordingly.

3   Mathematical Model

3.1   Steady Blood Flow Between the Rigid Boundaries

The simplest model of the blood flow in the stenosed artery in the presence of the guide catheter (Fig. 1) is the steady viscous flow between the rigid coaxial cylinders. According to the well-know solution of the problem the axial flow is

$$V(r) = \frac{\delta P}{4\mu L}\left(R_2^2 - r^2 + \frac{R_2^2 - R_1^2}{\ln(R_2/R_1)}\ln\left(\frac{r}{R_2}\right)\right) \qquad (1)$$

where R_2 is the radius of the artery, R_1 is the radius of the wire/catheter, μ is the blood viscosity, L is the distance between locations of the proximal and distal measurement sites, δP is the measured pressure drop.

The computed FFR in the straight part of the blood vessel is estimated on the CFD model, which in the limit of the rigid wall and the steady inflow tends to the Poiseuille solution

$$V_P(r) = \frac{\delta P}{4\mu L}\left(R_2^2 - r^2\right) \qquad (2)$$

From (1) and (2) the error in the FFR values computed basing on the measurement signals and CFD computations can be estimated.

3.2   Pulsatile Blood Flow in Compliant Vessels

Heart contraction produces oscillations of the pressure and flow that propagate along the vessels, and the speed of the pulse waves vary from $c = 5{-}8$ m/s in large elastic arteries to $c = 10{-}12$ m/s in small resistive blood vessels. In elderly individuals and in the case of atherosclerosis, hypertension and some other cardiovascular disorders the pulse wave velocity increases up to $c = 25$ m/s [19]. The wave propagation and reflection at the arterial branching, atherosclerotic plaques, lesions and other non-uniformities produce complex superposition of the propagated and reflected waves. Spectral and wave-intensity analysis of the registered signals can reveal novel features of hemodynamics of stenosis and diagnostic indexes.

In this paper the axisymmetric wave propagation between the coaxial cylinders is proposed as the model of the pulsatile blood flow and pressure wave propagation in the compliant artery when the guiding catheter is inserted (Fig. 1).

Fluid flow is governed by incompressible Navier-Stokes equations

$$\nabla \cdot \vec{v} = 0,$$
$$\rho \left(\frac{\partial \vec{v}}{\partial t} + (\vec{v} \cdot \nabla) \vec{v} \right) = -\nabla p + \mu \nabla^2 \vec{v}, \tag{3}$$

the mass and momentum conservation equations for the incompressible vessel wall

$$\nabla \cdot \vec{u} = 0,$$
$$\rho_w \frac{\partial^2 \vec{u}}{\partial t^2} = -\nabla p_s + \nabla \cdot \hat{\sigma}, \tag{4}$$

where $\vec{v}$ is the flow velocity, $\vec{u}$ is the wall displacement, ρ and ρ_w are the mass densities for the blood and wall, μ is the fluid viscosity, p and p_s are the hydrostatic pressures in the fluid and solid, $\hat{\sigma}$ is the stress tensor for the vessel wall.

The viscoelastic Kelvin-Voight body has been used as rheological model for the layers:

$$\sigma_i + \tau_w \frac{\partial}{\partial t} \sigma_i = A_{ik} \varepsilon_k + \mu_w \frac{\partial}{\partial t} \varepsilon_k \tag{5}$$

where A_{ik} is the matrix of elasticity coefficients, μ_w is the wall viscosity, τ_w is the stress relaxation time, $\vec{\sigma}^T = \{\sigma_{11}, \sigma_{22}, \sigma_{33}, \sigma_{23}, \sigma_{13}, \sigma_{12}\}$ is the stress vector, $\vec{\varepsilon}$ is similar strain vector, $\varepsilon_{ik} = (\nabla_i u_k + \nabla_k u_i)/2$, T is transposition sign. For the isotropic material A_{ik} contains Young modules E, Poisson ratio v and shear modulus G = E/(2 + 2v), while orthotropic and transversely isotropic materials can also be considered in (5) [21, 22].

The boundary conditions include the no-slip flow condition at the inner rigid surface; continuity conditions for the fluid and solid velocities and the stress components at the fluid-wall interface:

$$r = R_1 \quad : \quad \vec{v} = 0 \tag{6}$$

$$r = R_2 \quad : \quad \vec{v} = \frac{d\vec{u}}{dt}, \quad \vec{\sigma}_n = \vec{\sigma}_n \tag{7}$$

At the outer surface of the blood vessel the no displacement or no stress boundary conditions can be taken in the form

$$r = R_2 + h \quad : \quad \vec{\sigma}_n = 0 \text{ or } \vec{u} = 0 \tag{8}$$

where h is the thickness of the arterial wall, n and τ denotes the normal and tangential components.

At the ends of the tube the fastening conditions for the tube

$$z = 0; L \quad : \quad \vec{u} = 0, \tag{9}$$

the input wave at the inlet and the wave reflection condition at the outlet of the tube

$$z = 0 \quad : \quad p(t,0) = p_0(t), \tag{10}$$

$$z = L \quad : \quad p(t, L) = \Gamma p_0(t), \tag{11}$$

where Γ is the complex reflection coefficient equal to the ratio of the amplitudes of the reflected and propagates waves [19, 20], $\mathrm{Re}(\Gamma) \in [0, 1]$ and $\mathrm{Im}(\Gamma)$ accordingly corresponds to resistivity and capacity of the downstream vasculature [21, 22] are considered.

The solutions of the problem (3) and problem (4)–(5) which are coupled via the boundary conditions (8)–(11) have been found as a superposition of the steady solution and small axisymmetric disturbance in the form of the normal mode:

$$\{\vec{v}, p\} = \{\vec{v}^*, p^*\} + \{\vec{v}^\circ, p^\circ\} \cdot e^{st + ikz}$$
$$\{\vec{u}, p_s\} = \{\vec{u}^*, p_s^*\} + \{\vec{u}^\circ, p_s^\circ\} \cdot e^{st + ikz}$$

where $\vec{v}^\circ$, $\vec{u}^\circ$, p°, p_s° are the amplitudes of the corresponding disturbances, $k = k_r + ik_i$, $s = s_r + is_i$, s_i is the wave frequency, k_r is the wave number, s_r and k_i are spatial and temporal amplification rates, z is the axial coordinate. The steady part $\{\vec{v}^*, p^*\}$ is identified with Poiseuille flow (1) between the rigid surfaces.

The amplitudes $\vec{v}^\circ$, $\vec{u}^\circ$, p° p_s° can be obtained from (3)–(4) as Fourier expansions

$$p = \sum_{j=0}^{n} C_{1j} J_0(i\gamma_j r) e^{i(\omega_j t - \gamma_j x)},$$

$$p_s = \sum_{j=0}^{n} (C_{8j} J_0(i\gamma_j r) + C_{9j} Y_0(i\gamma_j r)) e^{i(\omega_j t - \gamma_j x)},$$

$$V_r = \sum_{j=0}^{n} i\gamma_j (C_{2j} J_1(i\gamma_j r) + C_{3j} J_1(i\beta_j r) + C_{10j} K_0(i\gamma_j r) + C_{11j} K_0(\kappa_j r)) e^{i(\omega_j t - \gamma_j x)},$$

$$V_x = \sum_{j=0}^{n} i(C_{2j} \gamma_j J_1(i\gamma_j r) + C_{3j} \beta_j J_1(i\beta_j r) + C_{10j} K_0(i\gamma_j r) + C_{11j} K_0(\kappa_j r)) e^{i(\omega_j t - \gamma_j x)}, \tag{12}$$

$$U_r = \sum_{j=0}^{n} i\gamma_j (C_{4j} J_1(\kappa_j r) + C_{5j} Y_1(\kappa_j r) + C_{6j} J_1(i\gamma_j r) + C_{7j} Y_1(i\gamma_j r) +$$
$$+ C_{12j} K_1(i\gamma_j r) + C_{13j} K_1(\kappa_j r)) e^{i(\omega_j t - \gamma_j x)},$$

$$U_x = \sum_{j=0}^{n} (C_{4j} \kappa_j J_0(\kappa_j r) + C_{5j} \kappa_j Y_0(\kappa_j r) + C_{6j} i\gamma_j J_0(i\gamma_j r) + C_{7j} i\gamma_j Y_0(i\gamma_j r) +$$
$$+ C_{12j} K_0(i\gamma_j r) + C_{13j} K_0(\kappa_j r)) e^{i(\omega_j t - \gamma_j x)}.$$

where $\beta_j^2 = \gamma_j^2 + i\omega_j/i\omega_j\nu.\nu$, $\kappa_j^2 = \omega_j^2\rho_w/\omega_j^2\rho_w\mu_w.\mu_w - \gamma_j^2$, $\gamma_j = \omega_j/c_j$, c_j is the speed of the jth harmonics, C_{kj} are unknown constants, $J_{0,1}$, $Y_{0,1}$ are Bessel and $K_{0,1}$ are modified Bessel functions of the 1st and 2nd kind.

The difference of the obtained solution (12) and the well-known Womersley solution at different boundary conditions [23, 24] is the modified Bessel functions $K_{0,1}$ in the expressions of the fluid velocities and wall displacements which become infinite at $r = 0$ and, therefore, are absent in the Womersley solution for the hollow tube (at $R_1 \rightarrow 0$). The constants C_{kj} can be obtained by substitution of (12) into the boundary conditions (6)–(11). The resulting expressions are not present here because of their complexity.

4   Results and Discussions

The pressure and flow distributions in the pulsatile flow between the coaxial rigid (guiding catheter/wire) and compliant viscoelastic surfaces have been computed on (9) using the following physiological parameters: $\rho = 1050$ kg/m^3, $\rho_s = 1000$–1300 kg/m^3, $\mu = 3.5 \cdot 10^{-3}$ Pa·s, $\mu_s = 1$ Pa·s, $\tau_s = 0.01$–0.1 s, $R_1 = 0.18$–1.25 mm, $R_2 = 0.75$–2.5 mm, $\mathrm{Re}(\Gamma) = 0; 0.5; 0.9$, $\mathrm{Im}(\Gamma) = 1 \pm i$. The computed $p(t, r, x)$ and $\vec{v}(t,r,x)$ distributions have been averaged over the cross-sectional area between the two surfaces and then compared to the solutions of the same problem formulation (3)–(11) at $R_1 = 0$ [20]. The aim of the study was to check whether the pressure signals measured for the pulsatile blood flow between two surfaces and in some cases in quite a narrow gap between them (($R_2 - R_1$)/$R_2 \sim 0.5 - 0.75$) are consistent with the CFD computations for the flows in rigid tubes without the axial obstacles [11–13]. The input pressure waveforms $p_0(t)$ and the wave reflection coefficients Γ have been taken in the same form for both geometries.

The non-dimensional axial flow profiles $V_x(r^\circ)$ computed at the same pressure gradient $\delta P/L = $ const and different relative size of the guiding catheter/wire $R_1/R_2 = 0{,}1 \div 0{,}5$, where $r^\circ = r/R_2$ are presented in Fig. 7. The flow profiles are built at $r \in [R_1/R_2, 1]$, non-dimensioned by the maximal Poiseuille velocity, and the axial obstacle is plotted at

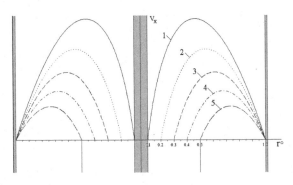

Fig. 7. Axial flow profiles $V_x(r^\circ)$ at different values $R_1/R_2 = 0{,}1; 0{,}2; 0{,}3; 0{,}4; 0{,}5$ (curves 1–5 accordingly).

$r = \pm 0.1$. The non-dimensional WSS at the inner and outer surfaces are presented in Fig. 8. In the presence of the catheter/wire the total energy dissipation due to the viscous drag is bigger than in the hollow tube (Poiseuille flow). The dissipation is bigger for the thin wires located in the centre of the blood vessel in the region of the maximal blood velocity, because thinner wires produce bigger shear rates.

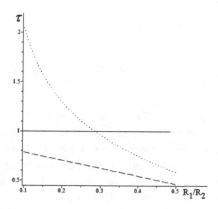

Fig. 8. WSS at the inner rigid (dotted line) and outer compliant (dashed line) walls at $R_1/R_2 = 0,1 \div 0,5$. The solid line corresponds to the Poiseuille flow.

When the constant flow rate regime $Q = $ const between the cylinders is maintained by different pressure gradients or active response of the vessel wall to the changed wall shear rate, the velocity profiles have different shapes produced by the main harmonics presented by the Bessel function $J_0(r)$ (Fig. 9).

The FFR values have been computed for different sets of the material parameters and for the individual geometries of the 45 segments of the coronary arteries examined in this study. The corresponding distributions are shown in Fig. 10. The standard model

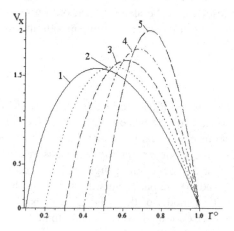

Fig. 9. Axial flow profiles $V_x(r^\circ)$ for the case $Q = $ const. The labels are the same as in Fig. 7.

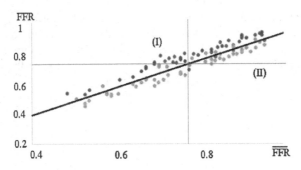

Fig. 10. Measured FFR (vertical axis) versus the FFR computed on the standard (I) and developed (II) models.

is based on the Womersley solution for the pulsatile flow in the hollow tube, while the developed model is based on the expressions (12) for the pulsatile flow between two coaxial cylinders. The coronary segment supposed to be straight and its inner radius, lengths and wall thickness estimated on the angiographic images (R_2, L and h).

In spite of possible patient-specific variations in the blood viscosity and wall rheology, the numerical computations on the developed model are closer to the FFR values measured via the CathLab, than the one computed for the flows in cylindrical geometries.

It was shown the developed model estimates stenosis severity more correctly. The vertical and horizontal lines in Fig. 10 correspond to the critical values FFR = 0.75; the cases with FFR >0.75 are strongly recommended for immediate stenting [6]. In the-most of the considered cases the measured and computed FFR values correlate for the both models, while the standard model underestimates the stenosis severity more frequently than the developed model. Since complexity of the numerical computations for the realistic vessel geometry with and without the catheter do not differ significantly, the developed model is more preferable for the virtual FFR technologies [25]. Besides, neglecting the high frequency components by smoothing of the measured signals leads to lower mean values for P_a but not P_d which results in overestimation of the stenosis severity. The obtained results must be also checked out on more complex geometries like curved/twisted tubes and in presence of long smooth and irregular stenoses.

5   Conclusions

Pressure signals registered before $P_a(t)$ and behind $P_d(t)$ the stenosis possess different oscillatory behaviour, because of the wave reflections at the site of the stenosis. The important diagnostic parameters crucial for decision making on surgery of the stenosis (stenting, bypass, grafts) are made on the signals measured in the presence of the guiding catheter and wire with the pressure gauge, while the computational approaches for estimation of the hemodynamic parameters are based on the simplified models. It was shown the mathematical model of the pulsatile flow between the rigid and compliant cylinders is more precise for the FFR estimation than the model of the flow in the hollow rigid tube without any obstacles along the axis.

Is was shown the mathematical model of the steady and pulsatile flow between the rigid and compliant surfaces predicts more accurate results for the diagnostic index $<P_d(t)>/<P_a(t)>$. It was also shown the pulsatile high frequency component gives complementary information on the stenosis severity.

References

1. Vlodaver, Z., Wilson, R.F., Garry, D.J.: Coronary Heart Disease: Clinical, Pathological, Imaging, and Molecular Profiles. Springer, US (2012)
2. Layek, G.C., Mukhopadhyay, S., Gorla, R.S.R.: Unsteady viscous flow with variable viscosity in a vascular tube with an overlapping constriction. Intern. J. Engin. Sci. 47, 649–659 (2009)
3. Caussin, C., Larchez, C., Ghostine, S., et al.: Comparison of coronary minimal lumen area quantification by sixty-four–slice computed tomography versus intravascular ultrasound for intermediate stenosis. Am. J. Cardiol. 98, 871–876 (2006)
4. Li, Y., Zhanga, J., Lub, Z., Panb, J.: Discrepant findings of computed tomography quantification of minimal lumen area of coronary artery stenosis: correlation with intravascular ultrasound. Eur. J. Radiol. 81, 3270–3275 (2012)
5. Silber, S., Albertsson, P., Aviles, F.F., et al.: Guidelines for percutaneous coronary interventions. The task force for percutaneous coronary interventions of the european society of cardiology. Eur. Heart J. 26, 804–847 (2005)
6. Pijls, N.H.: Is it time to measure fractional flow reserve in all patients? J. Am. Coll. Cardiol. 41, 1122–1124 (2003)
7. Tonino, P.A.L., Fearon, W.F., De Bruyne, B., et al.: Angiographic versus functional severity of coronary artery stenoses in the FAME study fractional flow reserve versus angiography in multivessel evaluation. J. Am. Coll. Cardiol. 55, 2816–2821 (2010)
8. Fihn, S.D., Gardin, J.M., Abrams, J., et al.: ACCF/AHA/ACP/AATS/PCNA/SCAI/STS guideline for the diagnosis and management of patients with stable ischemic heart disease. Circulation 126, 354–471 (2012)
9. Hinds, M.T., Park, Y.J., Jones, S.A., Giddens, D.P., Alevriadou, B.R.: Local hemodynamics affect monocytic cell adhesion to a three-dimensional flow model coated with E-selectin. J. Biomech. 34, 95–103 (2001)
10. Xiong, G., Choi, G., Taylor, C.: Virtual interventions for image-based blood flow computation. Comput. Aided Des. 44, 3–14 (2012)
11. Taylor, C.A., Fonte, T.A., Min, J.K.: Computational fluid dynamics applied to cardiac computed tomography for noninvasive quantification of fractional flow reserve. J. Am. Coll. Cardiol. 61, 2233–2241 (2013)
12. Qi, X., Lv, H., Zhou, F., et al.: A novel noninvasive method for measuring fractional flow reserve through three-dimensional modeling. Arch. Med. Sci. 9(3), 581–583 (2013)
13. Rajani, R., Wang, Y., Uss, A., et al.: Virtual fractional flow reserve by coronary computed tomography – hope or hype? EuroIntervention 9(2), 277–284 (2013)
14. Dodge, J.T., Brown, B.G., Bolson, E.L., Dodge, H.T.: Lumen diameter of normal human coronary arteries. Influence of age, sex, anatomic variation, and left ventricular hypertrophy or dilation. Circulation 86, 232–246 (1992)
15. Kizilova, N.: Blood flow in arteries: regular and chaotic dynamics. In: Awrejcewicz, J., Kazmierczak, M., Olejnik, P., Mrozowski, K. (eds) Dynamical Systems. Applications. pp. 69–80. Lodz Politechnical University Press (2013)

16. Barclay, K.D., Klassen, G.A., Young, Ch.: A method for detecting chaos in canine myocardial microcirculatory red cell flux. Microcirculation **7**(5), 335–346 (2000)
17. Trzeciakowski, J., Chilian, W.: Chaotic behavior of the coronary circulation. Med. Biol. Eng. Comp. **46**(5), 433–442 (2008)
18. Canic, S., Hartley, C.J., Rosenstrauch, D., Tambaca, J., Guidoboni, G., Mikelic, A.: Blood flow in compliant arteries: an effective viscoelastic reduced model, numerics, and experimental validation. Ann. Biomed. Engin. **34**(4), 575–592 (2006)
19. Nichols, W., O'Rourke, M., Vlachopoulos, Ch. (eds.): McDonald's blood flow in arteries: theoretical, experimental and clinical principles, 6th edn. Hodder Arnold, London (2011)
20. Lighthill, J.: Waves in Fluids. Cambridge University Press, Cambridge (2001)
21. Kizilova, N., Hamadiche, M., Gad-el-Hak, M.: Mathematical models of biofluid flows in compliant ducts: a review. Arch. Mech. **64**, 1–30 (2012)
22. Kizilova, N., Hamadiche, M., Gad-el-Hak, M.: Flow in compliant tubes: control and stabilization by multilayered coatings. Intern. J. Flow Control. **1**, 199–211 (2009)
23. Cox, R.H.: Wave propagation through a Newtonian fluid contained within a thick-walled, viscoelastic tube. Biophys. J. **8**, 691–709 (1968)
24. Milnor, W.R.: Hemodynamics. Williams & Wilkins, Baltimore (1989)
25. Min, J.K., Berman, D.S., Budoff, M.J., et al.: Rationale and design of the DeFACTO (Determination of Fractional Flow Reserve by Anatomic Computed Tomographic AngiOgraphy) study. J. Cardiovasc. Comput. Tomogr. **5**, 301–309 (2011)

Health Informatics

Ontological Analysis of Meaningful Use of Healthcare Information Systems (MUHIS) Requirements and Practice

Arkalgud Ramaprasad[1(✉)], Thant Syn[2], and Mohanraj Thirumalai[1]

[1] College of Business Administration, University of Illinois at Chicago,
601 S Morgan Street, Chicago, IL, USA
{prasad,mthiru1}@uic.edu
[2] School of Business Administration, University of Miami, 5250 University Drive,
Coral Gables, FL, USA
thant@miami.edu

Abstract. We present an ontology of meaningful use of healthcare information systems (MUHIS), and an analysis of its requirements and practices using the ontology. We map (a) the Stages 1 and 2 meaningful use requirements set by the Centers for Medicaid & Medicare Services (CMS) for Electronic Health Records (EHR), and (b) the current literature on meaningful use, to derive the ontological map of the two respectively. Both maps are fragmented and incomplete. We high-light the gaps (a) in the requirements, (b) in practices, and (c) between require-ments and practices, and highlight the bright, light, blank, and blind spots in MUHIS. We discuss why these gaps should be (a) bridged if they are important, (b) ignored if they are unimportant, or (c) reconsidered if they have been overlooked. Thus ontological analysis can provide systemic feedback for continuous improve-ment of MUHIS though systematic changes in policies and practice.

Keywords: Ontology · Meaningful use · Healthcare Information Systems · Electronic health records

1 Introduction

Meaningful Use of Healthcare Information Systems (MUHIS) is a work-in-progress at the national level in the USA and other countries [1–4], at the local level in many states and cities, and at the enterprise level in many hospitals, physician practices, and other healthcare providers. Its requirements and practices are evolving in tandem, and along different paths depending on the initial conditions, incentives, and the environment. It is seen as an instrument for addressing the national (USA, for example) concerns about the cost, quality, and safety of healthcare. Consequently, there is a constant pressure to continuously and rapidly improve MUHIS. To catalyze the evolution, the Centers for Medicaid & Medicare Services (CMS) in the USA has set Stages 1 and 2 meaningful use requirements for Electronic Health Records (EHR) [5]. The requirements specify the outcomes, associated objectives, and corresponding measures. There are incentives for meeting the objectives. Fulfilling the requirements will be necessary but not sufficient

© Springer International Publishing Switzerland 2015
G. Plantier et al. (Eds.): BIOSTEC 2014, CCIS 511, pp. 315–330, 2015.
DOI: 10.1007/978-3-319-26129-4_21

for harnessing the full potential of MUHIS; it has far greater potential than envisioned in the present requirements. The requirements and MUHIS have to evolve quickly in tandem to meet the rapidly increasing global demands on healthcare. It would be a challenge to make the MUHIS 'elephant' dance.

MUHIS is a large, complex, and ill-structured problem. It is a 'wicked' problem [6]. We have to manage its 'wickedness' through feedback and learning to help it evolve rapidly. To do so, we have to (a) abstract from the diverse, often contradictory, and heterogeneous requirements and practices of MUHIS, and (b) apply it to the reformulation of requirements and practices [7–9]. We need a clear framework and method for abstraction and application [9, 10] to avoid replaying the proverbial story of the five blind men each of whom imagined an elephant as a rock, an arrow, a fan, a rope, and a tree trunk after touching its body, tusk, ear, tail, and leg, respectively [11, 12]. A wise man settles their argument about the ontic nature of the elephant by piecing together the picture for them. Fortuitously, the wise man in the story could see and recognize the elephant; without him the blind men's argument would likely have continued ad infinitum. Analogously we need wise men and women who can see and recognize the 'big picture' of MUHIS. The framework should guide the abstraction, inform the application, and structure the visualization of the MUHIS. It should thus help (a) to limit the fragmentation of the requirements and practices, (b) to make the system greater than the sum of its parts, and (c) to evolve MUHIS systemically and systematically. The framework itself should be adaptable to the evolution of requirements and practices through scaling, extension, reduction, refinement, and magnification of its components. It should be extensible to the Stage 3 requirements in the US, for example.

In the following, we will present an ontology [12] for MUHIS and discuss a method of mapping it using the framework. Thus, we will present a map of the "knowledge structure" [13] of requirements and practices of MUHIS as an ontological map. The ontological map represents a "virtual knowledge landscape" [14, p. 505] based on textual empirical data about the requirements and practices. It will help visually recognize the coherence and lack of it in the cumulative domain knowledge, and therefore help correct the lacuna when appropriate [15, 16]. Thus, it will provide "support for navigating the knowledge landscape." [17, p. 1] Further, "[i]ncrementally computed information landscapes are an effective means to visualize longitudinal changes in large document repositories…" [18, p. 352] such as the requirements and practices of MUHIS. It will aid the continuous improvement of MUHIS.

First, we will describe an ontology of MUHIS. We will explain the conceptual foundations of the framework and its bases in MUHIS requirements and practice. We will also discuss the face, content [19], semantic [20], and systemic validity [21] of the framework.

Second, we will describe the method for mapping the requirements specifications and practice literature onto the ontology and explain the mapping process. We will discuss the reliability and validity of the mapping. We will demonstrate the visualization of the ontological map from the mapped data based on the map.

Third, we will describe the gaps within requirements, within practice, and between requirements and practice using the ontological map. We will discuss the importance of these gaps and their implications for future requirements specifications and practice.

Fourth, and last, we will describe how the method can be used to develop incremental maps [18] over time to generate feedback and facilitate learning in the evolution of MUHIS. We expect that continuous assessment and improvement of MUHIS using the proposed method will eventually lead to the realization of the guiding vision.

2   Ontology of MUHIS

Ontologies "… provide a shared and common understanding of a domain that can be communicated between people and heterogeneous and widely spread application systems." [22, p. 1] They "… make it possible to understand, analyze, exchange or share knowledge of a specific domain and therefore they are becoming popular in various communities. However, ontologies can be very complex and therefore visualizations can support users to understand the ontology easier. Moreover, graphical representations make ontologies with their structure more manageable. For an effective visualization, it is necessary to consider the domain for which the ontology is developed and its users with their needs and expectations." [23, p. 123]

Ontology is the study of being in contrast to epistemology which is the study of knowing. Its focus is on objects, their categories, and the relationships between them. Ontologies represent the conceptualization of a domain [24]; they organize the terminologies and taxonomies of a domain. An ontology is an "explicit specification of a conceptualization." [25, p. 908] It is used to systematize the description of a complex system [26]. "Our acceptance of an ontology is… similar in principle to our acceptance of a scientific theory, say a system of physics; we adopt, at least insofar as we are reasonable, the simplest conceptual scheme into which the disordered fragments of raw experience can be fitted and arranged." [27, p. 16]

There are potentially many ways of representing a domain ranging from a natural-language narrative to a formal mathematical formulation (when possible). The ontology is a structured natural-language representation, more formal than a narrative but less formal than a mathematical formulation. It is particularly suited for 'wicked' problems such as MUHIS. It is easy to understand and apply the ontology.

The ontology for MUHIS is shown in Fig. 1. It encapsulates the logic of MUHIS. It has been formulated manually by the authors from the meaningful use outcomes, objectives, and measures [5] and their knowledge of the structure and functions of an information systems. There is no computerized method for extracting such an ontology (a) at this level of granularity, (b) which is parsimonious (fits a letter size page with legible font), and (c) has high semantic validity [20] (each combination is a natural English sentence as explained below). During the formulation two of the authors iterated between abstraction of the framework from and its application [10] to the requirements until the model (a) was logically complete, and (b) covered all the objectives, requirements, and criteria. It is similar to the process described by Ramaprasad & Mitroff [10] and Ramaprasad [9] for the formulation of strategic problems.

The ontology has five columns representing the five dimensions of MUHIS; two of the dimensions together comprise the Health Information Systems. Each dimension is defined by a one- or two-level taxonomy. The dimensions are linked by words/phrases

		Healthcare Information Systems			
Management		**Structure**	**Function**	**Stakeholders**	**Outcome**
Analysis	[of]	Technology [for]	Acquisition	Recipients	Efficiency
Specification		Hardware	Analysis	Patients	Quality
Design		Software	Interpretation	Families	Safety
Implementation		Networks	Application	Population	Disparities
Maintenance		Processes	Distribution	Providers	
Assessment		Policies		Physicians	
		Personnel		Nurses	
				Pharmacists	
				Payers	
				Employers	
				Insurers	
				Regulators	
				Government	

(column connectors: [of information by/to] · [to meaningfully manage] · [of/in healthcare])

Four Illustrative components of meaningful use of HIS from 3360 (6x4x5x7x4) level-1 components:
1. Specification of technology for analysis of information by providers to meaningfully manage cost of healthcare.
   Examples: electronic health records software, data mining software
2. Design of processes for acquisition of information by patients to meaningfully manage quality of healthcare.
   Examples: access to online lab results, formation of social networks
3. Implementation of policies for application of information by government to meaningfully manage disparities in healthcare.
   Examples: wellness education policies, Medicaid reimbursement policies
4. Implementation (deployment) of personnel for interpretation of information by insurers to meaningfully manage safety of healthcare.
   Examples: data mining specialists

Fig. 1. Ontology for Meaningful Use of Healthcare Information Systems (MUHIS).

interleaved between the respective columns. The columns are ordered left to right such that the concatenation of a word from each column with the interleaved words/phrases results in a meaningful natural English sentence. Four such concatenated sentences are shown, with examples, at the bottom of Fig. 1. In the following we will discuss the dimensions, the taxonomies, and the concatenations in greater detail.

2.1 Dimensions of the Ontology

The rightmost column is 'Outcome' and it lists the four critical healthcare outcomes which need to be meaningfully managed using HIS. They are efficiency, quality, and safety of healthcare and disparities in healthcare – a Core and Menu Set outcome [5]. There are many other Core and Menu Set outcomes. We interpret them as means to the four outcomes in the ontology. For example, consider the Core outcome to 'Engage patients and families in their healthcare'. In the ontology patients and families are stakeholders in achieving the desired healthcare outcomes using the Health Information

System. Similarly, consider the Menu Set outcome to 'Improve population and public health' – the Population is a stakeholder receiving healthcare to achieve the desired outcomes. We have been able to relate all the present Stages 1 and 2 outcomes to the four outcomes. In the future, additional outcomes or subcategories of outcomes can be added, or some of the outcomes deleted for application to a particular context.

The second column from the right (Stakeholders) is a taxonomy of stakeholders in HIS. They are the recipients of healthcare (patients, families, and the population as a whole), the providers of healthcare (physicians, nurses, and pharmacists), payers for healthcare, employers of recipients, insurers of recipients, regulators of healthcare, and the government. The categories of stakeholders are not mutually exclusive – an entity may have multiple roles. For example, a recipient may also be a payer, and a self-insured employer may also be the insurer. The categories may not also be exhaustive – they may need to be extended or reduced. The present taxonomy is a generic, parsimonious list of stakeholders whose interests and roles in meaningfully managing the healthcare outcomes need to be considered.

The third and fourth columns (Structure, Function) from the right are the common structural and functional components of an information system. They have been adapted to the CMS terminology [5]. The structural components of HIS are the technology (hardware, software, and networks), processes, policies, and personnel. The functional components are acquisition, analysis, interpretation, application, and distribution of information.

The leftmost column (Management) lists the functions necessary to manage HIS to assure their meaningful use. These are common functions in the analysis, design, and assessment of any information system. They are analysis, specification, design, implementation, maintenance, and assessment; they have been derived from the HealthIT terminology [28].

2.2 Illustrative Components

Each concatenation of words/phrases across the framework is a potential component of MUHIS. There are 3360 ($6 \times 4 \times 5 \times 7 \times 4$) level-1 and 7920 ($6 \times 6 \times 5 \times 11 \times 4$) level-2 components. We will focus our discussion on the level-1 components and subsume within them the details of the second level. Four level-1 components are listed at the bottom of Fig. 1 with an example for each; they are discussed below.

First, consider 'Specification of technology for analysis of information by providers to meaningfully manage cost of healthcare.' This could include specification of cost-effective electronic health records software to provide the type of clinical decision support required by CMS for meaningful use. It could also include data mining software to be used by a large regional Health Maintenance Organization (HMO) to determine the most efficacious drugs for a commonly occurring chronic condition.

Second, consider 'Design of processes for acquisition of information by patients to meaningfully manage quality of healthcare.' It could include design of processes for online access of lab results (required by CMS in Stages 1 and 2), or processes to foster formation of social networks of cancer patients to acquire information from each other and form support groups.

Third, consider 'Implementation of policies for application of information by government to meaningfully manage disparities in healthcare.' It could include policies to transmit health data to government agencies, wellness education policies for those living in the 'food deserts', and reimbursement policies which help counter the imbalance due to socio-economic status.

Fourth, and last, consider 'Implementation (deployment) of personnel for interpretation of information by insurers to meaningfully manage safety of healthcare.' It could entail deployment of data mining specialists to discover early warnings about new potentially unsafe drugs.

Further, as shown in the examples, each component may be instantiated in multiple ways. The many instantiations constitute the MUHIS. We note that some components may not be instantiated at all in a given context. For example, without health insurance the fourth illustrative component above may be irrelevant. In general, the absence of instantiation may reflect either an error of omission (blind spot) or an irrelevant component (blank spot) in that context.

2.3 Validity of the Ontology

In assessing the validity of the framework we note that it is an ontology not the ontology for MUHIS; we recognize that there can be other equally valid frameworks. Each framework can be a lens to study the domain; each lens can offer different insights about the domain. Given that the MUHIS problem is complex and ill-structured, 'wicked' [6], a singular ontology is unlikely. We offer a framework and its associated insights. It derives its validity from its (a) logical construction, (b) comprehensiveness, (c) interpretability, and (d) completeness.

First, the logic of the MUHIS ontology's dimensions can be deconstructed as follows:

- Meaningful Use of Healthcare Information Systems = Meaningful Use + Healthcare Information Systems
- Meaningful Use = Management + Stakeholders + Outcome
- Healthcare Information Systems = Structure + Function

Thus, the dimensions comprehensively cover the connotation of MUHIS. They can be easily interpreted by a user.

Second, the categories of the taxonomy for each dimension are logical and generally accepted in the respective disciplines. Moreover, should a category or subcategory be missing from a taxonomy, it can be easily added. By the same token, a redundant category or subcategory can be easily removed. These corrections of potential errors of omission and commission will not invalidate the rest of the framework. Thus, the taxonomies of the dimensions are comprehensive and interpretable.

Third, the ordering of the dimensions fits the rules of English grammar – thus rendering the concatenations in natural English and making them meaningful and hence interpretable. Further, all the components (concatenations) encapsulated in the framework taken together provide a complete, closed description of MUHIS.

Fourth, and last, the parsimonious representation of the ontology provides a panoptic view of MUHIS which can be analyzed with minimal cognitive strain. A user can conveniently and meaningfully explore its dimensions, elements, and components at different levels of granularity.

Thus, we believe that the framework's face validity [19], content validity [19], systemic validity [21], and semantic validity [20] are high. It parsimoniously encapsulates the complexities of the system; it makes the MUHIS 'elephant' known and visible and hence can be used to map MUHIS systemically and systematically. It is a simple, powerful tool to synthesize and visualize the MUHIS knowledge domain, to analyze the accumulation of knowledge over time, and visualize its trajectory. It provides a holistic approach to visualize the map and guide the progress of a domain, for example, to answer the question: How can we continuously improve MUHIS? We explore these possibilities in the following.

3   Method

3.1   Mapping MUHIS Requirements

We mapped all the Stages 1 and 2 requirements onto the ontology through consensus mapping. The requirements were obtained from the CMS website [5]. Each objective was mapped individually, considering it in the context of the associated outcome and measures. The total number of objectives mapped = 51. All the objectives were first mapped by one author, reviewed and modified by the other, and the discrepancies between the two discussed and resolved in the final mapping. The mapping does not distinguish between the core and menu objectives, and those for eligible professionals, eligible hospitals, and CAHs (Critical Access Hospitals). We provide two examples of mapping in the following.

Consider the Stage 1 core objective of "Implement drug-drug and drug-allergy interaction checks." It is one of a set of objectives with the stated outcome of "Improving quality, safety, efficiency, and reducing health disparities." We mapped the objective for quality, safety, and efficiency outcomes but not for disparities; we could not see a direct link from the discussion of the objective and its measures to managing disparities. We mapped it to the ontology as: "Implementation of technology/processes for application of information by providers to meaningfully manage efficiency/quality/safety." We note that the objective corresponds to six components of the ontology, not just one.

Consider the Stage 2 core objective of "Provide patients the ability to view online, download and transmit their health information within four business days of the information being available to the EP." Although the stated outcome of the objective is "Patient Electronic Access", we inferred the ultimate outcome to be primarily quality. It could be efficiency and safety too, but we did not find sufficient evidence to justify them. We mapped the objective to the ontology as: "Implementation of technology for distribution of information by/to recipients/providers to meaningfully manage quality." Again, we note that the objective corresponds to two components of the ontology.

Mapping the meaningful use objectives was straightforward in most cases. It required little interpretation except in the mapping the outcomes of a few objectives as illustrated above. The mapping was recorded on an Excel spreadsheet using one row per objective and a column per element of the ontology. All but 2 of the 51 objectives were mapped to all the five dimensions of the framework for a total of 65 full and 10 partial components.

3.2 Mapping MUHIS Practice

We mapped all the articles indexed in PubMed that contain the term "Meaningful Use" in the title/abstract and belong to the following MeSH major topics: "Medical Informatics", "Medical Records Systems, Computerized", "Electronic Prescribing", and "Computer Communication Networks". In addition, we also included articles specifically designated to MeSH major topic "Meaningful Use". The combined result was filtered by date (2009 – March 1, 2013) as well as the availability of abstract. We obtained a total of 200 articles. Of these, 43 were announcements, editorials, etc. and 7 were non-US. They were excluded from the study. The remaining 150 articles were mapped by the authors onto the ontology based on their titles and abstracts. Each article was mapped by one author and validated by the other. Differences in mapping between the two were resolved through discussion. As with the requirements an article could be coded (a) on all or some of the dimensions, and (b) into a single or multiple components of the framework. Of the 150 articles, 63 were coded on all the dimensions and 87 on a subset, for a total of 214 components and 1964 partial components. All the data were maintained and mapped on spreadsheets (Google Docs and Excel).

4 Results

4.1 Ontological Map of MUHIS Requirements

The ontological map of MUHIS requirements is shown in Fig. 2. The elements correspond to the first level of the ontology. The number in parenthesis adjacent to each element is the frequency of its occurrence in the set of objectives. The bar below the element is proportional to the frequency using the total number of objectives (51) as the denominator. The profile is very similar for Stages 1 and 2 requirements and hence they are not shown separately. The total frequency for elements in a column may exceed the total number of CMS objectives due to one to many mapping of objectives to components as illustrated and explained earlier.

The mosaic of the MUHIS requirements as a whole is evident from the ontological map. It has many bright spots (high frequency elements), light spots (low frequency elements), and blank/blind spots (no frequency elements). The no frequency elements may be 'blank' by choice or 'blind' by oversight – it cannot be resolved based on the data. The mosaic may be summarized in a complex sentence with higher frequency elements in bold and decreasing frequency left to right as follows:

Implementation/maintenance of **technology/processes**/policies for **distribution/acquisition/** application of information by/to **providers/recipients**/government to meaningfully manage **quality/efficiency/safety**/disparities of healthcare.

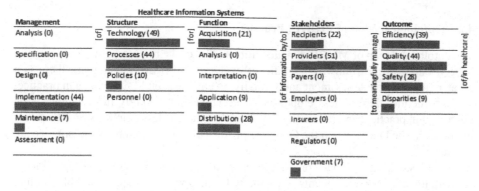

Fig. 2. Ontological map of Stages 1 and 2 meaningful use requirements.

The partial histogram of Stages 1 and 2 meaningful use requirements shown in Fig. 3 highlights the most common components of the requirements using the structured construction of the ontology – the bright spots. On the left is the synthetic requirement based on the ontology, and on the right the total frequency of its occurrence and a proportional bar. As we have noted earlier, a CMS requirement may be deconstructed into multiple synthetic requirements. The full histogram (not shown due to space constraint) portrays the bright, light, and blank/blind spots at the component level, in contrast to the element level visualization in the ontological map (Fig. 4).

Fig. 3. Partial histogram of Stages 1 and 2 meaningful use requirements.

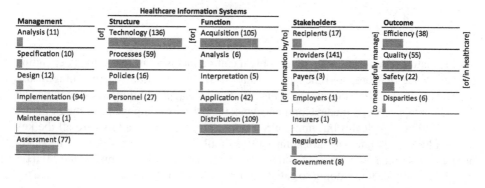

Fig. 4. Ontological map of meaningful use practice.

4.2 Ontological Map of MUHIS Practice

The ontological map of MUHIS practice mirrors that of the requirements, as one would expect, but with the following significant exceptions: (a) in Management there is a greater emphasis on Assessment and virtually no emphasis on Maintenance in practice; (b) in Structure, there is less emphasis on Processes and slightly more emphasis on Personnel in practice; and (c) among Stake holders there is less emphasis on Recipients in practice. The Function and the Outcome profiles of the requirements and practice are similar.

The mosaic of the MUHIS practice as a whole is evident from the ontological map. As with the map of requirements it has many bright spots, light spots, and blank/blind spots. The mosaic may be summarized in a complex sentence with higher frequency elements in bold and decreasing frequency left to right as follows:

> **Implementation/assessment**/design/analysis/specification/maintenance   of   **technology/ processes**/personnel/policies for **distribution/acquisition**/application/analysis/interpretation of information by/to **providers**/recipients/government/regulators/payers/insurers to meaningfully manage **quality/efficiency/safety**/disparities of healthcare.

The partial histogram of meaningful use practice shown in Fig. 5 highlights the bright spots using the synthetic components of the ontology. Its construction is similar to Fig. 3.

Assessment of Technology for Acquisition of information by/to Providers to meaningfully manage Quality of/in healthcare   20
Assessment of Technology for Distribution of information by/to Providers to meaningfully manage Quality of/in healthcare   20
Implementation of Technology for Distribution of information by/to Providers to meaningfully manage Quality of/in healthcare   20
Implementation of Technology for Acquisition of information by/to Providers to meaningfully manage Quality of/in healthcare   18
Implementation of Technology for Distribution of information by/to Providers to meaningfully manage Efficiency of/in healthcare   18
Implementation of Technology for Acquisition of information by/to Providers to meaningfully manage Efficiency of/in healthcare   17
Implementation of Technology for Acquisition of information by/to Providers to meaningfully manage Safety of/in healthcare   14
Implementation of Technology for Distribution of information by/to Providers to meaningfully manage Safety of/in healthcare   14
Assessment of Technology for Acquisition of information by/to Providers to meaningfully manage Efficiency of/in healthcare   13
Assessment of Technology for Distribution of information by/to Providers to meaningfully manage Efficiency of/in healthcare   13

Fig. 5. Partial histogram of meaningful use practice.

5 Discussions

Words matter. The formulation of a problem can be inclusive or restrictive, depending on the choice of words and their connotations. We have formulated meaningful use inclusively in the MUHIS ontology (Fig. 1). The Management dimension includes all the major steps of a system development cycle; the primary components of the Structure and Function of a Health Information System are incorporated; and so are all the key Stakeholders and Outcomes. Meaningful use should be 'meaningful' for all the stakeholders for all the key 'uses' (outcomes). The inclusive formulation makes the MUHIS 'elephant' fully visible – doing so can diminish the costs of fragmentation and drive the benefits of integration.

CMS has formulated meaningful use narrowly in its Stages 1 and 2 requirements, as shown in the ontological map in Fig. 2. The narrow formulation may be driven by its mission – their primary site for MUHIS is 'HealthIT.gov' not 'HealthIS.gov', emphasizing technology not the system. It may be motivated by the strategy for implementation – to start where there may be greatest leverage and to proceed in stages. It may also be determined by their decisions about their role. They may see motivating recipients and providers as their role but not motivating payers and employers. Similarly, they may see motivating implementation as part of their role but not analysis, specification, and design – the latter could be the EMR vendors' role. Last, the similarity of Stages 1 and 2 maps suggest a continuity of focus. We cannot adduce the reasons for the map shown in Fig. 2 and its continuity but we can assert that the Stages 1 and 2 requirements by themselves are unlikely to result in MUHIS in its panoptic connotation expressed in the ontology. The narrow formulation is likely to be suboptimal if not dysfunctional.

Consider the Stage 1 objective: 'Implement drug-drug and drug-allergy interaction checks'. These checks will directly affect the Quality and Safety [29–31] Outcomes of healthcare [32]. Their effectiveness will depend upon the providers' response to the alerts issued based on the checks. Recent Assessment shows that more than 90 % of the alerts are overridden due to alert fatigue [29, 33, 34], information overload [35], poor user interface Design [30, 36, 37], poor Specification of the critical interactions [37], and inadequate Analysis [38, 39] of the interactions. It will be necessary to include most of the blank elements in the map of Stages 1 and 2 (Fig. 2) to improve the effectiveness of the checks. First, it would be necessary to Assess [40, 41] the current system to provide feedback [33] for Analysis [38, 39], Specification, and Design of the system. Second, the Assessment could be done internally by a provider, locally, or by a conference of all the Stakeholders [34, 38, 42]. Third, any Assessment and feedback will entail extensive Analysis [38, 39] and Interpretation [43] of empirical data [44]. Thus, the success of a large number of components encapsulated in the ontology will be essential for effectively implementing the 'drug-drug and drug-allergy interaction checks'. In absence of a systematic systemic [40] perspective, the checks may be implemented but they may be meaningless, especially if they are overridden constantly [45].

In addition to the above, the fragmentation in policies and practice may be simply a consequence of their development being distributed across many policymaking groups. These groups are designed to represent different stakeholders and their interests. The diversity is intended to make the policies inclusive; but by the same token it makes it difficult to integrate them systematically and systemically. Constraints of time and other resources in the operation of these groups often result in their devolving their recommendations to the lowest common denominator of agreement. Moreover, lacking a synthesizing tool such as the ontology, it is easier to settle on the lowest common denominator than to discuss the logic of inclusion and exclusion of each component. Metaphorically, in the absence of a good map it is easier and safer to stick to the familiar pathways.

Similarly, part of the fragmentation in practice is likely a reflection of the policy which guides it. In addition, its fragmentation is likely amplified by its spatial and temporal distribution across many organizations, operating at different levels of technological sophistication, and implementing the system incrementally. If the fragmentation can be made visible, as we have using the ontology, the visibility can provide feedback to promote greater integration.

6 Conclusion

The ontological maps and histograms provide clear visualizations of the gaps within each and between them. Some of these gaps definitely need to be bridged, as in the case of decision support for drug-drug and drug-allergy interactions. The policy makers and practitioners have to assess the importance of the other gaps and change requirements and practices to bridge them. This process of feedback and change has to be ongoing for continuous improvement of MUHIS. Ontological maps such as the ones presented in this paper can provide the foundations for visualizing the domain, monitoring the incremental changes, and making it complete and integrated.

In summary, we present an ontological meta-analysis and synthesis of MUHIS requirements and practice [46]. It highlights the domain's bright spots which are heavily emphasized, the light spots which are lightly emphasized, the blank spots which are not emphasized, and the blind spots which have been overlooked. It also highlights the biases and asymmetries in MUHIS requirements and practice; they can be realigned to make them stronger and more effective.

As we have emphasized earlier our ontology is one lens through which one can study MUHIS. There can be other equally valid frameworks. Each lens will likely yield a different map and thus different insights into the bright, light, and blank/blind spots. Each of these sets of insights will be a product of observing the phenomenon systematically through a systemic framework, of a different way of making the 'elephant' visible. Reconciling these differences, in addition to changing the map of each will advance knowledge of MUHIS and can set the research/practice agenda for the domain.

The ontology is extensible and reducible, and hence the method is adaptable to the developments in MUHIS. Should a new Function or Stakeholder of MUHIS emerge in the future, they can be added to the framework. Or, should a new subcategory of Providers becomes a key Stakeholder, the framework can be extended to accommodate the change. By the same token, if a category becomes irrelevant, it could be eliminated from consideration. The extensibility and reducibility will also help trace the evolution of the constructs in and the logic of MUHIS.

Last, but not the least, visualization is key to making sense of and interpreting 'big text data' like the emerging requirements and practice of MUHIS. The ontology provides an easy and intuitively understandable vehicle for visualization. Note, for example, the ontological maps can be used to study the evolution of MUHIS over time by creating maps for different cross-sections of time. It can also be used to study the map at different levels of granularity using more refined/coarsened taxonomies. These are works in progress. Feedback based on incremental ontological maps will help to continuously improve MUHIS. With the current ontological map of MUHIS requirements and practice it is unlikely that the full vision of meaningful use will be realized – they have to evolve a lot.

The evolution has to balance the emphasis on the categories, dimensions, and components of the ontology. It has to balance the bright, light, blank, and blind spots.

Following are three examples:

- The emphasis on the Stakeholders has to be balanced. All the stakeholders, individually and in interaction with each other, collectively affect the outcome.
- With the increasing role of 'big data' and data mining in healthcare the low emphasis on Analysis and Interpretation will likely have to be increased significantly
- Meaningful Use is itself a dynamic concept which will evolve with time. MUHIS too has to be equally dynamic. To do so the emphasis on Analysis, Specification, Design, and Maintenance (in Management) will have to be increased considerably.

The ontological map can guide the evolution now and in the future. Since the analysis for this paper, the Stage 3 requirements in the US have been announced. They can be easily mapped to the ontology. More important, the concept of MUHIS itself is likely to be transformed with new disruptive technologies (for example, mobile health or mHealth technologies) which in turn will necessitate a fundamental rethinking of the requirements and practice. Ontological analysis and synthesis can help guide this transformation systematically and systemically. If such transformations are not managed effectively, there is as danger of the disruptive technologies becoming destructive, and for the healthcare system to be unable to harness its value quickly.

A major strength of the method is the synoptic view of the domain it provides based on the population of requirements and research articles. By the same token, its strength is also critically dependent upon the inclusiveness of the search. Another major strength is its use of natural English to model the problem; again the semantic variability of the natural language could also be a weakness by introducing errors in coding. Last, while the parsimony of the ontology is a significant strength, significant errors of omission for the sake of parsimony will weaken the method. The omission of a key category can make the ontology, and consequently the study, 'blind' to the associated requirements and practice.

The relatively small number of requirements and research articles allowed the authors to code them manually. As the domain grows it will be necessary to automate or at least semi-automate the coding. Qualitative analysis tools such as NVivo can be used to partially automate the coding. It will be incorporated in the future.

Last, the method is new and in its early stages of development. In the future we propose to systematically address some of the weaknesses mentioned above. We also propose to benchmark the method with other methods for synthesizing the knowledge in a domain.

References

1. Dermer, M., Morgan, M.: Certification of primary care electronic medical records: lessons learned from Canada. J. Healthc. Inf. Manage.: JHIM **24**, 49–55 (2010)
2. Ke, W.C., Hsieh, Y.C., Chen, Y.C., Lin, E.T., Chiu, H.W.: Trend analysis and future development of Taiwan electronic medical records. Stud. Health Technol. Inf. **180**, 1230–1232 (2012)

3. Kim, H., Kim, S.: Legislation direction for implementation of health information exchange in Korea. Asia-Pac. J. Public Health/Asia-Pacif. Acad. Consortium Public Health **24**, 880–886 (2012)
4. Varroud-Vial, M.: Improving diabetes management with electronic medical records. Diab. Metab. **37**(suppl. 4), S48–S52 (2011)
5. https://www.cms.gov/Regulations-and-Guidance/Legislation/EHRIncentivePrograms/Meaningful_Use.html
6. Churchman, C.W.: Wicked problems. Manage. Sci. **14**, B-141 (1967)
7. Ramaprasad, A.: Role of feedback in organizational-change - review and redefinition. Cybernetica **22**, 105–113 (1979)
8. Ramaprasad, A.: On the Definition of Feedback. Behav. Sci. **28**, 4–13 (1983)
9. Ramaprasad, A.: Cognitive process as a basis for MIS and DSS design. Manage. Sci. **33**, 139–148 (1987)
10. Ramaprasad, A., Mitroff, I.I.: On formulating strategic problems. Acad. Manage. Rev. **9**, 597–605 (1984)
11. Börner, K., Chen, C., Boyack, K.W.: Visualizing knowledge domains. Ann. Rev. Inf. Sci. Technol. **37**, 179–255 (2003)
12. Ramaprasad, A., Valenta, A.L., Brooks, I.: Clinical and translational science informatics: translating information to transform health care. In: Azevedo, L., Londral, A.R. (eds.) Proceedings of HEALTHINF 2009 – Second International Conference on Health Informatics, pp. 135–141. INSTICC Press, Porto (2009)
13. Zhang, J., Xie, J., Hou, W., Tu, X., Xu, J., Song, F., Wang, Z., Lu, Z.: Mapping the knowledge structure of research on patient adherence: knowledge domain visualization based co-word analysis and social network analysis. PLoS ONE **7**, e34497 (2012)
14. Scharnhorst, A.: Constructing knowledge landscapes within the framework of geometrically oriented evolutionary theories. In: Mathies, M., Malchow, H., Kriz, J. (eds.) Inegrative Systems Approaches to Natural Social Dynamics, pp. 505–515. Springer, Heidelberg (2001). http://www.virtualknowledgestudio.nl/staff/andrea-scharnhorst/documents/constructing-knowledge-landscapes.pdf
15. Hoeffner, L., Smiraglia, R.: Visualizing domain coherence: social informatics as a case study. Adv. Classif. Res. Online **23**, 49–51 (2013)
16. Noar, S.M., Zimmerman, R.S.: Health Behavior Theory and cumulative knowledge regarding health behaviors: are we moving in the right direction? Health Educ. Res. **20**, 275–290 (2005)
17. Kazimierczak, K.A., Skea, Z.C., Dixon-Woods, M., Entwistle, V.A., Feldman-Stewart, D., N'Dow, J.M.O., MacLennan, S.J.: Provision of cancer information as a "support for navigating the knowledge landscape": findings from a critical interpretive literature synthesis. Eur. J. Oncol. Nurs., 1–10 (2012)
18. Syed, K.A.A., Kröll, M., Sabol, V., Scharl, A., Gindl, S., Granitzer, M., Weichselbraun, A.: Dynamic topography information landscapes – an incremental approach to visual knowledge discovery. In: Cuzzocrea, A., Dayal, U. (eds.) DaWaK 2012. LNCS, vol. 7448, pp. 352–363. Springer, Heidelberg (2012)
19. Brennan, L., Voros, J., Brady, E.: Paradigms at play and implications for validity in social marketing research. J. Soc. Mark. **1**, 3 (2011)
20. Kotis, K., Vouros, G.: Human-centered ontology engineering: the HCOME methodology. Knowl. Inf. Syst. **10**, 109–131 (2006)
21. Horn, B.R., Lee, I.H.: Toward integrated interdisciplinary information and communication sciences: a general systems perspective. In: Proceedings of the Hawaii International Conference on System Sciences, vol. 244, pp. 244–255. IEEE (1989)

22. Fensel, D.: Ontologies: A Silver Bullet for Knowledge Management and Electronic Commerce. Springer, Heidelberg (2003)
23. Kriglstein, S., Wallner, G.: Human centered design in practice: a case study with the ontology visualization tool knoocks. In: Kraus, M., Mestetskiy, L., Richard, P., Braz, J., Csurka, G. (eds.) VISIGRAPP 2011. CCIS, vol. 274, pp. 123–141. Springer, Heidelberg (2013)
24. Gruber, T.R.: Ontology. In: Liu, L., Ozsu, M.T. (eds.) Encyclopedia of Database Systems. Springer, Heidelberg (2008)
25. Gruber, T.R.: Toward principles for the design of ontologies used for knowledge sharing. Int. J. Hum. Comput. Stud. **43**, 907–928 (1995)
26. Cimino, J.J.: In defense of the Desiderata. J. Biomed. Inf. **39**, 299–306 (2006)
27. Quine, W.V.O.: From a Logical Point of View. Harvard University Press, Boston (1961)
28. http://www.healthit.gov/providers-professionals
29. Crosson, J.C., Schueth, A.J., Isaacson, N., Bell, D.S.: Early adopters of electronic prescribing struggle to make meaningful use of formulary checks and medication history documentation. J. Am. Board Fam. Med. **25**, 24–32 (2012)
30. Rahmner, P.B., Eiermann, B., Korkmaz, S., Gustafsson, L.L., Gruvén, M., Maxwell, S., Eichle, H.-G., Vég, A.: Physicians' reported needs of drug information at point of care in Sweden. Br. J. Clin. Pharmacol. **73**, 115–125 (2012)
31. Spina, J.R., Glassman, P.A., Simon, B., Lanto, A., Lee, M., Cunningham, F., Good, C.B.: Potential safety gaps in order entry and automated drug alerts: a nationwide survey of VA physician self-reported practices with computerized order entry. Med. Care **49**, 904–910 (2011)
32. Classen, D.C., Phansalkar, S., Bates, D.W.: Critical drug-drug interactions for use in electronic health records systems with computerized physician order entry: review of leading approaches. J. Patient Saf. **7**, 61–65 (2011)
33. Smithburger, P.L., Buckley, M.S., Bejian, S., Burenheide, K., Kane-Gill, S.L.: A critical evaluation of clinical decision support for the detection of drug-drug interactions. Expert Opin. Drug Saf. **10**, 871–882 (2011)
34. Phansalkar, S., van der Sijs, H., Tucker, A.D., Desai, A.A., Bell, D.S., Teich, J.M., Middleton, B., Bates, D.W.: Drug–drug interactions that should be non-interruptive in order to reduce alert fatigue in electronic health records. J. Am. Med. Inf. Assoc. (2012)
35. Callen, J.L., Westbrook, J.I., Georgiou, A., Li, J.: Failure to follow-up test results for ambulatory patients: a systematic review. J. Gen. Intern. Med. **27**, 1334–1348 (2011)
36. Seidling, H.M., Phansalkar, S., Seger, D.L., Paterno, M.D., Shaykevich, S., Haefeli, W.E., Bates, D.W.: Factors influencing alert acceptance: a novel approach for predicting the success of clinical decision support. J. Am. Med. Inf. Assoc. **18**, 479–484 (2011)
37. Gaikwad, R., Sketris, I., Shepherd, M., Duffy, J.: Evaluation of accuracy of drug interaction alerts triggered by two electronic medical record systems in primary healthcare. Health Inf. J. **13**, 163–177 (2007)
38. Phansalkar, S., Desai, A.A., Bell, D., Yoshida, E., Doole, J., Czochanski, M., Middleton, B., Bates, D.W.: High-priority drug–drug interactions for use in electronic health records. J. Am. Med. Inf. Assoc. **19**, 735–743 (2012)
39. Takarabe, M., Shigemizu, D., Kotera, M., Goto, S., Kanehisa, M.: Network-based analysis and characterization of adverse drug-drug interactions. J. Chem. Inf. Model. **51**, 2977–2985 (2011)
40. Saverno, K.R., Hines, L.E., Warholak, T.L., Grizzle, A.J., Babits, L., Clark, C., Taylor, A.M., Malone, D.C.: Ability of pharmacy clinical decision-support software to alert users about clinically important drug–drug interactions. J. Am. Med. Inf. Assoc. **18**, 32–37 (2011)

41. Warholak, T.L., Hines, L.E., Saverno, K.R., Grizzle, A.J., Malone, D.C.: Assessment tool for pharmacy drug–drug interaction software. J. Am. Pharmacists Assoc. **51**, 418–424 (2011)
42. Hines, L.E., Malone, D.C., Murphy, J.E.: Recommendations for generating, evaluating, and implementing drug-drug interaction evidence. Pharmacother. J. Hum. Pharmacol. Drug Ther. **32**, 304–313 (2012)
43. Dhabali, A.A.H., Awang, R., Zyoud, S.H.: Clinically important drug–drug interactions in primary care. J. Clin. Pharm. Ther. **37**, 426–430 (2012)
44. Haueis, P., Greil, W., Huber, M., Grohmann, R., Kullak-Ublick, G.A., Russmann, S.: Evaluation of drug interactions in a large sample of psychiatric inpatients: a data interface for mass analysis with clinical decision support software. Clin. Pharmacol. Ther. **90**, 588–596 (2011)
45. Yu, D.T., Seger, D.L., Lasser, K.E., Karson, A.S., Fiskio, J.M., Seger, A.C., Bates, D.W.: Impact of implementing alerts about medication black-box warnings in electronic health records. Pharmacoepidemiol. Drug Saf. **20**, 192–202 (2011)
46. Ramaprasad, A., Syn, T.: Ontological meta-analysis and synthesis. In: Proceedings of the Nineteenth Americas Conference on Information Systems, Chicago, Illinois, 15–17 August 2013

Using Digital Pens for Maternal Labor Monitoring: Evaluating the PartoPen in Kenya

Heather Underwood[1]([⊠]), John Ong'ech[2], Maya Appley[1], Sara Rosenblum[1], Addie Crawley[1], S. Revi Sterling[1], and John K. Bennett[1]

[1] ATLAS Institute, University of Colorado Boulder, Boulder, CO, USA
{heather.underwood,maya.appley,sara.rosenblum,addie.crawley,
revi.sterling,jkb}@colorado.edu
[2] Kenyatta National Hospital, Nairobi, Kenya
J.Ong'ech@knh.or.ke

Abstract. The goal of the PartoPen system is to enhance the partograph, a paper-based labor monitoring tool intended to promote timely delivery of quality care by birth attendants in developing countries. The PartoPen digital system provides audio instructions for measuring and recording labor progress indicators, real-time decision support based on recorded measurements, and time-based patient-specific reminders. Previous studies of the PartoPen system showed improved partograph completion rates among students in nursing classrooms at the University of Nairobi (UoN) in Kenya. This paper presents the results of two continuation studies conducted in the maternity ward of Kenyatta National Hospital (KNH) in Nairobi. In this paper we identify and discuss the interrelated factors impacting PartoPen adoption and use in the labor ward at KNH, and review the challenges and opportunities likely to face digital pen deployments in other healthcare settings.

Keywords: Digital pens · Maternal health · ICTD · Kenya · Partograph · Clinical Decision-Support System (CDSS)

1 Introduction

The World Health Organization (WHO) estimates that 300,000 women die every year due to pregnancy-related complications, most of which occur in developing countries [1]. Timely and informed labor monitoring by a skilled attendant can help prevent many of the main causes of maternal death – hemorrhage, infection, unsafe abortion, eclampsia, and obstructed labor [2]. Globally, the WHO promotes the paper partograph as an effective and cost-efficient tool for monitoring labor, and preventing obstructed labor and resulting complications. Used correctly, the partograph provides decision support that assists in early detection of maternal and fetal complications during labor. Especially in rural clinics, early detection allows transport decisions to be made in time for a woman to reach a regional facility capable of performing emergency obstetric procedures.

Despite the positive reports of improved maternal outcomes resulting from correct partograph use [3, 4, 5], several recent studies in Kenya have reported underuse and

© Springer International Publishing Switzerland 2015
G. Plantier et al. (Eds.): BIOSTEC 2014, CCIS 511, pp. 331–345, 2015.
DOI: 10.1007/978-3-319-26129-4_22

incorrect use of the partograph at all levels of maternity care [6, 7, 8]. The well-documented barriers to partograph use include partograph shortage, staff shortage, low partograph knowledge and training, and the perspective that the partograph is time consuming and redundant [6]. The goal of the PartoPen project is to mitigate some of the barriers preventing correct and widespread partograph adoption using an interactive digital pen, dedicated pen software, and partograph forms printed with a background dot pattern that is recognized by the pen. Using only the digital pen and the existing paper form, the PartoPen addresses training and resource barriers by providing audio-based decision support, patient-specific reminders, and partograph use instructions. Prior PartoPen work at the University of Nairobi [9] suggests that the PartoPen is effective in multiple healthcare settings: initial training, training reinforcement, and use with actual patients. These results motivated two of the studies described in this paper. These studies focus on populations at two ends of the healthcare spectrum: nursing students with little training or clinical experience using the partograph, and nurse midwives at KNH, who are well-trained and generally have many years of experience using the partograph in the labor ward.

Our previous work examined the effect of PartoPen use on partograph completion in nursing classrooms with third and fourth year nursing students. The results of follow-on maternity ward studies are presented here. In addition, we discuss various environmental factors that led to different outcomes in the two studies.

We first summarize the results from the PartoPen nursing student study, and give an overview of the technological components of the PartoPen system. The remainder of the paper discusses the results of maternity ward studies conducted at KNH in July and August 2012, with follow-up in July 2013.

2   Background

2.1   PartoPen Software System

Over the past fifty years, a large body of work on pen-and-paper computing and pen-and-paper user interfaces (PPUIs) has been developed. More recently, digital pens have been used and evaluated for usability and efficiency in the healthcare context both by patients [10, 11], and by nurses [12, 13]. The key findings in these studies is that digital pen technology is an intuitive and usable technology with great potential, but healthcare-specific digital pen applications must be designed to meet the actual needs of the user. The Livescribe (LS) digital pen technology [14] used by the PartoPen system captures pen input and digitizes paper content by using a unique location tracking and page identification technique patented by the Anoto AB group [15].

The current implementation of the PartoPen system uses the LS Echo digital pen running the custom PartoPen software that allows the pen to meaningfully interact with the paper partograph. The Echo pen has a built in microphone, speaker, and OLED display. The pen relies on a rechargeable lithium ion battery, which is advertised to last about 36 h during normal use. A battery life of 20 to 26 h has been observed, depending upon the amount of audio played during use. Pens can store between 200 and 800 h of audio, or the equivalent amount of text data, depending on the pen model, and all stored data can be downloaded to a desktop computer using a standard micro-USB cable.

One of the goals of the PartoPen system is to enforce birth attendant training on correct use of the partograph, as this has been cited as a significant barrier to consistent use of the form. The WHO partograph user manual, and a local partograph manual issued to clinics by the Kenyan Ministry of Health, are the primary resources for partograph instruction in Kenya. The PartoPen system makes the instructions found in these manuals accessible directly from the partograph itself. The PartoPen uses fixed print "button" regions around the partograph text to provide verbatim audio recordings of the instructions found in the partograph use manuals. Thus, by tapping on these "buttons," nurses and nursing students can get short informational prompts on how to use each section of the form correctly.

One of the most commonly cited barriers to partograph use is the inability to interpret the data plotted on the partograph and take appropriate action. Nursing students and less-experienced nurses often plot the data correctly on the partograph, but fail to derive the meaning of the plotted data, or do not remember what actions to take based on the data that they have plotted. The decision support functionality of the PartoPen addresses these issues by interpreting plotted data based upon page location, and providing real-time feedback on the appropriate actions to take. Currently, the PartoPen provides decision support in three of the partograph sections: cervical dilation, liquor/amniotic fluid, and fetal heart rate.

The labor ward at KNH delivers approximately 1000 babies during the "busy" months from October to March, or roughly 34 babies every day. On average there are 4–6 nurses working at a time, and based on survey data collected at the end of the PartoPen study, nurses on average are responsible for 5–7 patients during a day shift and 7–10 patients during a night shift. The WHO recommends a maximum ratio of one nurse to three patients to ensure compliance with partograph completion protocols. In the survey, nurses nearly unanimously reported that staff shortage is the most common reason for low partograph completion rates. While the PartoPen does not replace nurses or supplement the shortage of nurses in the labor ward, it does provide a reminder system intended to help busy and tired nurses keep track of when patients need measurements taken.

2.2   PartoPen Nursing Student Study

Ninety-five nursing students in their third and fourth years of study at the UoN School of Nursing Sciences participated in the study. Students were asked to complete a partograph worksheet, which consisted of two patient case studies and two blank partograph forms printed with the dot pattern. The students recorded the patient data on the blank partograph forms as if they were actively monitoring that patient during labor. In each worksheet, students received two of three possible patient case studies. The three case studies represent three possible labor outcomes. Mrs. A's data represents an uncomplicated, timely labor that progresses without medical intervention. Mrs. B's data illustrates a case of prolonged or obstructed labor. Mrs. C's labor progression data illustrates an increasing number of complications, including fetal distress, and ultimately results in a cesarean section.

The students were first divided into three groups. Group 1 was the control group, and Groups 2 and 3 were the intervention groups. Group 1 students completed a partograph worksheet task with a PartoPen in "silent logging mode," and received no instructions on how to use the technology. Group 2 completed the same worksheet task, but used a fully functional PartoPen in "use" mode. The PartoPen software in "use" mode for the student pilot has two main components: instructions and decision support. Group 2 received no training on how to use the technology. Group 3 received a fully functional PartoPen in "use" mode and a 15-min introduction and demonstration of the PartoPen system before completing the partograph worksheet task.

Using an unpaired t-test, the difference between Group 1 (M = .520, SD = .141) and Group 3 (M = .722, SD = .089) for the patient case study Mrs. C, a prolonged labor resulting in a CS, was found to be significant; $t(8) = 2.709$, $p = 0.0267$. These data suggest that for more challenging or complex labor cases, the availability and utilization of the PartoPen instruction prompts promotes more accurate form completion.

After each group completed the worksheet task, students were asked to participate in a short focus group session. Students unanimously reported that plotting contractions was one of the most difficult sections of the partograph, because both duration and frequency are plotted together using a combination of bar charts and coloring patterns. Students also reported unanimously that plotting descent of the fetal head was particularly challenging. Difficulties plotting descent of the fetal head can also be attributed to having to plot on the same graph as another measurement (cervical dilation), but may also be due in part to the nursing school transitioning to a different partograph version that requires users to plot the descent in increments of one instead of two, and on the left side of the graph instead of the right.

The completion results of the 'contractions' section of the partograph show improvements in all three case studies (Mrs. A, B, and C) between groups that did and did not use the PartoPen. There was a statistically significant improvement in contraction plotting on the Mrs. C case study between Group 1 (M = .513, SD, .232) and Group 3 (M = .803, SD = .139); $t(8) = 2.399$, $p = 0.0433$.

'Descent of fetal head' measurements also showed a significant improvement on the Mrs. C case study between Group 1 (M = .337, SD = .152) and Group 2 (M = .585, SD = .162); $t(10) = 2.699$, $p = 0.0223$. In addition, there was a very significant improvement on descent plotting on the Mrs. C case study between Group 1 and Group 3 (M = .705, SD = .137); $t(8) = 4.028$, $p = 0.0038$.

The UoN PartoPen study indicated that PartoPen use in classrooms can improve students' ability to correctly complete a partograph form. The study results also suggest that significant PartoPen training is not required to achieve these benefits. A significant increase in partograph completion and accuracy was observed with minimal prior training, due to the intuitive design, push-based functionality, and the enhancement – rather than replacement – of the current paper-based system.

3   Maternity Ward Study

The follow-on PartoPen studies in the maternity ward at KNH examined the impact of the digital pen software system on partograph completion by nurse-midwives monitoring patients during actual labor. The partograph used in this study is pictured in Appendix A. The first study was conducted in July-August of 2012 at Kenyatta National Hospital. The study was designed as a pre- and post study, which compared partograph completion rates for partographs completed in June (without the PartoPen system) and in August when the PartoPen system was in use by nurses.

3.1 Methodology

Currently KNH evaluates partograph completion using a rubric with four options – "complete," "incomplete," "correct," or "incorrect" – boxes for each partograph category (fetal heart rate, moulding, cervical dilation, etc.) Due to the wide range of variation in how partographs are used and completed, this basic evaluation rubric does not correctly capture the actual completeness of the partograph, or the real usefulness the data recorded on the partograph. Therefore, we created a new rubric that would better assess these measures. This rubric is built upon the basic tenants of the evaluation tool used by KNH. The rubric has grading criteria for each partograph category, including a separate set of grading criteria for the labor summary printed at the bottom of each partograph. For each partograph category there are three grading criteria: (1) measurements recorded, (2) symbols correct, and (3) spacing correct. The total possible value for each of these grading criteria is determined by the time between patient admission and delivery.

Previous efforts to evaluate partograph completion required researchers to continuously observe nurses during labor monitoring to assess partograph completion [16], or researchers were required to use a coarse-grained categorization scheme similar to the original KNH rubric [17].

The new rubric also has some limitations. Chief among these is the fact that the time of admission is used to determine how many measurements are expected on the partograph form. The time of admission, however, does not always accurately reflect when the woman went into active labor or when the partograph was started.

All of the partographs collected during the study were first categorized by delivery mechanism – spontaneous vaginal delivery (SVD) and cesarean section (CS). The CS deliveries were further categorized into emergency CS (EmCS) and "other", which includes voluntary CS and CS due to previous CS scars. Deliveries of twins, triplets, or deliveries lasting less than one hour were noted among the SVD partographs, but not included in the data analysis because partographs (a) are not designed to monitor multiple births, and (b) do not provide beneficial monitoring for labors that are less than one hour in length.

3.2 Quantitative Results

Initially all of the collected partographs from June (369) were compared to all of the collected partographs from August (457). This blanket analysis of partograph completion rates between June and August attempted to capture any broad improvements that may have occurred due to researcher presence, or a general increase in interest and attention to the partograph because of the PartoPen study taking place. These results are captured in a previous paper [18], and briefly summarized below.

Phase 1 Data Analysis Summary. In the initial data analysis process, the collected partographs were graded and checked by two pairs of research assistants according to the new evaluation rubric. Each partograph received two scores: a composite completion score and a summary score. The composite score was calculated by dividing the number of points received by the total number of points possible for all three grading criteria (mark existence, correct mark symbol, and correct mark spacing) for each partograph section (fetal heart rate, cervical dilation, etc.). The summary score reflects the completion percentage for the partograph summary section at the bottom of the form, which summarizes the labor (and is usually completed after a patient delivers).

Based on this initial analysis, there were no statistically significant improvements in the composite scores between June and August partographs. However, there were improvements in the summary scores for both SVDs and CSs. This result can be attributed to several possible factors. First, the partographs used in June were slanted and blurred due to frequent photocopying, whereas the partographs used in August were printed individually (to assure the unique dot pattern on each form). This made the August partographs significantly easier to read and, presumably easier to complete. Second, the improvement in summary scores was likely a result of the increased awareness and underscored importance of the partograph that occurred during the PartoPen study.

The lack of improvement in completion rates for the overall composite partograph scores in the presence of the increased focus on the partograph is likely due to the impact of understaffing. Understaffing thwarts completing the graphical portion of the partograph because the ratio of nurses to patients (often between 1:5 and 1:10) does not allow for regular half-hour measurements to be taken for each patient. The PartoPen system cannot replace trained staff members, and does not directly address the understaffing barrier to partograph completion.

The data from the broad comparison of June and August partographs suggest that the PartoPen does not have an overall impact on partograph completion, at least not in facilities like KNH, which have highly trained and experienced, but critically overworked, staff.

Phase 2 Data Analysis. After the initial broad data analysis, a more fine-grained analysis was performed on the PartoPen data to ascertain if and how the PartoPen functionality impacted partograph completion rates. Partographs completed in June were compared to August partographs that were *actually completed* with the PartoPen. The PartoPen was used to complete 48 of these partograph forms. PartoPens were only given to nurses at KNH during the study, which excluded the nursing students who were actively working in the labor ward as part of their clinical rotation. Student-completed partographs in August, which were not completed with a PartoPen, were excluded from Phase 2 analysis. In addition, many partographs were only partially completed with the PartoPen, due to nurse rotations and patient handoffs. These partially completed partographs were also excluded from Phase 2 analysis.

The comparison of all the partographs completed with PartoPens versus the August partographs not completed with the PartoPens versus all of the June partographs is represented in the Fig. 1. This histogram illustrates that August partographs completed with the PartoPens never received a completion score lower that 25 %, whereas both June and August partographs completed without the PartoPen did. Additionally, the August partographs completed with the PartoPen had the highest percentage of partographs in the 75–100 % completion range.

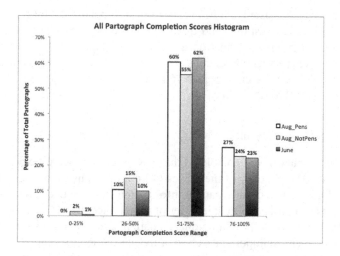

Fig. 1. A histogram of partograph completion scores for August partographs completed with a PartoPen, August partographs not completed with a PartoPen, and June partographs.

4   2013 Follow-up Study

At the conclusion of the PartoPen maternity ward study, the nurses and hospital staff had successfully incorporated the PartoPen system into the daily operations of the labor ward. The system in place at KNH at the end of the study included 20 PartoPens, a printer capable of printing the dot pattern partographs, PartoPen chargers, and extra pen caps and ink replacements. Nine months after the completion of the 2012 PartoPen maternity ward study, a follow-up study was conducted with KNH nurses to assess the performance and impact of the PartoPen system.

Upon returning to the KNH labor ward in May 2013, researchers found that all 20 of the PartoPens were accounted for, 19 out of the 20 PartoPen were functional, and over 600 digital partograph records were present on the PartoPens ranging from September 2012 to April of 2013 (the printer used to print dot-patterned partographs failed in April 2013, and replacement parts were not readily available in Nairobi). During the May 2013 visit, PartoPen researchers identified a local printer model alternative and worked with hospital IT staff to establish a recurring printer toner order, so as to remove this responsibility from the already busy nurses and record-office staff in the labor ward.

4.1 Methodology

During the May 2013 visit, PartoPen researchers surveyed KNH labor ward nurses regarding their impressions of PartoPen deployment and use. Twenty-six nurses completed a paper survey about the PartoPen and its affect on labor ward operations and patient care.

The paper survey consisted of nine questions (see Appendix B); four YES/NO questions, two Likert scale questions, and three free response questions. The survey also included basic demographic information, and a rank-order question where nurses ranked the importance of the partograph sections. The surveys were completed during morning patient handoff. Nurses took an average of 15 min to complete the survey, and the nurses were not compensated for their time.

4.2 Follow-up Survey Results

Thirteen of the 26 nurses who completed the survey felt they were 'experts' using the PartoPen system. The majority of the nurses (19 out of 26) used some combination of partograph information and other patient information to make decisions about patient care. The nurses were asked to rank in order of importance the different sections of the partograph as they relate to providing quality patient care. Nine nurses ranked patient name and age as the most important section of the partograph to complete. Eight nurses ranked fetal heart rate as the most important section of the partograph, and seven nurses ranked the partograph sections sequentially (i.e., the most important section is the topmost section of the form, and the least important is the bottommost portion of the form). One nurse ranked contraction frequency as the most important, and one nurse ranked cervical dilation as the most important section of the partograph to complete. The responses from the survey suggest that certain information on the partograph is more useful for making critical decisions about patient care, which may indicate that a simplified and restructured form that highlights these sections (and makes them easier to complete) could be useful in this setting. Nurses largely prioritized patient information and fetal heart rate as the most important portions of the form. In the PartoPen study, some of the qualitative feedback received by nurses indicated that using larger boxes for information entry for these sections considerably improved the usability and readability of these critical pieces of information.

The survey also asked nurses to identify if there are certain kinds of labor or patients who do not need a partograph. Twenty of the 26 nurses said that there were patients who do not need a partograph during labor. Elective cesarean sections, false labors, and patients who arrive already in the second stage of labor were the most common responses for labors that do not require a partograph to monitor labor progress. Elective cesarean sections are scheduled in advance and are categorized separately from emergency C-sections that happen as a result of complications during labor. Additionally, Kenyatta National Hospital, as the leading referral hospital, receives a very high volume of patients who are in the second stage of labor. Although KNH administrative policies require that a partograph be used during all labors without exception, staff shortages make prioritization necessary when deciding to begin or continue a partograph for a patient. Since KNH is a referral hospital, many patients arrive late in labor in poor condition, and completing paperwork or a partograph is not the highest priority of hospital staff. The result is blank or retroactively completed partographs.

Nurses were also asked to identify patients and labor types that benefit the most from being monitored with a partograph. Nurses were allowed to circle more than one labor type out of SVD, CS, IUFD, Referral, and 'Other'. Twenty-three out of 26 nurses said that spontaneous vaginal deliveries (SVD), which are often categorized as 'normal' labors, benefit the most from correct partograph use. Eleven out of 26 nurses circled CS, emergency cesarean sections, as benefiting the most from partograph use, and 7 out of 26 nurses circled 'Referral'.

The survey asked several PartoPen-specific questions, including whether the nurses had observed any changes in the labor ward because of the PartoPen. This question was included in the survey to follow up on qualitative observations and discussions at the end of the 2012 studies that suggested labor ward nurses were feeling an increased sense of pride in their job because of the interest of senior hospital staff, and reliance on labor ward nurses to explain the project and demonstrate its functionality. Additionally, only labor ward nurses were given PartoPens, and this sense of privilege was mentioned several times by nurses as rewarding. Twenty-four of the 26 nurses said 'yes', there had been changes in the labor ward because of the PartoPen. The majority of the changes nurses described related to the reminder functionality of the PartoPen. Nurses frequently noted the reminders being effective for providing more timely care and making patient care more efficient. Better decisions and easier chart interpretation were also noted as significant changes resulting from PartoPen use in the labor ward.

4.3 Secondary Data Analysis

Based on the data from the 2013 surveys, the data from the 2012 maternity ward study was re-examined, as follows: First, only the SVD partographs were included, as the majority of nurses indicated that SVD patients benefit most from partograph use. In addition, partograph sections that nurses deemed most important (i.e., patient information and fetal heart rate) were examined individually.

The SVD partographs were analyzed in three categories: August SVDs completed with the PartoPens, August SVDs completed without the PartoPens, and all of the SVDs from June. Using the same grading and evaluation rubric, these partographs were analyzed with respect to completion. The results of this analysis are shown in Fig. 2. Frequency in this histogram is represented as a percentage of the total number of partographs present in the sample (37 August partographs completed with the PartoPens, 206 August partographs completed without the PartoPens, and 153 partographs completed in June). The histogram illustrates that August partographs completed with the PartoPens never received below 25 % completion, and this set had the highest percentage of partographs in the 75–100 % range.

The same set of SVD partographs was then analyzed, looking specifically at the completion of the 'patient information' and 'fetal heart rate' sections. While fetal heart rate completion did not change significantly between the three groups, a significant difference was observed in patient information completion between August PartoPen

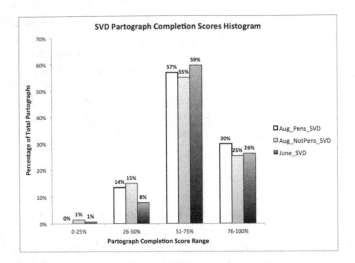

Fig. 2. A histogram of SVD-only partograph completion scores (for August partographs completed with a PartoPen for spontaneous vaginal deliveries only, August partographs completed without a PartoPen for SVDs only, and June partographs for SVDs only).

SVDs (M = .949, SD = .086) and June SVDs (M = .882, SD = .152) using a paired t-test (t(188) = 2.6178, p = .0096). This difference may be attributable to several factors, including the improved readability and larger space for the patient information fields on the PartoPen version of the partograph form.

5   Discussion

The results from the maternity ward studies do not generally exhibit significant differences in partograph completion rates between partographs completed with the PartoPens and those not completed with the PartoPens. In retrospect, this result is not surprising. The PartoPen system was designed to address training barriers that have been cited as significant obstacles to correct partograph use. However, the PartoPen system was deployed at Kenyatta National Hospital, one of the leading training and teaching facilities in Kenya. KNH has a highly trained and knowledgeable staff who are less likely to benefit from the training re-enforcement aspects of PartoPen use. The other cited barriers to partograph use, including staff shortages and lack of supplies, are not directly addressed by the PartoPen system, thus at KNH, any training reinforcement benefit the PartoPen provided was overshadowed by other barriers.

The positive results in the nursing student study demonstrate that the PartoPen is beneficial for partograph training for less-trained staff or for students learning how to use the partograph. In a controlled environment like a classroom where the primary focus is on the task of completing a form rather than delivering a baby, the PartoPen's training reinforcement and decision-support functionality are fully utilized. In the chaotic and understaffed environment of the labor ward at KNH, the primary focus is on patients, not on paperwork, thus the design objectives of the PartoPen system did not align well with the primary focus of the KNH nurses.

The next iteration of the PartoPen project will be deployment at more rural and local levels of maternity care, where nurse training, rather than staff numbers and supplies, is the more problematic issue. The primary contributions of the PartoPen study in the maternity ward at KNH include nurses' reflections on PartoPen usability, nurses' perceptions of useful versus complete partographs, and initial data on the durability and infrastructure requirements of the PartoPen system, which can be used in future deployments of the platform in other labor wards.

From interviews with the nurses and researcher observations, the reminders issued by the PartoPen had the most impact on nurse behavior, although this impact did not translate into increased partograph completion, for the reasons described below. The partograph used in the study was supplemented with PartoPen Reminder ID boxes at the very bottom of the form. Nurses were instructed to use these boxes to record a memorable patient code, such as a patient's initials or the room number where the patient was located. This patient code would be displayed on the OLED display on the PartoPen when the reminder for that patient sounded. The goal of the reminder system was to ensure timely patient checkups by nurses who are busy, distracted, or simply have forgotten to check on one of their many patients. However, when the ratio of nurses to patients is between 1:7 and 1:10, even if a nurse has correctly recorded a patient code and receives the patient's reminders, she may be assisting with another labor, checking on another patient, etc. Many of the nurses reported receiving the reminders but being unable to act on them because they were already involved with a different patient. Additionally, the design of the system was not as helpful to nurses who had their hands busy, as the patient code was displayed textually on the screen, and nurses were often unable to stop what they were doing to look at the pen and read the patient reminder ID.

6   Conclusions

The initial objective of the PartoPen maternity ward studies was to examine the impact of digital pen technology on *partograph completion*. This objective assumed that a primary barrier to partograph completion was a lack of training and knowledge on how to complete and interpret the form. However, the highly skilled staff at KNH did not lack in training or knowledge, but rather, suffered from staff and resource shortages, which the PartoPen was not designed to address. Despite the disparity between the study goals and observed study site realities, several important observations were made that may contribute to future work in this area.

First, every clinic or hospital has a unique set of problems, personnel and procedures, which have to be identified and addressed during both study design and implementation. The PartoPen maternity ward study design did not adequately account for the myriad confounding factors present at KNH, including under-staffing issues, different birth rates between months compared, and the presence of (different groups of) nursing students in the labor ward during the intervention month, but not the control month. Unlike the PartoPen nursing student study design, the maternity ward study was not designed such that *only* the affect of the PartoPens on partograph completion could be measured. In one analysis, study results were evaluated assuming an experimental study where nurses were given the intervention (the PartoPen) and the nursing students present in the labor ward were the controls. This was not the ideal study design, as the experimental and control groups were not well matched in terms of training, background, or experience. A more appropriate study design for this environment would be a paired comparison of individual nurses' performance on partographs for similar labor types with and without the PartoPen during comparably busy shifts.

The study design that was used – a combination of qualitative and quantitative data collection – illustrates a disparity between the data from nurse surveys and research observation and the data from the partograph completion evaluation. When surveyed, all of the nurses reported that they considered partograph information to be important, and that they relied upon this information. Interviews with nurses also revealed that nurses considered the partograph is an essential tool in the labor ward. However, the low partograph completion scores, regardless of the PartoPen intervention, suggest that the partograph was often under-utilized, filed out retroactively, or filled out incompletely. This result is not indicative of a lack of diligence or aptitude, rather a lack of adequate staffing. Thus, partograph completion rates should not be routinely equated with quality of care, particularly at a short-staffed referral facility. It would therefore be premature to promote the partograph universally without conducting large-scale studies on the direct association between partograph use and maternal and child outcomes, which account for environmental and social circumstances unique to the study site.

Second, health informatics interventions, especially in developing countries, are often consumed by the technological aspects of the project. We sometimes fail to recognize the benefit of addressing immediate and simple issues, which do not necessarily require technological intervention. The qualitative feedback received by nurses indicated that the cleaner PartoPen form with larger boxes for information entry considerably improved the usability and readability of the form. The cleaner form was simple to produce within the existing workflow and with existing equipment, and could have been done independently of the PartoPen project.

Finally, the PartoPens deployed at KNH were successfully used and sustained for over nine-months of continuous hospital use. This illustrates the robustness of the system, as well as a willingness among nurses to use the PartoPens on a daily basis. The PartoPen maternity ward study helped identify the environmental and physical challenges present in the KNH labor ward, and illustrated both the challenges and opportunities that arise when deploying a digital pen software system in a maternity ward setting. The results of this study are encouraging for the continued and expanded use of digital pen systems in healthcare, and stress the need for more in-depth and well-designed studies in this area.

Acknowledgements. We would like to express our sincere appreciation to the staff and nurse-midwives at Kenyatta National Hospital, without whose support and participation this work would not have been possible. This research was funded in part by the ATLAS Institute at the University of Colorado Boulder, a National Science Foundation Graduate Research Fellowship, and a Bill and Melinda Gates Foundation Grand Challenges Exploration Grant (OPP1061309).

Appendix A

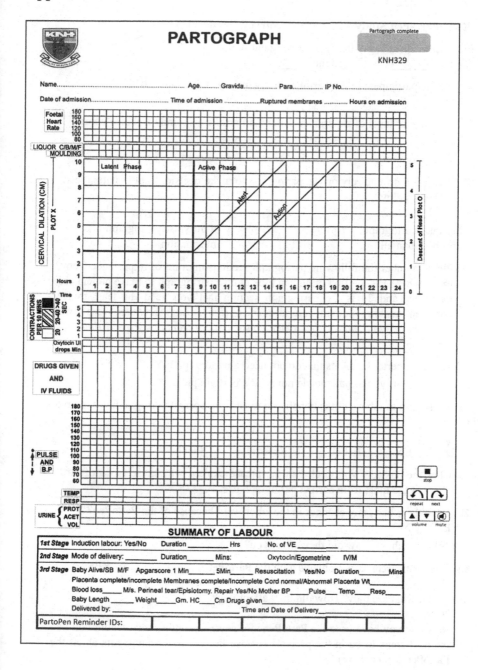

Appendix B

KNH PartoPen Study – Participant Survey – June 2013

Please provide answers for the following questions:

Age: _____   Gender: _____

1) Did you participate in the PartoPen study in July and August 2012? YES
NO

2) What is your level of experience using the PartoPens? (circle a choice below)

      1   2   3   4   5

No experience   Expert

3) How much do you rely on the partograph to make decisions about patient care? (circle a choice below)

      1   2   3   4   5

I don't use partograph   I **only** use partograph
information at all   information

4) Please number the following partograph sections in order of importance from 1 to 24, where 1 is the first thing you look at on a partograph to make patient care decisions, and 24 is the partograph section that you need the least to feel confident making patient care decisions. *(Partograph sections were listed below in original survey, but space constraints prevents us from listing them here.)*

5a) Are there certain patients that do not need a partograph? YES   NO

5b) If yes, what type of patients do not need a partograph?

6a) What kind of labors/births benefit the most from correct partograph use? (circle all that apply)

SVD   CS   IUFD   Referral   Other?

6b) For the answers you circled in 6a, please explain why these types of births benefit the most from correct partograph use.

7a) Have there been any changes in the labor ward because of the PartoPen?
YES   NO

7b) What are they? How did they affect you?

8a) Have there been any problems with the PartoPen?   YES
NO

8b) If yes, what are they? How did they affect you?

9) Now that you have used the PartoPens for several months, what would you like the next steps to be in the project? (Please list any other comments about the PartoPen project here).

References

1. World Health Organization, Maternal mortality (2013). http://www.who.int/mediacentre/factsheets/fs348/en/ (accessed 26 September 2013)
2. United Nations. The Millennium Development Goals Report, pp. 1–80 (2010)
3. Kwast, B.E., Lennox, C.E., Farley, T.M.M.: World Health Organization partograph in management of labor. Lancet **343**, 1399–1404 (1994)
4. Mathai, M.: The partograph for the prevention of obstructed labor. Clin. Obstet. Gynecol. **52**(2), 256–269 (2009)
5. Lavender, T., Hart, A., Smyth, R.M.D.: Effect of partogram use on outcomes for women in spontaneous labor at term. The Cochrane database of systematic reviews (2013)
6. Opiah, M.M., et al.: Knowledge and utilization of the partograph among midwives in the Niger Delta Region of Nigeria. Afr. J. Reprod. Health **16**(1), 125–132 (2012)
7. Qureshi, Z.P., Sekadde-Kigondu, C., Mutiso, S.M.: Rapid assessment of partograph utilisation in selected maternity units in Kenya. East Afr. Med. J. **87**(6), 235–241 (2010)
8. Lavender, T., et al.: Students' experiences of using the partograph in Kenyan labor wards. Afr. J. Midwifery Womens Health **5**(3), 117–122 (2011)
9. Underwood, H., Sterling, S.R. Bennett, J.: The PartoPen in training and clinical use: two preliminary studies in Kenya. In: proceedings 6th International Conference on Health Informatics/HealthINF 2013 (2013b)
10. Lind, L., Karlsson, D., Fridlund, B.: Digital pens and pain diaries in palliative home health care: professional caregivers' experiences. Med. Inform. Internet Medicine **32**(4), 287–296 (2007)
11. Lind, L., Karlsson, D., Fridlund, B.: Patients' use of digital pens for pain assessment in advanced palliative home healthcare. Int. J. Med. Informatics **77**(2), 129–136 (2008)
12. Procuniar, M. Murphy, S.: Intuitive information technology: enhancing clinician efficiency. In: AMIA Annual Symposium proceedings/AMIA Symposium pp. 1247–1248 (2008)
13. Estellat, C., et al.: Data capture by digital pen in clinical trials: a qualitative and quantitative study. Contemp. Clin. Trials **29**(3), 314–323 (2008)
14. Livescribe Inc., Livescribe Echo Smartpen Specs(2013). http://livescribe.com (accessed 26 September 2013)
15. Anoto, Anoto Products (2013). http://anoto.com (accessed 26 September 2013)
16. Rotich, E., et al.: Evaluating partograph use at two main referral hospitals in Kenya. Afr. J. Midwifery Womens Health **5**(1), 21–24 (2011)
17. Khonje, M.: Use and documentation of Partograph in urban hospitals in Lilongwe - Malawi: Health workers perspective (2012)
18. Underwood, H., Sterling, S.R., Bennett, J.: The PartoPen in practice: evaluating the impact of digital pen technology on maternal health in Kenya. In: ICTD 2013 proceedings (2013a)

An Integrated Virtual Group Training System for COPD Patients at Home

Jonathan B.J. Dikken[1]([✉]), Bert-Jan F. van Beijnum[1], Dennis H.W. Hofs[1],
Mike P.L. Botman[1], Miriam M. Vollenbroek-Hutten[1,2], and Hermie J. Hermens[1,2]

[1] Institute for Biomedical Technology and Technical Medicine (MIRA), University of Twente,
P.O. Box 217 7500AE Enschede, The Netherlands
{j.b.j.dikken,b.j.f.vanbeijnum}@utwente.nl,
{d.hofs,m.vollenbroek,h.hermens}@rrd.nl
[2] Roessingh Research and Development, Roessinghsbleekweg 33b, 7522AH Enschede,
The Netherlands

Abstract. Chronic Obstructive Pulmonary Disease (COPD) patients often experience a downward spiral of fear for breathlessness, inactivity and social isolation which leads to a bad physical condition. Motivation to keep patients compliant to their training scheme is a key factor in home-based exercise training. This paper presents the Integrated Training System for COPD patients; a home based virtual group exercise system to facilitate improvement of the exercise capacity safely at home using a virtual group environment. The four components of the system are the Home Trainer, the Virtual Exercise Environment, the Web Portal and the Controller. These components are implemented in a prototype, in which as much as possible existing components are used. An in-training evaluation was performed to evaluate the subsystems used during a training exercise. All subsystems are working correctly during the evaluation. In this paper the focus for the Integrated Training System is on COPD patients, but the system might be used for other groups such as Chronic Heart Failure patients or elderly people in general.

Keywords: Physical condition · Virtual group training · Chronico obstructive pulmonary disease · Home-based exercise training · Integrated training system · Design · Evaluation · Exergaming

1 Introduction

Chronic Obstructive Pulmonary Disease, a common disease characterized by persistent airflow limitation, is one of the leading diseases in many countries which will grow to the 4th largest cause of death in 2030 [1, 2]. In 2007 in the Netherlands 323.600 people, about 2 % of the population, were diagnosed with COPD. These figures are comparable to the surrounding countries [3].

The downward spiral of breathlessness fear, inactivity and social isolation leads to a bad physical condition [2]. To overcome this downward spiral patients can be enrolled in a pulmonary rehabilitation program (PRP) which improve the exercise

© Springer International Publishing Switzerland 2015
G. Plantier et al. (Eds.): BIOSTEC 2014, CCIS 511, pp. 346–359, 2015.
DOI: 10.1007/978-3-319-26129-4_23

capacity [4–6]. However, most benefits deteriorate after the rehabilitation program is finished [5, 7–10].

Maintenance strategies can retain the effects of a pulmonary rehabilitation program. Du Moulin et al. shows that home-based exercise training is effective as maintenance of the exercise capacity [11]. Also Beauchamp et al. showed a significant improvement of the exercise capacity with a community based maintenance exercise program [12]. Motivation to keep patients compliant to their training scheme is a key factor in home-based exercise training.

This paper presents a home-based virtual group exercise system to facilitate improvement and maintenance of the physical condition of COPD patients. In this paper we focus on the technical design of the system and the medical case it should cover. The system should cover all important aspects of home-based exercise training: means to do the training, motivational support and professional guidance. Therefore we call it the *integrated* training system (ITS). The goal of the system is to facilitate improvement and maintenance of the physical condition of COPD patients safely at home using a (virtual) group environment. This will reduce disabilities in activities of daily living [13, 14].

In *Backgrounds* relevant training aspects, motivation aspects, adherence aspects and existing exergames will be given. After the *Design Considerations* are explained, the *Architecture* of the ITS will be drawn. The *Implementation* will be tested in the *Evaluation*. With *Discussion and Conclusion* this paper will be finalized.

2 Backgrounds

2.1 Training

Different opinions exist about the use of either power training or endurance training as the most suitable method to improve the physical capacity [15]. Studies have conflicting outcomes on which training intensity and method gain the best results, however all studies suggest an improvement of the physical capacity by physical training [2, 3, 9, 16, 17]. The type of exercise should correspond as much as possible with the activities of daily living. Cycling, walking and walking stairs are the most suitable exercise forms.

To be effective, a training session should be intense enough. The optimal heart rate is between 60 and 80 % of the maximal heart rate [2, 18]. With 60–80 % of the maximal heart rate a patient will train in the aerobic zone. Staying in the aerobic zone for the whole training is not mandatory, but the total amount of time in the aerobic zone determines the efficiency of the training.

2.2 Adherence and Motivation

The above results regarding physical training for COPD patients have been used by the KNGF, the Dutch physical therapist association, to develop the therapeutic guideline for COPD patients [19].

The lack of therapy adherence of COPD patients is a known problem with physical exercises [5]. Therapy adherence can be increased by enjoyment and social interaction [20]. Burke et al conducted a meta-analysis of 44 studies to qualify the effect of the

348   J.B.J. Dikken et al.

setting of the training [21]. Four categories where defined: home-based training without involvement of third parties, home based training with consultation (e.g. by phone), center-based training and center based training with additional attention for group dynamics. A superior result was found in groups with a high social cohesion among the participants in comparison to normal center-based training and home-based training with consultation. The latter two had a superior result in regard to individual training without involvement of third parties. One can conclude that both good group dynamics as professional consultation result in a better therapy outcome.

Social motivation theories can help in increasing therapy adherence. One social motivation theory is social support, which is associated with how networking helps people cope with stressful events and enhance psychological well-being and can be categorized in appraisal, companionship, emotional, instrumental and informational support [22, 23]. Another theory is the social comparison theory which includes competition, cooperation and normative comparison between members of a groups [18]. These theories will be used in the implementation of the Integrated Training System.

2.3 Existing Exergaming

Several professional and consumer exergames are used for improving the physical condition of patients. Professional products include the Cybex Trazer, LightSpace, and Sportwall. Consumer products include the Sony PlayStation (with Dance Dance Revolution), Nintendo Wii and Xavix (with J-Mat). The energy expenditure with these 6 systems are comparable with the energy expenditure of walking [24]. The Cybex Trazer, LightSpace, Sony PlayStation with DDR and Xavix are based on moving to specific positions. These four systems require non-continuous dynamic movements. Such movements are unsuitable for COPD patients because injuries can occur. Measuring and controlling the intensity of the non-continuous dynamic movements is difficult. The Sony PlayStation with DDR supports multiplayer sessions at distinct locations. The other systems don't support virtual groups where users play at distinct locations and can see each other. Social interaction is limited when players are at distinct locations.

With Sportwall a player should hit specific positions on a wall with the hand or a ball. This system requires non-continuous dynamic movements as well. Because it is a professional product it is unsuitable to put at patients' homes. This system lacks the possibility to train in virtual groups and thus the possibility for social interaction.

Wii Sports include five simulation games which can be controlled by arm movements: baseball, boxing, bowling, golf and tennis. A precise motor system is important to use the described systems. Also this system has the disadvantages of non-continuous dynamic movements.

A system more tailored to the need of elderly people is the Espresso Bike in combination with the NetAthlon riding software. This system was used in a randomized clinical trial in older adult cognition [25]. It has a simple user interface and doesn't require very fast responses from the users. The training intensity is accurately controllable by changing the resistance of the bike. The intensity can be measured accurately. The system focus on cognition aspects and the improvement of the physical condition is a side issue.

For COPD patients a system is needed that is safe, has a low risk for injuries, can measure and control the intensity accurately and support social interaction.

None of these systems are built for – and suitable for accessible physical exercises at home for elderly people. None of the systems use a virtual group environment to support social interaction for the enhancement of therapy adherence.

3   Design Considerations

- Cycling on a home trainer was chosen, because a home trainer is well known for COPD patient from rehabilitation programs and can measure and set the training intensity. Using the home trainer at home increases the accessibility of the system. The ability to control the power by changing the resistance contributes to the safety of the system.
- A virtual environment will be used to be able to have a group setting at home. In the virtual environment patient can train and interact with each other.
- The oxygen saturation level should be above 90% during training sessions to prevent desaturations.

Requirements were elicitated from a literature review, observations of PRPs and interviews with physiotherapists and movement therapists using a scenario and the People-Activity-Context-Technology (PACT) framework [26]. A detailed description of the design process is given by Dikken et al. [27, 28].

4   Architecture

The ITS (Integrated Training System) is divided into four components: the *home trainer* (HT), the *virtual exercise environment* (VEE), the *controller* (C) and the *web portal* (WP). The home trainer is an ergo bike with some additional sensors, the virtual exercise environment is a computer game in which a patient can cycle together with other patients and the controller ensure a safe and efficient training by collecting and analyzing data from the home trainer and giving feedback. The controller sends the physiological and exercise data to the web portal. On the web portal this data is shown. The overview of the architecture is shown in Fig. 1.

Users of the ITS are COPD patients and their supervisors, the physiotherapist. Patients use the whole system, while the physiotherapists only use the web portal component.

4.1   Home Trainer

With the home trainer patient can perform physical activities. On the home trainer the training intensity can be controlled and patient data is acquired. Patients will experience force feedback from the home trainer.

The home trainer component contains all sensors and actuators of the system. Quantities that need to be measured are power, cadence, heart rate and oxygen saturation level. This component consists of a home trainer, a heart rate belt and a pulse oximeter.

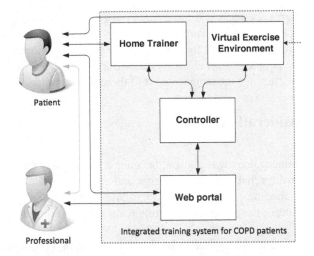

Fig. 1. High level architecture of the integrated training system.

Fig. 2. Virtual exercise environment.

4.2   Virtual Exercise Environment

The virtual exercise environment (VEE), showed in Fig. 2, provides motivation during a physical training. The VEE is essentially a computer game in which multiple players can cycle in the same virtual environment. The avatar of a patient cycles with a speed that reflects the performance of the patient in the virtual environment, but the users are kept close to each other to up keep the group spirit during the whole training.

Several motivation theories are implemented, which is explained in the next section. Each session is a game in which the patient who cycles with the best performance will win the game (social comparison – competition). The performance is calculated by how close a patient cycles to his individual goal. Players are motivated when they cycle together in the virtual environment and can see each other (social support – companionship). When a patient fails to cycle with a similar performance of the other group members, he will slow down a get in the back. To prevent a player to drop out because he is too far behind, that player will get a boost to keep up with the other group members.

The boost will stop when the player gets close to the other group members to prevent disturbance of the competition.

During training all interaction with the user is provided by the VEE, except for the force feedback by the home trainer. The interaction of the VEE includes a user interface with an overview of the important measured values, such as power and heart rate and an overview with the current performance.

4.3   Web Portal

The web portal (WP) provides after-exercise motivation: patient can retrieve their progress, set and monitor personal and group goals and give feedback on training results of group mates. Also the portal is used to give the physiotherapist insight in the training progress and configure the system.

The Web Portal supports functionalities like patient enrollment, creating exercise schemes and templates, monitoring training results and messaging. Training scheme templates can be specified by the therapist and may be further personalized by the therapist or the patient into training schemes. A training scheme is a training prescription applicable to a specific known patient and that is valid for a certain period (over time a training scheme may change). A training scheme consists of a sequence of training scheme items, each such item specifies a certain phase in the training session, example phases are warming-up, intense, cooling down. Each phase has a specified duration and cycling intensity goal.

Given a group of patients, the structure of a training session will be quite similar for each patient, however there may be individual differences in terms of durations or intensities. From the therapists point of view: exercises have a certain structure common for patients with a similar chronic disease and the same phase of the rehabilitation process. These similarities are captured by the concept of training session template. Once such a template has been specified (by the therapist), concrete patient training schemes are created through a mechanism of instantiation, and such an instance may be tuned for the individual patient. When an actual training session is started, the appropriate training scheme is selected and used to configure the controller. System and patient data is gathered during the session and transferred to the portal and stored.

After a training session both the patient and the professional can see a summary of training on the web portal. This includes a good indication of the performance with a mark and important events such as over performance, underperformance and possible desaturation events during the training. A screenshot is shown in Fig. 3. Patients can review each other's performance (Fig. 4; social comparison – competition and cooperation) and give feedback through messages to motivate each other (social support – companionship and emotional support). The professional will use the training summary to give advices.

A group of patients has the goal to reach each individual goal (social comparison – cooperation). The patients can compare themselves to a patient who is performing well (social comparison – normative comparison). Informational support (social support) is given by sharing patient stories with how they cope with the disease.

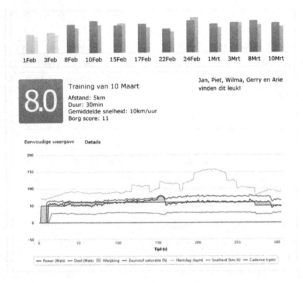

Fig. 3. Web portal - training result page.

Fig. 4. Web portal - patient profile page.

A detailed description of the web portal including design rationale and the design process is given by Botman [29].

4.4 Controller

The controller monitors and guides the safety and performance during a training session and facilitates data exchange between all components. For the safety and performance three control modules are used: the *performance loop*, the *safety loop* and the *positioning system*. The controller has several interface modules to connect to the other components. A design overview is shown in Fig. 5.

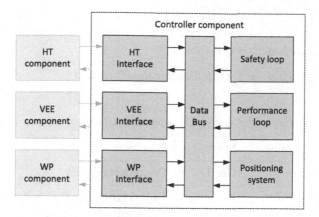

Fig. 5. Modules of the controller component. HT: home trainer, VEE: virtual exercise environment, WP: web portal.

Important measured data, such as power, heart rate, oxygen saturation level and performance, is send to the web portal by the controller.

All modules, both interface modules as control modules, use a data bus to exchange real-time data. With a data bus the controller is highly flexible. Modules are not aware of each other's existence and can be added, removed, updated and replaced easily. Different execution frequencies are possible, enabling usage of sensors with different sample.

Performance Loop. To facilitate improvement of the physical condition a training session needs to be effective. This is handled by the performance module. The performance is calculated to be able to give visual feedback on the performance and to adapt the resistance of the home trainer.

Based on the current power (P_c) and the target power (P_{target}), which is set in the web portal, the performance is calculated as shown in Eq. (1). The result is a dimensionless value between 0 and 1, where 1 indicates the best possible performance and 0 the worst possible performance.

$$performance = \begin{cases} \dfrac{P_c}{P_{target}} & if \quad P_{target} > P_c \\ \dfrac{P_{target}}{P_c} & if \quad P_{target} \leq P_c \end{cases} \tag{1}$$

The speed for the VEE is calculated based on the performance and the configured maximal speed constant (v_{max}) as shown in Eq. (2).

$$v_{VEE} = v_{max} * performance \tag{2}$$

Safety Loop. The safety is handled by the safety module. The safety module has three different states based on the measured oxygen saturation level (SpO_2). The SpO_2 is compared with a desaturation threshold as shown in Eq. (3). The desaturation threshold is set in the web portal. The status is stored for the configured interval, for example 60 s. Based on the status values the saturation state is determined and the corresponding feedback is executed, as shown in Table 1.

$$Status = SpO_2 > DesaturationThreshold \qquad (3)$$

Table 1. Saturation state in safety module based on saturation status shown in Eq. (3).

Status values	State	Feedback
All true	Good	Continue normally
Some true	Warning	Patient is urged to slow down and intensity is decreased
All false	Bad	Training session is terminated

The target power is decreased in case of the warning state. As a result the performance loop will lower the training intensity by lowering the resistance of the home trainer.

Positioning System. Part of the motivational support is handled by the positioning module. As mentioned above patient are kept together in the virtual exercise environment. When a patient gets too far behind the player in front he will get a boost to prevent the patient from getting farther behind. For this Eq. (2) is extended with the speed correction factor (SCF). The new formula is shown in Eq. (4). The SCF is a value between 1 and 20. The farther a patient gets behind, the larger the SCF value.

$$v_{VEE} = v_{max} * performance * SCF \qquad (4)$$

For the Speed Correction Factor different algorithms can be used with specific advantages and disadvantages. A linear algorithm will influence the speed more than an exponential algorithm for small distances. However, the exponential function will result in very high speeds when a patient is getting further behind.

Another considered algorithm determined the distance in the virtual exercise environment between the player in front and the current player. This results in a non-zero speed, even when the player doesn't cycle at all. Such an algorithm prevents a player from getting behind at all costs.

5   Implementation

The four components described architecture is implemented in a prototype:

HT: Bremshey BE5i home trainer with a polar T31 heart rate belt.
VEE: WebAthletics cycling game running on an Asus ME301T Android tablet with a 22" LG 22EA53VQ monitor placed in front of the home trainer.
C: Developed in Java and is running on the same tablet as the VEE.
WP: Developed on top of the Liferay Portal, running on a dedicated server.

The Bremshey BE5i has a 32-step servo motor to control the resistance, the gear. To support patients with cycling at the right performance a controller module is built to set the gear based on the performance. When $P_c/P_{target} < 0.6$ for 5 s the gear is shifted up. When $P_c/P_{target} > .25$ for 5 s the gear is shifted down.

The web portal receives the training data during a training session by a web service. This data is stored in the using the EDF format. Beside the training data, the data model contains training schedules, training schedule templates, personal identification information, personal messages and group messages.

6   Evaluation

The evaluation was performed with 4 healthy subjects, cycling in separated rooms in the same multiplayer session. Each subject was instructed to cycle as close as possible to the given target power. During the training session all relevant parameters were recorded (time, power, cadence, heart rate, speed in VEE, gear, distance, relative distance, target power, performance and SCF). The relative distance for each player is the distance between the player and player in front. Performance, relative distance and cadence are shown in Fig. 6. After the training session subjects were asked to fill in a short questionnaire.

Results from the evaluation are shown in Table 2. The distance between all players is small (min: 0.0 m, mean: 13.1 m, max: 51.0 m), while the performance for subject 4 was suboptimal. The performance was calculated from the recorded power and target power. This is compared with the recorded performance. The average deviation is less than 0.1 %.

Table 2. Evaluation results.

Subject	S1	S2	S3	S4	All
Relative distance (m)					
Min	0	0	0	0	0
Mean	4.3	14.2	5.4	28.3	13.1
Max	43	42	44	51	51
Performance					
Mean	0.95	0.91	0.95	0.80	0.90
Performance deviation					
Mean	0.003	0.000	0.002	-0.001	0.001
Standard deviation	0.058	0.064	0.050	0.068	0.060

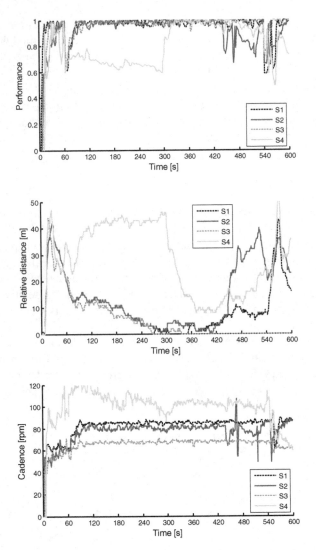

Fig. 6. Performance, relative distance and cadence of the four subjects during the exercise.

6.1 Control Modules

With a suboptimal performance of one of the subjects the distance control module managed to keep all subjects close to each other in the virtual environment. The distance control module worked correctly during the training. Also the performance control module worked correctly. A small deviation exists between real-time calculated performance and the calculation afterwards. This might be caused by a small delay in the power value used in the real-time calculation.

6.2 Other Issues

The subjects reported some issues about the automatic gear control, feedback on the performance and feedback on the position of other users. The automatic gear control caused the suboptimal performance of one of the subjects. The subject was cycling with a low power and high cadence, but the gear didn't shift up. Manual gear control can be added to prevent this situation and give the users more control. Feedback about the performance is currently given by displaying the current power and the target power in the virtual exercise environment. Users have to compare these numbers themselves to get an indication about their performance, while the performance is an important parameter during the training session. Suggested is to use a graphical performance indicator.

Finally remarks were given about the position of other users in the virtual environment. A user can see other players in front of him with a limited range. Players who are too far in front or behind the user are not visible. A rear view or a third person view, a map of the environment with indicators of the other players, will overcome this problem.

7 Discussion and Conclusion

The goal of the Integrated Training System is to facilitate improvement and maintenance of the physical condition of COPD patients safely at home using a (virtual) group environment. The current prototype satisfies to the goal of the system, but leaves room for improvement. Further evaluation is recommended.

The algorithms used in controller modules need to be validated. For each algorithm several aspects of the algorithm can be varied. For example with the performance module the performance increases linear with an increasing power, when the power is lower than the target power. When the power is higher than the target power the performance decreases hyperbolic. This could be replaced by a linear function as well. Further evaluation will determine which alternative is the best indicator for the performance. The gear algorithm can be improved when the cadence is taken into account. With a high cadence the threshold to shift up can be decreased, while with a low cadence the threshold to shift down can be decreased.

In this paper the technical design of the Integrated Training System is described. Further research into the economic and legal aspects is needed. A sound business case should be created. Motivation theories are implemented in the system. The next step is to evaluate the system in a clinical trial with the objective to evaluate the system functionality with respect to the motivational strength and to investigate the effectiveness of the system with respect to improvements of adherence to the therapy.

The focus is on COPD patients in this study. However the Integrated Training System can be used for other groups with no or only limited adaptations to the system. It can be used for Chronic Heart Failure patients or even elderly people in common to facilitate improvement of the physical condition to improve the quality of life.

References

1. Mathers, C.D., Loncar, D.: Projections of global mortality and burden of disease from 2002 to 2030. PLoS Med. **3**(11), 2011–2030 (2006)
2. Global initiative for chronic obstructive lung disease, global strategy for the diagnosis, management, and prevention of chronic obstructive pulmonary disease (2013)
3. Rijksinstituut voor Volksgezondheid en Milieu, Gezondheid en determinanten - Deelrapport van de VTV 2010 Van gezond naar beter. Houten: Bohn Stafleu Van Loghum (2010)
4. Croitoru, A., et al.: Benefits of a 7-week outpatient pulmonary rehabilitation program in COPD patients. Pneumologia **62**(2), 94–101 (2013)
5. Nici, L., et al.: American thoracic society/European respiratory society statement on pulmonary rehabilitation. Am. J. Respir. Crit. Care Med. **173**(12), 1390–1413 (2006)
6. Shahin, B., et al.: Outpatient pulmonary rehabilitation in patients with chronic obstructive pulmonary disease. Int. J. COPD **3**(1), 155–162 (2008)
7. Karapolat, H., et al.: Do the benefits gained using a short-term pulmonary rehabilitation program remain in COPD patients after participation? Lung **185**(4), 221–225 (2007)
8. Egan, C., et al.: Short term and long term effects of pulmonary rehabilitation on physical activity in COPD. Respir. Med. **106**(12), 1671–1679 (2012)
9. Gosselink, R.: Respiratory rehabilitation: improvement of short-and long-term outcome. Eur. Respir. J. **20**(1), 4–5 (2002)
10. Spruit, M.A., et al.: Exercise training during rehabilitation of patients with COPD: a current perspective. Patient Educ. Couns. **52**(3), 243–248 (2004)
11. Du Moulin, M., et al.: Home-based exercise training as maintenance after outpatient pulmonary rehabilitation. Respiration **77**(2), 139–145 (2009)
12. Beauchamp, M.K., et al.: A novel approach to long-term respiratory care: results of a community-based post-rehabilitation maintenance program in COPD. Respir. Med. **107**(8), 1210–1216 (2013)
13. Garcia-Aymerich, J., et al.: Risk factors of readmission to hospital for a COPD exacerbation: a prospective study. Thorax **58**(2), 100–105 (2003)
14. Tak, E., et al.: Prevention of onset and progression of basic ADL disability by physical activity in community dwelling older adults: a meta-analysis. Ageing Res. Rev. **12**(1), 329–338 (2013)
15. Puhan, M.A., et al.: How should COPD patients exercise during respiratory rehabilitation? comparison of exercise modalities and intensities to treat skeletal muscle dysfunction. Thorax **60**, 367–375 (2005)
16. Korczak, D., et al.: Outpatient pulmonary rehabilitation-rehabilitation models and shortcomings in outpatient aftercare. GMS Health Technol. Assess. **6**, Doc11 (2010)
17. Puente-Maestu, L., et al.: Comparison of effects of supervised versus self-monitored training programmes in patients with chronic obstructive pulmonary disease. Eur. Respir. J. **15**(3), 517–525 (2000)
18. Janssen, P.: Lactate Threshold Training. Human Kinetics Publishers, Champaign (2001)
19. Gosselink, R., et al.: KNGF-richtlijn Chronisch obstructieve longziekten, Koninklijk Nederlands Genootschap voor Fysiotherapie (2008)
20. Ryan, R.M., et al.: Intrinsic motivation and exercise adherence. Int. J. Sport Psychol. **28**(4), 335–354 (1997)
21. Burke, S.M., et al.: Group versus individual approach? a meta-analysis of the effectiveness of interventions to promote physical activity. Sport Exerc. Psychol. Rev. **2**(1), 19–35 (2006)
22. House, J.S.: Work Stress and Social Support. Addison-Wesley Pub. Co., Reading (1981)

23. Sonderen, L.P.: Het meten van sociale steun. Groningen: University Library Groningen (1991)
24. Bailey, B.W., McInnis, K.: Energy cost of exergaming: a comparison of the energy cost of 6 forms of exergaming. Arch. Pediatr. Adolesc. Med. **165**(7), 597–602 (2011)
25. Anderson-Hanley, C., et al.: Exergaming and older adult cognition: a cluster randomized clinical trial. Am. J. Prev. Med. **42**(2), 109–119 (2012)
26. Huis in't Veld, R.M.H.A., et al.: Exergaming and older adult cognition: a cluster randomized clinical trial. J. Telemed. Telecare **42**(2), 109–119 (2012)
27. Dikken, J.B.J.: Design and evaluation of an integrated training system for COPD patients. In: Biomedical Signals and Systems, University of Twente: Enschede, p. 105 (2012)
28. Dikken, J.B.J., van Beijnum, B.J.F., Hermens, H.J.: Integrated training system for the improvement of the physical condition of COPD patients. In: Fourth Dutch Conference on Bio-Medical Engineering Egmond aan Zee, The Netherlands, BME (2013)
29. Botman, M.P.L.: Design and evaluate a webportal for physiotherapists and COPD patients for virtual group training at home. In: Biomedical Signals and Systems, University of Twente, Enschede, p. 105 (2013)

A Modular Android-Based Multi-sensor mHealth System

Avval Gupta$^{(\boxtimes)}$, Anju Kansal, Kolin Paul, and Sanjiva Prasad

Indian Institute of Technology Delhi, New Delhi, India
avval.07@gmail.com

Abstract. This paper describes the design and methodology for building a versatile, portable and user-friendly mHealth solution for capturing data using multiple electronic sensors. The paper identifies the various dimensions of information that need to be captured to provide a degree of trustworthiness in the system, and describes the protocols and mechanism for its realization. The application uses a set of basic electronic biomedical sensors to capture vital medical information of the patient, combines these with personal, biometric and contextual information and generates an XML based encounter report. The biomedical sensors are connected to a micro-controller that interfaces to a smartphone via one of many possible connectivity modes. The applications running on the smartphone can then analyse the data and metadata and provide feedback to the health care provider and patient, or upload the XML report to a server on demand.

Keywords: mHealth · Android · *mDroid* · Multiple sensors · Data acquisition · Aggregation · Metadata validation · Protocols

1 Introduction

The proliferation of smart mobile phones [2], better internet services as well as the development of cheaper and portable sensors provide a unique opportunity for altering the modes of health care delivery. *mHealth* describes the use of mobile phone technology in the delivery of health care to users in a timely and effective manner, improving the quality of patient experience, and lowering the cost of health care [3]. Even the most basic mobile phones can act as powerful tools when deployed in health care by delivering real-time, critical information through phone calls and text messages [4]. mHealth solutions can support continuous health monitoring at both the individual level (personal monitor) and population level [5]. However there remain several critical challenges in the development and adoption of mHealth solutions, particularly in the developing world [6,7].

Most scenarios involve the collection of various medical data of patients during a clearly identified period, which we call an *encounter*. Medical data forms the primary information that is collected in any mHealth encounter. The raw data collected are usually later collated and entered into a computer, and can

© Springer International Publishing Switzerland 2015
G. Plantier et al. (Eds.): BIOSTEC 2014, CCIS 511, pp. 360–377, 2015.
DOI: 10.1007/978-3-319-26129-4_24

then be uploaded to servers. Manual entry at any stage makes the entire system vulnerable to human errors and will require constant scrutiny to ensure reliability [8]. Another important aspect of any mHealth implementation is ensuring that the data are exchangeable. Existing systems frequently overlook the fundamental requirements of interoperable data standards [9]. Also, medical data by themselves will not be of much use unless accompanied by certain metadata that add a degree of trustworthiness to the system.

Thus we would desire an mHealth solution that: (1) has an automated system for collection of medical data by health workers with minimal training, (2) consists of all basic medical sensors and is yet portable, (3) is affordable for mass usage in under-developed countries, (4) collects data in a format that can be easily exchanged and usable within heterogeneous systems, and (5) provide metadata to make the information complete and useful having an acceptable degree of veracity.

mDroid [1] is an effort which makes use of an Android based device to capture medical data through various electronic sensors with ease, combines them with personal and biometric information of the patient to form an XML based medical report, which can be uploaded to a server on request.

Currently there are a variety of sensors that can be interfaced to computing and communication devices, thus reducing the possibility of errors while entering medical data. However, each such sensor comes with its own peculiarities on how it interfaces with smartphones or servers. Configuring each sensor to operate together with a smartphone app is cumbersome. Each sensor may present different interfaces or have different formats for communication and drivers for synchronisation. In other words, there are few if any standard protocols for sensor abstraction, other than those at the level of physical communication (e.g. Bluetooth HDP).

This paper presents the *design and prototype implementation of mDroid*, an mHealth system that comprises middleware residing on a smartphone and its counterpart on a micro-controller board that permits the connection of a variety of standard sensors to an Android application. The *mDroid* design accommodates a large class of sensors, and can work over different communication modes (Bluetooth, Wi-fi, USB). *mDroid* allows the application developer to build the system in a modular fashion, separating device connection and data communication issues from the application logic. Thus it provides an abstraction from the connection/communication mechanisms as well as from the particulars of the sensors used. Furthermore, by collecting and tagging medical data with metadata, *mDroid* is able to render them in a standard logical form which can be easily amenable to a variety of analyses or conversion to formats supported in common medical repositories.

The prototype implementation of *mDroid* which we describe here is capable of capturing medical data such as Body Temperature, Pulse Rate, Oxygen Saturation, Galvanic Skin Response and ECG, personal and biometric information of the patient, sensor metadata, information about the health worker along with date, time and GPS information.

Fig. 1. End-to-end data flow of the *mDroid* system.

Figure 1 shows a bird's eye view of the *mDroid* system for data acquisition from a collection of sensors to data repositories. These sensors have a wired connection with a micro-controller board which, in turn, fetches samples from the sensors and transmits them to a smartphone. The smartphone app forms the heart of the whole system driving the encounter between the health worker and the patient. It acts as an aggregating device and generates an XML based report consisting of the various parameters captured. The encounter report is then uploaded through a delay tolerant implementation on availability of network to the server, from which it can be pushed to various repositories as required.

Having presented the schematic structure of the *mDroid* system above, we now outline the structure of the rest of the paper. In Sect. 2, we first present the different types of data that are captured (Sect. 2.1), followed by how data acquired from the sensors is transferred to the smartphone (Sect. 2.2). In particular, we detail the logical and physical structure of the communication packets (Sect. 2.3), and then describe the validation of the data on the smartphone (Sect. 2.4). Next we describe the prototype implementation of *mDroid* in Sect. 3, detailing both the sensor hardware (Sect. 3.1) as well as the Android app (Sect. 3.2). Section 4 presents how *mDroid* can be used in a variety of scenarios. We briefly review the related work in Sect. 5, including a comparison with *mDroid*. Finally we conclude in Sect. 6, outlining various lines of future work.

2   Data Collection in mHealth

The *mDroid* system is designed to capture data efficiently from multiple sensors simultaneously, rather than individually, which can be time-consuming, inefficient, error-prone and cumbersome. Concurrent capture is achieved by using a micro-controller board which acts as an intermediate platform that is capable of connecting to multiple sensors at the same time and transmitting data to the smartphone through one of the connectivity options (Bluetooth, Wi-fi, etc.) supported by the two communicating devices. The smartphone acts as an aggregating device where further data and metadata collection and processing takes place before finally sending the data (possibly tagged with metadata) downstream to various data repositories as required by the application.

Table 1. Summary of information dimensions captured.

Component	Description
Medical Data	Captured from sensors
	Validated in the app
Personal Information	Basic patient information e.g. name, age, gender, contact information etc.
	Manually entered
Biometric Information	Patient Image captured from integrated camera
	Used for tagging the data
Sensor Metadata	Information about sensors e.g. serial no., batch no., make, model etc.
	Fetched from sensors
Health worker Information	Health worker Id, name etc.
	Manually entered
Date, Timestamp, GPS Location	Associated with data collection encounter
	Captured from smartphone OS/middleware
	Used for tagging the data

2.1   Types of Data Captured

We first identify the different kinds of data that need to be captured in a reliable mHealth system. *mDroid* is designed to be a complete system that is easy to use and which covers all important aspects needed to be captured in a reliable and useful mHealth solution. Medical data form the primary pieces of information to be collected in any mHealth encounter. However, medical data alone will not be of much use unless they are tagged with certain crucial metadata. These metadata are required to attest to the provenance and quality of the data collected, so that the medical data collected can carry a degree of assurance of authenticity and trustworthiness.

We propose the following classification of information that need to be captured for a good mHealth system, as shown in the Table 1. For simplicity, we have listed only a limited set of metadata classes in our prototype implementation. This set of data and metadata can be enlarged in future extensions and adaptations of the *mDroid* design.

Medical Data form the principal pieces of information collected in the mHealth scenario. These data comprise readings from multiple biomedical sensors (e.g., temperature from a thermometer, etc.) collected together. The design of the system allows fetching multiple sensors' readings together within a particular time frame. This allows for the various physiological parameters to be correlated for any medical/research analyses.

Personal Information comprises the basic personal information about the patient, such as Name, ID, Gender, Date of Birth etc. It also includes basic contact information (for contacting the subject in case of any emergency or abnormality detected during analysis). In the *mDroid* system, we collect personal information by manual entry into the smartphone app.

Biometric Information forms another dimension of the patient information. Including biometric information adds several advantages to the system. It can, for instance, be used to digitally identify the patient in the data repository. It also induces confidence that the data collected by the health worker is in fact authentic and has been collected by visiting the patient. In our prototype *mDroid* system, a simple and straightforward way to include biometric information is to capture the patient's image by using the integrated camera of the smartphone. Alternative biometric techniques may also be explored, particularly in scenarios where confidentiality is paramount.

Sensor Metadata is fetched from external sensors through the micro-controller board. The sensors' specifications require them to be able to provide device details such as serial number, batch number, model, make etc. This information is captured so that any malfunctioning sensors can be identified and reported. It can be also used to add further verification of usage of assigned sensors by health workers.

Health Worker Information comprises information such as the name, ID etc. used to identify the user who is responsible for the mHealth encounter. It is entered manually by the user in the app and is helpful in tracking the performance of health workers assigned for the data collection job, e.g., for quality and training purposes.

Timestamp and GPS information is used to tag the data collected (especially those collected automatically) during an encounter. This information is captured through the smartphone operating system. It helps identify the date and geographical location of the mHealth encounter and is a useful parameter in verification and validation of the data collected.

2.2 Data Acquisition from Sensors to Smartphone

The *mDroid* system supports multiple medical sensors. These sensors provide two kinds of data (1) medical data, and (2) sensor metadata. The algorithm for obtaining a set of data samples from the sensors by the smartphone app is shown in Fig. 2.

   In general, any request for acquiring data involves the smartphone application sending a REQUEST byte to the micro-controller board, which in turn fetches data samples from the sensors and sends them back to the smartphone app bundled in a RESPONSE packet. Once a RESPONSE packet is received by the app, it undergoes a series of validations and integrity checks as explained in Sect. 2.4. Certain data values may be re-fetched in case of corrupt or unacceptable values until finally the request is completed and user is satisfied with the readings fetched.

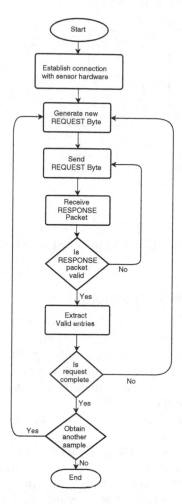

Fig. 2. Flow chart for obtaining data samples from the sensors by the smartphone app.

2.3 Structure of Data Exchanged Between Microcontroller and Smartphone

The data obtained from the sensors are categorised into *discrete data points* and *continuous data streams*. Discrete data comprises the discrete physiological parameters such as temperature, pulse rate, oxygen saturation in blood etc., whereas continuous data covers graphical medical parameters such as ECG, which are required to be plotted on a graph over an uninterrupted time interval. It should be noted here that the distinction between discrete and continuous sensors is only applicable for collection of medical data. No such distinction applies to the metadata collected across the sensors.

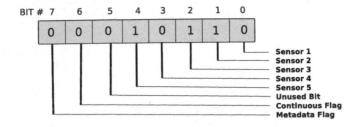

Fig. 3. Example of REQUEST byte. The request is for a combination of data from sensors 2, 3 and 5. The MSB is not set denoting that the request is for medical data from these sensors.

The fundamental messages exchanged between the micro-controller board and the smartphone are REQUEST byte, RESPONSE packet, data stream, and STOP byte.

2.3.1 Structure of REQUEST Byte

The fetching of all kinds of sensor data is controlled by the smartphone application. Based on the set of data required, the application generates a REQUEST byte and sends it to the sensor hardware. A REQUEST byte is used to encode *'what'* has to be fetched. It specifies the kind of data that has to be fetched – is it a request for medical data or sensor metadata? In the case of medical data, is it a request for a discrete data packet or for a continuous sensor data stream? Additionally, the REQUEST byte also encodes the specific sensors from which these data are requested.

The physical structure of the REQUEST byte is exemplified in Fig. 3. In the current version, the *mDroid* system is capable of handling 5 types of medical sensors. The 5 lower order bits (Bit 0 through Bit 4) of the REQUEST byte correspond to the five respective sensors while the highest bit (Bit 7), called Metadata Flag, is used distinguish between the type of data requested (medical data or metadata) for the specified sensor(s). Bit 6, called Continuous Flag, is used to differentiate between discrete data request and continuous stream request. One bit (Bit 5) is left unused for future extension. The design can be easily extended to support more sensors by increasing the size of the REQUEST word. The uniqueness of this design lies in the fact that it provides flexibility in using variable combinations of sensors at the same time as per the requirement of the mHealth application.

2.3.2 Structure of RESPONSE Packet

Whenever the micro-controller board receives a request for collecting sensor metadata or a discrete set of medical readings, it fetches the values from the sensors, couples them together into a data packet and transmits the packet to the connected smartphone. We call this data packet a RESPONSE packet.

This RESPONSE packet design allows all the sensor data points collected at a time to be transmitted together. Coupling readings from different sensors makes it possible to correlate the different physical parameters.

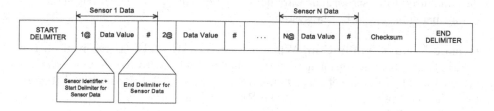

Fig. 4. Design of a RESPONSE packet.

The physical structure of a RESPONSE packet is shown in Fig. 4. Each data packet consists of a frame start delimiter, end delimiter, sensor identifiers, sensor data start and end delimiters, and actual sensor data (medical or metadata). Sensor identifiers are used to label particular sensors' data.

The design of the RESPONSE packets allows the *mDroid* system to form custom data packets depending on the specific sensors requested by the user application, supporting flexibility in the set of sensors used by an application and eliminating the need for maintaining a fixed sequence of sensors in the data packets.

2.3.3   Continuous Data Stream and STOP Byte

There are certain medical parameters which are required to be captured continuously over an uninterrupted time interval. Examples of such parameters are ECG, PPG etc. which make sense only when they are plotted over a continuous time interval. The continuous data stream is acquired separately without coupling the values with any other sensor data.

On receiving a request for a continuous stream, the micro-controller board sends the stream start byte followed by the sensor identifier to indicate the sensor whose data is going to be streamed. It starts fetching data points from the sensor and transmitting a stream of values to the smartphone device, until it receives a STOP request.

The STOP request is a special character represented as a byte, which tells the micro-controller to stop streaming continuous data.

2.4   Data Validation

An important design decision of the *mDroid* system has been to focus on making the data collected have a degree of veracity and trustworthiness. Keeping in mind this design decision, the system performs, on the smartphone, various validation checks on the data collected for every mHealth encounter.

The data which are entered manually into the app need to satisfy certain format validations depending on the field being entered. However, there is only so much trust that can be imposed on manually entered data and the system relies greatly on the trustworthiness of the user for entering correct details. The data acquired from sensor hardware goes through the following validation checks before being accepted by the system.

Packet Format Validation. When a RESPONSE packet is received on the smartphone from the sensor hardware, it is first checked to see that it conforms to the RESPONSE packet format illustrated in Sect. 2.3.2. It is verified to contain all the required delimiters. If the RESPONSE packet fails this check, it is discarded and the REQUEST byte is sent again to re-fetch the samples.

Integrity Check. Once the format validation is passed, the RESPONSE packet is checked to ensure the integrity of the data received. The micro-controller board calculates the checksum by taking exclusive-OR of all the bytes contained in the response and then attaches to the RESPONSE packet as shown in Fig. 4 before sending it to the smartphone. The checksum is again calculated using the same method on the smartphone end and crosschecked with the existing checksum value contained in the RESPONSE packet to confirm that the data packet is not compromised. Just as in the case of packet format validation, if the RESPONSE packet fails this check, it is discarded and the REQUEST byte is sent again to re-fetch the samples.

Content Validation. Once a packet has been validated to be complete and intact, the smartphone app extracts various data readings from the packet based on the packet delimiters. The system performs two kinds of content validation on these data values extracted. First, the data type is checked to match with the sensor specification, i.e., whether the data type should be integer, float etc. filtering out corrupt values that do not match the specifications. Second, a check is performed against the valid range of values defined for the specific sensor values. It may be noted here that this range needs to be specified by the user (or the application designer).

In addition to automatic validation checks, the system also facilitates user validation by giving control to the user to choose fetching of data samples from sensors as many times as needed till she gets satisfactory results.

2.5   Coupling All Data and Metadata Together

The final piece that completes data collection through the *mDroid* system is the report generation module which is responsible for combining and representing all kinds of data captured in a meaningful and useful format that can be easily transferred across different platforms and easily interpreted. This is achieved by generation of an XML based report, which links all the pieces of information together. The design of the medical record is based on XML because of its simplicity of representation, versatility and widespread acceptance as a vehicle for information exchange in the modern era.

3   Implementation

The components used in the implementation of the *mDroid* system were carefully selected keeping in mind portability and low power consumption. Also, special emphasis was paid to design the user interface of the mobile phone application such that an amateur user with minimal training can capture and upload data with ease. The *mDroid* implementation can be broadly divided into two parts (1) Sensor Hardware, and (2) Smartphone Application.

3.1   Sensor Hardware

Sensor hardware comprises all the biomedical sensors along with a micro-controller board to which these sensors have a wired connection. These biomedical sensors are basic sensors which have a simple interface with the micro-controller board and are available at low cost.

In our prototype implementation of the *mDroid* system, the main parameters captured are Body Temperature, Pulse Rate, Oxygen Saturation, Galvanic Skin Response and ECG.

Microcontroller Program. The medical sensors described above can be interfaced with a micro-controller board such as Arduino or Raspberry Pi and derive the power directly from the board. In our prototype implementation, we have used the Arduino UNO R3 board.

The code for interfacing the Arduino board to the medical sensors is written using the open source Arduino 1.0.4 software. The prototype implementation uses Bluetooth as the mode of connectivity between the board and the smartphone although the design does not depend on it. RN-42 Bluetooth Module is used for the Bluetooth transmission using the Serial Port Profile (SPP).

Algorithm 1. Pseudocode for the Microcontroller Program.

```
1  while connection exists do
2  |    read REQUEST byte;
3  |    decode REQUEST byte;
4  |    write generated RESPONSE to output port;
5  end
```

Once a connection has been established with the smartphone app, the micro-controller program keeps listening for a REQUST byte. Based on the request, it fetches the required values from the sensors, bundles them in a RESPONSE packet and transmits the packet to the smartphone. The pseudocode for the micro-controller program is shown in Algorithms 1 and 2.

Algorithm 2. Pseudocode for Decode REQUEST Byte Function.

Data: REQUEST byte

1 check MSB;
2 **if** *metadata flag is set* **then**
   | // Metadata is requested
3 | **for** *each sensor* **do**
4 | | **if** *bit is set* **then**
5 | | | fetch metadata;
6 | | | append to RESPONSE;
7 | | **end**
8 | **end**
9 **else**
10 | **if** *continuous flag is not set* **then**
   | | // Discrete data is requested
11 | | **for** *each sensor* **do**
12 | | | **if** *bit is set* **then**
13 | | | | fetch sensor data;
14 | | | | append to RESPONSE;
15 | | | **end**
16 | | **end**
17 | **else**
   | | // Continuous stream is requested
18 | | **while** *STOP byte is not received* **do**
19 | | | fetch continuous sensor data reading;
20 | | | write to output;
21 | | **end**
22 | | **return**;
23 | **end**
24 **end**
25 write RESPONSE to output;
26 **return**;

3.2 Android Application

The *mDroid* smartphone application is developed using the Android SDK 2.3.3 Java platform. Android was chosen for the implementation of the system owing to its widespread use, affordability, and extendibility. The application supports any device running Android version 3.0 and above with Bluetooth 2.0. The prototype implementation is done keeping in mind usage by rural health workers for generating an encounter report. The encounter workflow of the Android application comprises: (1) obtaining patient information and health worker information, (2) obtaining patient image, (3) obtaining medical data, and (4) combining all data into a medical report to be uploaded to server.

Obtaining Patient and Health Worker Data (Personal Information and Health Worker Information). The first component in the patient record system is obtaining the general information about the patient and health worker. The mHealth encounter starts with the user being asked to enter this information through a form based user interface. The module collects the name, ID, age, sex, and phone number of the patient along with name and ID of the health worker, and performs suitable validations on the data entered before allowing the user to proceed. The various validations imposed by the system are as follows[1]:

- All the entries must be filled. No field can be left blank.
- Name and Sex can take text values. Age and phone number should be numeric values.
- Sex can take only **M** or **F** as values.
- Phone number must be 10 digits.

Capturing Patient Image (Biometric Information). Once the patient and health worker information have been entered, the encounter workflow navigates the user to capture an image of the patient, which serves as biometric information about the patient. Once an image is captured, the user has the option of either capturing another image (if he/she is not satisfied with the current image) or proceeding to capture the medical data. The latest captured image is taken into account. The image name contains the timestamp when the image was captured and is of the form *IMG yyyymmdd hhmmss X.jpg*, where X is a random number. This allows the image to be uniquely identified on the server avoiding different image files having the same name.

Capturing Sensor Data (Medical Data and Sensor Metadata). The next step in the encounter workflow is to fetch readings from the sensors. In the implementation of our prototype, the user is prompted to select a combination of medical sensors whose data she wishes to collect.

The app then establishes a connection with the sensor hardware (via the micro-controller board), requests for metadata and medical data one after the other from those sensors using the algorithm described in Sect. 2.2. We have implemented a layered architecture for fetching data from the sensors to the Android device visualized as a communication stack. The details of the design of the stack merits a separate description and we intend to cover it in another paper.

A screenshot of the medical data collection is shown in Fig. 5. The user may choose to capture a 10 s ECG data, which is fetched as continuous stream separately and displayed to the user as a waveform. The ECG values get saved in a text file which can be later used in the report for uploading to the server.

Generating Encounter Report. Up till this point, we have patient information and health worker information input by the user, patient image captured using the smartphone integrated camera, and medical data along with sensor metadata fetched from the sensors. In the next step in the encounter workflow,

---

[1] The specific choices described here only illustrate that these data must be checked for conformance to a specified format.

Fig. 5. Screenshot of *mDroid* app.

the user proceeds to generate the encounter report. At this point, the app picks up the date, timestamp and GPS location from the Android device. Now, we have all the information dimensions captured. The app generates an XML based encounter report which contains all the information captured in a well-defined format. This report gets saved on the smartphone.

A sample medical record generated by the application is depicted in Fig. 6. If the user wishes, he/she can choose to upload this data record onto the server. The encounter report along with the captured image file and ECG text file is uploaded on the server, where the image and text files are stored in separate directories.

4   Application Scenarios

mDroid is a user-friendly, affordable and portable multi-sensor data collection system which has a straightforward encounter workflow for collecting, combining and uploading health information with ease without requiring any specialized or intensive training. The *mDroid* system has been designed in a fashion to make it a versatile mHealth data collection system that has ubiquitous usages, varying from data collection by health workers, usage by health professionals at their clinics or for personal health monitoring by individuals. To illustrate this point, let us consider the following scenarios.

Sushma is a moderately educated health worker in the Navli village of India working for a sub-health centre covering a population of about 500 people. Her job is to make door-to-door visits, note down various medical readings of the residents on a piece of paper, and bring them to the sub-health centre. Several such encounter records are stacked up in the sub-health centre from where they are manually entered into a central database on a monthly basis. This can be easily replaced with the *mDroid* system for data collection, which would not only make the life of health workers like Sushma a lot easier, but also get rid of

```xml
<?xml version='1.0' encoding='UTF-8' standalone='yes' ?>
<mdroid_encounter_record>
  <personal_data>
    <patient_name>Sumitha Reddy</patient_name>
    <patient_age>23</patient_age>
    <patient_sex>F</patient_sex>
    <patient_phone>9900099000</patient_phone>
    <patient_img>Images/IMG_20140502_052321_-1595727957.jpg</patient_img>
  </personal_data>
  <medical_data>
    <body_temperature sensor_id="TMP12345" batch="B1105">37.103002</body_temperature>
    <bpm sensor_id="OXY12345" batch="B2205">74</bpm>
    <spo2 sensor_id="OXY12345" batch="B2205">99</spo2>
    <skin_conductance sensor_id="GSR12345" batch="B3205">-0.082000004</skin_conductance>
    <skin_resistance sensor_id="GSR12345" batch="B3205">3860.0</skin_resistance>
    <ecg sensor_id="ECG12345" batch="B4305">ECG/ECG_20140502_052544_-45436273352.txt</ecg>
  </medical_data>
  <operator_data>
    <operator_name>Sushma Kishore</operator_name>
    <operator_id>AW_201415253</operator_id>
  </operator_data>
  <encounter_tags>
    <date>2014-05-02</date>
    <timestamp>05:26:52</timestamp>
    <gps>
        <latitude>27.234741</latitude>
        <longitude>77.487746</longitude>
    </gps>
  </encounter_tags>
</mdroid_encounter_record>
```

Fig. 6. A sample XML report generated by the *mDroid* app.

countless problems, errors and inaccuracies associated with maintaining paper-based records and manual entry to the databases.

In another scenario, a physician working at a clinic may need to monitor a patient having dengue where the he would need to keep track of patient's body temperature and pulse rate. The *mDroid* system combines various sensors into a single platform facilitating usage for monitoring patients in special situations that require monitoring more than one health parameter. The doctor can easily use the *mDroid* system by selecting only the required sensors, i.e. the temperature sensor and pulse rate sensor in this case.

In a third scenario, a patient suffering from anxiety can be monitored at home by doctors at a hospital by using the ECG and Galvanic Skin Response (GSR) parameters. The team of doctors need not continuously observe his parameters, but instead an application running on the smartphone can analyse the medical data and send an emergency notification if and when there is an indication of the onset of an anxiety attack.

The auto-generated encounter records are XML compliant, facilitating easy consolidation, transmission and retrieval of medical records from servers. Moreover, it allows existing protocols to be used for exchanging data among heterogeneous systems. The capability of XML to exchange data among different types of computing platforms enables disparate back-end systems in health care organisations to work together. The data records generated by the *mDroid* system can hence be easily made available to the care givers and researchers.

5   Related Work

Several technologies have emerged in the recent years for facilitating data capture in mHealth. Some of the related relevant efforts in this field are discussed in this section.

The Swasthya Slate [10] (Health Tablet) developed by Public Health Foundation of India (PHFI) is a useful system for collection of medical parameters from different sensors using Android based tablets. The ODK Sensors [11] framework facilitates development of sensor-based mobile applications by creating a common abstraction that enables all sensors to be accessed through a unified sensing interface. Other related frameworks for mobile health care are the SANA [12] technology developed at MIT and CommCare [13] developed by the Dimagi Corporation. The prime focus of the SANA tool is to transmit the medical information to the reach of various stakeholders in the health care domain. Comm-

Table 2. Comparison of *mDroid* with related software platforms.

Related platform/framework	Key aspects	*mDroid* in Comparison
Swasthya Slate [10]	- multi-sensor medical data collection system - no support for contextual data or XML report	- metadata collection makes system more reliable and trustworthy
ODK Sensors [11]	- framework for including multiple sensor data into mobile app - manages every connected sensor separately	- sensors connected to custom sensor hardware having single connection channel with phone - multi-sensor data collected can be correlated
SANA [12] and CommCare [13]	- support for pushing medical files in the form of plain text and multimedia to server - no support for external sensors	- combines form-based data collection with sensor data to generate complete encounter report
Funf [14]	- support for multiple built-in data probes including data collected by on-phone sensors and various kinds of data managed on the phone such as call-logs, media files etc. - no advertised support for external sensors	- provides support for collecting data from multiple external sensors simultaneously

Care is an open source mobile platform for CHWs and frontline workers, that seeks to boost workers' abilities through improved access, quality, experience and accountability of care. Funf [14] is a sensing and data processing framework for enabling the collection, uploading and configuration of a wide range of data signals accessible via a mobile phone.

The comparison of these software platforms with *mDroid* is summarized in Table 2. By including sensory data along with contextual data, *mDroid* proves to be a reliable and useful mHealth data collection tool.

BITalino [15] provides an interesting platform for data collection and management with an innovative, low cost, modular bio-signal sensor kit that facilitates making quick and easy-to-build medical devices and health tracker apps. It supports a host of sensors for data collection which can interface with computing platforms such as Arduino (and derivatives) and Raspberry Pi. We intend to embed this platform in our architecture for obtaining sensor data in our future experiments.

6   Conclusions

The fundamental concept of data collection in a variety of mHealth scenarios has been identified as an *encounter* between a patient and a user who is equipped with a collection of sensors connected to a micro-controller that communicates with a smartphone. The user collects not only medical data of the patient, but also various kinds of metadata that serve to contextualise the information, thus rendering it more complete and trustworthy. The smartphone acts as an aggregator of the sensor data (and metadata) which it packages into records that may be sent to medical repositories. The smartphone may also locally process the data and perform a variety of analyses that may improve the delivery of quality health care to the patient. The main contributions of this paper are to categorise the various kinds of data and metadata collected; to specify protocols by which data from multiple sensors may be combined, coupled and correlated with metadata; and to specify how these data and metadata may be validated – all within a simple, versatile and easy-to-use framework.

This paper extends [1] by clearly outlining the various kinds of data and metadata, and by performing enhanced validation checks to increase robustness and accuracy in the system. The logical and physical structures of all data exchanged in the system have been detailed highlighting the factors that make the system a flexible and scalable system in mobile health care. The notion of combining metadata information with clinical data increases confidence in the information collected and will help to a great extent in performing various kinds of analyses on the data collected. This work also details the algorithm implemented at the micro-controller board for decoding the REQUEST byte that unfolds how this design can be used to fetch various kinds of combinations of data from multiple external sensors.

The various kinds of metadata collected through the system provide a vision towards development of reliable systems in a variety of areas. In personal health monitoring, technologies like *mDroid* can be envisioned to be linked to a personal history database, wherein a health care professional can interpret the changes in patterns of the data collected over time, and help in analyses of personalised medical behaviours of a person. Correlating health worker information and accuracy of data collected can help in analysing and tracking the work of community health workers better, and serve as inputs for improving the quality and identifying areas of training required for them. Sensor metadata can help in identifying faulty sensors by observing the data collected using certain sensors over time. The information dimensions outlined in this paper have been limited to few parameters for simplicity. This design can be extended to have a host of other metadata information for building trustworthy and veracious mHealth solutions.

Although the design of the system architecture has been made targeting data collection in the health care domain, we note that the design of *mDroid* can be adopted in any sensor-based applications for capturing different dimensions of data in a simple and trustworthy manner.

In this paper we have confined ourselves to presenting the design of an encounter-based multi-sensor system. As noted above, a forthcoming paper will describe in greater detail the "Sensor Stack" for the implementation of the application-level communication protocol, as well as the realisation on an Android platform and micro-controller boards such as the Arduino or the Raspberry Pi.

There also is a lot of further work in the development and refinement of the *mDroid* project. Some of the directions for the future involve working on the security aspects including protection from network vulnerabilities; access control and authorisation aspects for the health data repositories; incorporating provenance analyses; distributed information flow analyses; and integration of XML based records with open source standard repositories for management of electronic medical records. Furthermore, we are interested in employing these concepts while outlining or developing a variety of applications for mHealth monitoring as well as other sensor-based data collection and analysis systems.

References

1. Kansal, A., Gupta, A., Paul, K., Prasad, S.: mDROID - an affordable android based mHealth system. In: Proceedings of the International Conference on Health Informatics, HEALTHINF 2014, pp. 109–116 (2014)
2. Cisco Visual Networking Index: Global Mobile Data Traffic Forecast Update, 2012–2017 (2013). http://www.cisco.com/en/US/solutions/collateral/ns341/ns525/ns537/ns705/ns827/white_paper_c11-520862.html
3. Wang, H., Liu, J.: Mobile phone based health care technology. Recent Pat. Biomed. Eng. **2**, 15–21 (2009)
4. Dglise, C., Suggs, L.S., Odermatt, P.: Short Message Service (SMS) applications for disease prevention in developing countries. J. Med. Internet Res. **14**, e3 (2012)
5. Kumar, S., et al.: Mobile health technology evaluation. Am. J. Prev. Med. **45**, 228–236 (2013)
6. WHO: mHealth: New Horizons for Health Through Mobile Technologies. In: Global Observatory for eHealth Series, vol. 3 (2011)

7. Mechael, P., Batavia, H., Kaonga, N., Searle, S., Kwan, A., Goldberger, A., Fu, L., Ossman, J.: Barriers and gaps affecting mHealth in low and middle income countries. In: A Policy White Paper commisioned by the mHealth Alliance (2010)
8. Otieno, C.F., Kaseje, D., Ochieng, B.M., Githae, M.N.: Reliability of community health worker collected data for planning and policy in a peri-urban area of Kisumu, Kenya. J. Community Health **37**, 48–53 (2012)
9. WHO: Management of patient information: Trends and challenges in Member States. In: Global Observatory for eHealth Series, vol. 6 (2012)
10. Swasthya Slate of Public health foundation of India. http://www.swasthyaslate.org
11. Brunette, W., Sodt, R., Chaudhri, R., Goel, M., Falcone, M., VanOrden, J., Borriello, G.: Open data kit sensors-a sensor integration framework for android at the application-level. In: Proceedings of the International Conference on Mobile Systems, Applications, and Services, MobiSys (2012)
12. The SANA Project (2010). http://sana.mit.edu/wiki/index.php?title=Overview
13. CommCare (2014). http://www.commcarehq.org/home/
14. Funf (2014). http://www.funf.org
15. BITalino (2013). http://www.bitalino.com

A Full Body Sensing System for Monitoring Stroke Patients in a Home Environment

Bart Klaassen[1]([✉]), Bert-Jan van Beijnum[1], Marcel Weusthof[1], Dennis Hofs[2],
Fokke van Meulen[1], Ed Droog[1], Henk Luinge[3], Laurens Slot[3], Alessandro Tognetti[4],
Federico Lorussi[4], Rita Paradiso[5], Jeremia Held[6], Andreas Luft[6], Jasper Reenalda[1,2],
Corien Nikamp[1,2], Jaap Buurke[1,2], Hermie Hermens[1,2], and Peter Veltink[1]

[1] Biomedical Signals and Systems Group, University of Twente, Enschede, The Netherlands
b.klaassen@utwente.nl
[2] Roessingh Research and Development B.V., Enschede, The Netherlands
[3] Xsens Technologies B.V., Enschede, The Netherlands
[4] Interdepartmental Center E. Piaggio, University of Pisa, Pisa, Italy
[5] Smartex SRL, Pisa, Italy
[6] Department of Neurology, University Hospital Zurich, Zurich, Switzerland

Abstract. Currently, the changes in functional capacity and performance of
stroke patients after returning home from a rehabilitation hospital is unknown to
a physician, having no objective information about the intensity and quality of a
patient's daily-life activities. Therefore, there is a need to develop and validate
an unobtrusive and modular system for objectively monitoring the stroke patient's
upper and lower extremity motor function in daily-life activities and in home
training. This is the main goal of the European FP7 project named "INTERAC-
TION". A complete full body sensing system is developed, whicj integrates Inertial Measurement Units (IMU), Knitted Piezoresistive Fabric (KPF) strain
sensors, KPF goniometers, EMG electrodes and force sensors into a modular
sensor suit designed for stroke patients. In this paper, we describe the complete
INTERACTION sensor system. Data from the sensors are captured wirelessly by
a software application and stored in a remote secure database for later access and
processing via portal technology. Data processing includes a 3D full body reconstruction by means of the Xsens MoCap Engine, providing position and orientation of each body segment (poses). In collaboration with clinicians and engineers,
clinical assessment measures were defined and the question of how to present the
data on the web portal was addressed. The complete sensing system is fully
implemented and is currently being validated. Patients measurements start in June
2014.

Keywords: Telemedicine · Architecture · Sensing system · Stroke · Home
environment · Daily-life activities · Monitoring · Performance · Capacity

1 Introduction

Currently, the changes in functional capacity and performance of stroke patients after
returning home from a rehabilitation hospital is unknown to a physician, having no
objective information about the intensity and quality of a patient's daily-life activities.

© Springer International Publishing Switzerland 2015
G. Plantier et al. (Eds.): BIOSTEC 2014, CCIS 511, pp. 378–393, 2015.
DOI: 10.1007/978-3-319-26129-4_25

As a consequence, the physician is unable to monitor the prescribed training program for sustaining or increasing the patient's motor capacity (what a patient is able to do) and performance (what a patient is doing in actual practice) and cannot give advice to the patient outside the hospital setting. Therefore, there is a need to develop and validate an unobtrusive and modular system for objective monitoring of daily-life activities and training of upper and lower extremity motor function in stroke patients. That is the main goal of the European FP7 project named INTERACTION [4]. A physician will be able to continuously evaluate the patient's performance in a home setting by using the INTERACTION system, allowing the physician to compare the patient's performance at home with the patient's capacity in the rehabilitation hospital. Thereby, the system will support the physician in making decisions on, for example, altering the prescribed training programs.

The INTERACTION sensor system is composed of Inertial Measurement Units (IMUs), Knitted Piezoresistive Fabric (KPF) strain sensors, KPF goniometers, EMG electrodes and force sensors. These sensors are integrated into a custom-made modular suit for stroke patients (e-textile), which consists of a shirt, a pair of trousers, shoes and gloves. The iterative design process for the sensor suit includes several usability tests as well as an extensive user requirements analysis with medical and technical experts. Data are captured wirelessly on a home-gateway, which transmits the data to a secure database. Portal technology can access and process the data. The results can be consulted by a clinician whenever necessary.

In this paper, we describe the complete sensing system, including the architecture and the requirements for presenting the assessment measures to clinicians. Specifically, in Sect. 2, the system requirements are given along with an overview of the whole system and a detailed description of each component. The design of the sensing suit is described in Sect. 3. In Sect. 4, the data processing aspects of the system will be explored in further detail. In Sect. 5, the design process of the data presentation is elaborated upon. In Sect. 6, the current implementation of the system is presented and finally, in Sect. 7, the conclusions and future work are described.

2 System Architecture

2.1 System Requirements

Five major requirements were set before the initial system development:

1. The system should compute and display motor capacity and performance measures to evaluate stroke patients during daily-life activities (for example: grasping an object) in a home setting.
2. The sensing system should be unobtrusive for patients to wear and easy to use.
3. The system should be divided into several modules: upper extremity (shirt), lower extremity (trousers), gloves and shoes. This will allow clinicians to assign different modules to different patients according to the clinicians specific interests.
4. Analysis of the sensor data will not be done in real time. The system should be able to store the computed data such that it can be accessed by a clinician when needed.

5. The system should present the performance information of the patient to the clinician, such that it optimally supports monitoring the progress of the patient and decisions about continued therapy. The clinician should be able to inspect the information in progressive detail from global performance parameters to details concerning the quality of specific movement tasks, according to his or her needs.

2.2 System Architecture Overview

The INTERACTION system's architecture is based upon a generic architectural approach described by Pawar et al. [9]. Figure 1 shows a general overview of the current system's architecture. The Body Area Network (BAN) is composed of several sensors listed in Table 1 and a home gateway. The Xsens wireless Awinda protocol is used to connect and synchronize the sensors to the home gateway, which captures the data and stores it in a European Data Format (EDF) [2]. The EDF file protocol was extended for the INTERACTION project by adding additional signal labels to the header of the file. The home gateway application uploads the EDF file to a secure and remote SQL database if an internet connection is detected. A server, installed at the University of Twente, runs Liferay portal software with Matlab [4, 5]. The portal obtains the data from the database and sends the results to Matlab for processing. The results are saved and visualized on the web-portal on request. Each component is explained in detail in the following subsections.

Table 1. Sensor overview.

Type of sensor	Number of sensors			
	Shirt	Trousers	Pair of shoes	Pair of gloves
IMU[a]	6	4	2	2
KPF Strain[b]	2			
KPF goniometer[b]	1	2		6
EMG electrode set[b]	2			
Force[c]			4	12

[a]Xsens MTw and MTw CE sensors [15].
[b]Developed by the University of Pisa.
[c]Tekscan FlexiForce® [12] and Interlink force sensors [5] for shoes and gloves.

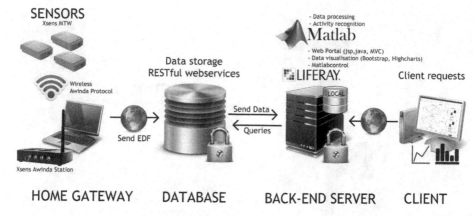

Fig. 1. System architecture. The home gateway captures sensor data and sends it in a EDF to a secure database. Portal technology can access and process the data and visualize the results to clinicians on a web-portal.

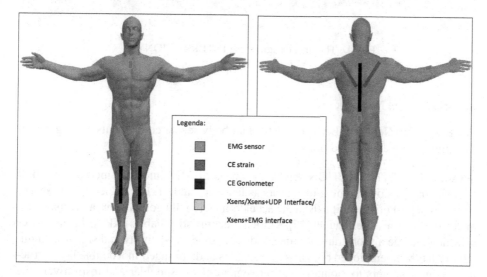

Fig. 2. Sensing system overview. The CE goniometers over the knee and spine are removable and only one EMG sensor set, located on the affected side of the patient, is connected.

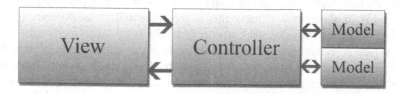

Fig. 3. A spring Model-View-Controller structure of a portlet.

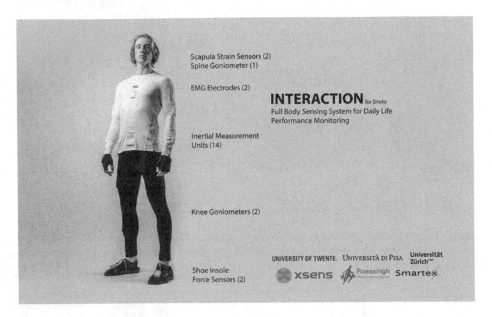

Fig. 4. The final design of the INTERACTION suit.

2.3   Body Area Network

The Body Area Network (BAN) consists of all body sensor components and a gateway to capture, store and upload sensor data.

Sensors. The INTERACTION sensor system is divided into four modules, each of which comprises of a number of sensors as listed in Table 1. Each Xsens MTw sensor box includes 10 primary signals: a 3D accelerometer, a 3D goniometer, a 3D magnetometer and one Pressure channel [15]. Knitted Piezoresistive Fabric (KPF), the properties of which include a short transient time, reduced aging, washability and signal reproducibility has been employed both as strain sensors and, arranged in a double layer structure, as goniometers to monitor joint movement of the shoulder and respectively the thoracic spine, knees and fingers. The KPF sensors are developed by the University of Pisa; the strains are fully integrated in the e-textile suit, but the goniometers for the knees and spine are removable. Two sets of EMG electrodes are integrated into the shirt (on the left and right shoulder) and the signal is pre-processed by an on-body front-end system into a smooth rectified signal. Only the EMG electrode set, located on the affected side of the patient, is connected. The choice of realizing an unobtrusive, minimal EMG sensor set led to the use of the electrodes on the deltoid only, due to its anti-gravitational function, to detect its activity and discriminate the presence of compensatory movements in the shoulder.

The kinaesthetic and kinetic glove module was designed to identify reaching and grasping activities. KPF goniometers were integrated in the metacarpal-phalangeal area of thumb, index and medium fingers. Moreover, force sensitive resistors were integrated in the palm and lateral side of the glove to measure contact pressures and give an indication of hand loading in stroke patients. An additional force sensor was integrated on the lateral side of the middle phalanx of the forefinger to complete the information derived from the goniometers and to improve the discrimination between hand positions that have similar joint angular values (Fig. 5). Force insoles were made for different shoe sizes to be fitted into the patients shoes. Each insole comprises of two pressure sensors: one under the heel and the other under the forefoot, such that they measure the pressures at the main pressure points under the feet (Fig. 6).

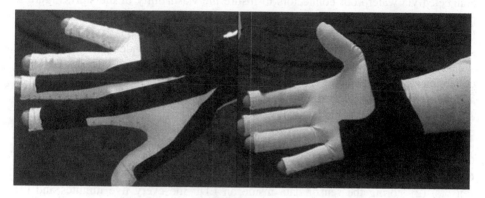

Fig. 5. Glove design. KPF goniometers were integrated in the metacarpal-phalangeal area of thumb, index and medium fingers. Force sensitive resistors were integrated in the palm, lateral side of the glove and the lateral side of the middle phalanx of the forefinger.

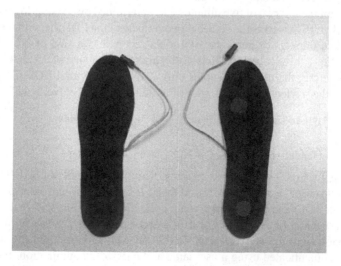

Fig. 6. Force shoe insoles. Each insole comprises of two pressure sensors: one under the heel and the other under the forefoot. The right insole shows the sensor locations.

The Xsens Awinda Auxiliary Data Functionality enables users to combine data coming from a wide range of devices with the inertial data from the MTw sensors and use the wireless link of the MTw as a means of transporting this data. The KPF strain sensors, KPF goniometers, EMG electrode set on the patient's affected side and force sensors on the gloves and shoes are each physically linked to a modified MTw sensor box by passing the data to the MTw's pressure channel. The strain sensors and force sensors on the shoes and the force and goniometers on the gloves are multiplexed to be sent over one channel. As a result, the wireless capabilities of the MTw's for data transmission from the BAN to the gateway are preserved. The integrated sensors are connected via waterproof connectors to insure the washability of the shirt. Figure 2 provides a global overview of the sensing system for the upper and lower extremity. Each MTw sensor unit outputs 10 primary signals, each of which is assigned a unique sensor label within the EDF file. The data collection rate is dependent on the number of sensors. In the INTERACTION project, the collection rate is set to 20 Hertz. This data collection rate has been assessed to be appropriate, since 3D kinematics is analyzed at a higher frequency (1800 Hertz) inside the MTw sensor units before transmission to the Awinda base station. This local analysis provides a more accurate estimation of acceleration and angular velocity values. 20 Hz is an appropriate rate for transmission of 3D orientation as well as the other quantities measured by the sensors, as specified in Table 1.

Gateway. The home gateway has three main functions: (1) collecting the data from the sensors, (2) storing the sensor data inside an EDF file every five minutes and (3) uploading the EDF file to the database. The sensor data is first logged using an Xsens mtb file structure to handle sensor packet loss and data retransmissions and then converted to EDF. The software automatically tries to reconnect with the sensors in case one or multiple sensors are out of range.

The data storage interval of five minutes was determined by considering the available network bandwidth as well as the decreasing overhead of the EDF file with measurement time. With five minutes of data, the EDF data record has a size of 1.6 MB in total according to Eq. 1 (also called the "payload") and a header size of 35.25 kB according to Eq. 2. Therefore, the header occupies only 2.15 % of the total EDF file space. The ratio between the payload and the header of a file is called the "overhead".

$$\text{Payload size} = M_t * F_s * N_{imu} * N_{signals} * 2 \text{ bytes} \tag{1}$$

$$\text{Header size} = \left(N_{imu} * N_{labels} + 1\right) * 256 \text{ bytes} \tag{2}$$

The inputs for Eqs. 1 and 2 are as follows: Nimu: 14 (Number of IMU's), Nsignals: 10 (Number of sensor signals per IMU), Nlabels: 10 (number of sensor labels), Mt: 300 s (Measurement time) and Fs: 20 Hertz (data collection rate). The 256 bytes in Eq. 2 is the size of the general EDF header. The data is uploaded to the database with SSL secure data encryption over the network using RESTful web services, and the users of the database are authenticated using a username and password combination. Furthermore, within the EDF file, only a device ID is used to identify each sensor suit, so no patient names are exchanged.

The home gateway software is used by clinicians to setup the INTERACTION system for collecting patient data at home or in the clinic. The user interface was designed in an iterative process throughout the INTERACTION project. Several usability tests were done during software development. Clinicians were monitored (with video cameras) while performing several measurement scenarios with the hardware and software while thinking out loud. After the scenarios were done, interviews were conducted to get the clinician's opinion. This cycle was repeated several times while developing the gateway software. Options in the interface include: choose a patient, choose the type of measurement (a calibration measurement, a performance measurement or a capacity measurement), switch a sensor (in case of a malfunction) and view sensor data. The interface provides visual feedback on the duration of the measurement, when sensors get out of range or when data is uploaded to the server.

2.4 Database

An SQL database was configured at the Roessingh Research and Development centre (RRD) by reason of their technical experience in secure databases. Dedicated API's were constructed for communication between the home gateway and portal using SSL. For obtaining the data, a correct combination of username and password is required to authenticate the user, and a separate authorization model is in use which determines the access rights of the user, including his or her reading and writing rights. A query engine is developed based on RESTful web services to obtain EDF sensor data from the RRD database on receiving a request from the web portal with a start and end time.

2.5 Portal

The web-portal is responsible for controlling and visualizing the data. We chose the Liferay portal framework as it provides a flexible working environment to develop portlets in a Spring Model-View-Controller (MVC) structure using Java, JavaScript, CSS and JSP. This structure is shown in Fig. 3. Liferay includes a dedicated Content Management System, which allows the portal to be personalized for different users by means of a detailed access-control scheme for assigning different rights to different users. The View component is responsible for displaying the processed data to the user and includes two visual libraries: the Highchart library [3] for graphs and the Bootstrap library [13] for a responsive layout and website elements. The front end Controller component is connected with the View and initiates the Model(s). The Model components obtain the data from the database by use of multiple queries and subsequently send the data to Matlab via a Matlab-Java bridge [8]. Users are able to send requests for different types of measurement data in a specific portlet (by pressing, for instance, a button on the web-portal). These requests are directly forwarded to the Controller component associated with that portlet. This Controller initiates several Models accordingly. With this MVC structure, we are able to process and visualize large amounts of data in an organized way. Multiple portlets can be constructed, each having a different function to show different types of data on the same web-page or on separate web-pages.

3   E-Textile Suit Design

The INTERACTION sensor suit combines a shirt, trousers, shoe insoles and gloves. These fabrics are characterised by a different amount of elastic component and different weight. A heavier fabric with higher elasticity was used for the sensing part of the shirt and the trousers to guarantee a good fitting with the body shape, whilst a fabric based on ceramic components was used to realise the rest of the suit. These mineral components, that are introduced at fibre level, affect the thermo-regulation of the body and improve blood micro-circulation when in contact with the skin for more than six hours. The properties of the fibres do not deteriorate after repeated washing cycles. Several prototypes were realized in a joint effort between designers, engineers, clinicians and stroke patients. To evaluate the functionality of the KPF strain and goniometer sensors, a series of testing prototypes have been designed. These prototypes offer the possibility to evaluate different configuration of the system by varying the location, the dimension and the number of textile sensors. The prototypes were accessorised with velcro® strips to facilitate this process. The positions of the strips were varied to test the functionality of the corresponding sensors that can be attached with the velcro® on the garments. Decisions on the materials, dimensions, shapes and locations of the textile EMG electrodes were made after a round of experiments, performed to determine the best solutions that optimized the functionality of the garments in terms of their sensing properties [11].

The sensing glove has to accurately fit the hand in order to provide an adherence similar to a second skin. Furthermore, thermal comfort is another fundamental requirement, solved by the selection of a suitable material in term of breathability and elasticity. A patented fabric has been used for the basic prototype of the glove, which combines two types of fibres: a polyamide microfiber and LYCRA® elastomer. In Fig. 5, the glove is shown in detail. The force shoe insoles were designed based on Regular shoe insoles. Two layers of fabrics were merged, where the force sensors were placed in between to make a tight fit. The force insoles are shown in Fig. 6.

The final designs resulted in a system that balances the wearability properties of the prototypes with the requirements in terms of the positions and mechanical constraints of the sensors. Patients tested the final designs, in which solutions were added to increase the easiness of wearing and removing the sweatshirt and the leggings (for example, zippers on the side of the shirt and sleeves). Different designs were made for male and female. In total, four complete sensing systems and one back-up system were made, each with several e-textile clothing sizes ranging from S to XL. In Fig. 4, the INTERACTION sensor system is shown.

4   Data Processing

Within INTERACTION, sensor data is captured from up to 14 IMU's. The data is stored in a secure database and will be processed over night. This ensures a short waiting time during the day for displaying patient reports to clinicians on a website. Matlab [7] software is used for data processing, which also includes several external libraries from Xsens. The data processing flow is shown in Fig. 7.

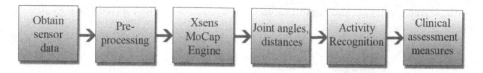

Fig. 7. Data processing flow.

The first step in processing the data is obtaining sensor data in EDF from the database. This step is realized by the use of RESTful web services to access the database and implemented via a custom-made Java portlet in the Liferay web-portal software. The second step is to initiate Matlab and pre-process the EDF sensor data for the later steps. The pre-processing includes matching the correct calibration data with the measurement data, estimating the orientation of the sensors and structuring the data. Furthermore, Additional sensors from the INTERACTION suit, such as the KPF strain sensors on the shoulders, goniometers and force sensors inside the glove and force sensors in the shoe insoles, are demultiplexed.

Step three consists of inserting the pre-processed sensor data into the Xsens MoCap Engine (XME). The XME computes poses of all body segments, where a pose is defined as an orientation and position of a body segment [10]. A full body 3D reconstruction can be made by using the XME. From these poses, joint angles according to the ISB standards and several kinematic distances (like the hand-sternum distance) are calculated as part of step four.

Table 2. Examples of clinical assessment measures in the INTERACTION system.

1	Arm usage of the affected and non-affected arm
2	Maximum reach of the affected and non-affected arm
3	Range of Motion of the elbow and shoulder of the affected and non-affected arm
4	Range of Motion of the trunk
5	Maximum grasping force of the affected and non-affected arm
6	Number of grasps of the affected and non-affected arm
7	Number of steps, step length and step time
8	Weight support by affected and non-affected leg

In step five, basic activities of the patient are detected based on the results of step three and four. A number of daily-life activities were classified and these activities are shown in Fig. 8. The activity algorithms were developed with the goal of getting a high specificity in identifying the activities. In the final step, the INTERACTION clinical assessment measures are computed. These measures are presented to the clinicians in the form of a report and should provide valuable insight into the patient's capacity and performance during daily-life activities. These measures are being determined in a joint effort among clinicians and engineers. Several examples of these measures are listed in Table 2.

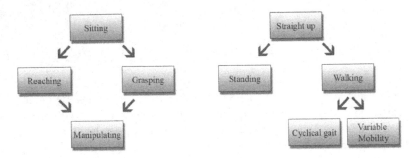

Fig. 8. Activity classification.

5   User Interface

Designing a graphical user interface for clinicians to access the web-portal and determining which clinical assessment parameters to present on the web-portal is one of the major challenges in the INTERACTION project. The INTERACTION system will be collecting data that clinicians are not familiar with in current practice and the data has to be presented in a format that clinicians can understand and evaluate within a few minutes. Therefore, in close collaboration with clinicians, we investigated which clinical outcome measures are relevant and how to present the data in such a way that the capacity and performance of a patient can be easily evaluated and compared over time. We first collaborate with clinicians directly involved in the project only and, when final decisions have been made, will evaluate the results with clinicians not related to the project. At the start, interviews were conducted with clinicians and engineers from the Netherlands (Roessingh Research and Development centre, Enschede) and Switzerland (University hospital in Zürich and Cereneo Rehabilitation centre, Vitznau). We concluded the following: Clinicians can have as many as 40 stroke patients in treatment at a given moment, all of whom have to be evaluated within one hour by the end of the week. This amounts to only a few minutes per week to analyze the performance of each patient. Hence, there is a need for a basic overview of all patients on the web-portal with an option to successively drilldown to a particular data set for a particular patient. As soon as the patient data is processed, it will be available in a report format on the website. This report includes the assessment measures of INTERACTION for the upper extremity and lower extremity. A basic overview of all the subjects is available. The clinician is then able to choose a subject and a measurement day and ask for the report for that day. In this report, the patients capacity measurements in the clinical setting (consisting of multiple clinical measures such as the Berg Balance Scale [1]) are compared with his or her performance measurements in a home environment and can also be compared with healthy subject measurements. An example of the upper extremity webpage is shown in Fig. 9.

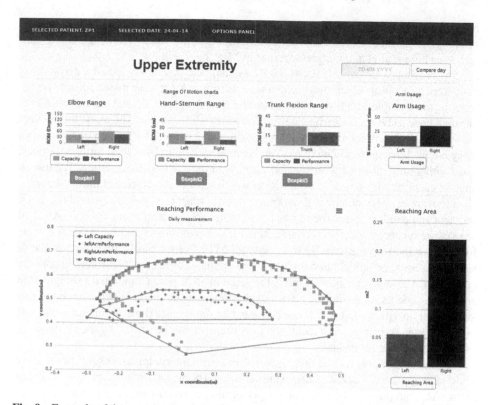

Fig. 9. Example of the upper extremity webpage. The patients capacity is compared with his or her performance during reaching activities for the left and right arm. Several range of motion graphs are shown on top. A plot of the hand-sternum distance in the transversal plane (x,y) is shown on the bottom. The outline is visualized and the area in this outline is computed.

Different Range of Motion charts for the elbow angle, hand-sternum distance and trunk flexion angle are shown for comparing capacity with performance during reaching activities. Box plots are available as well to show the distribution. Furthermore, the reaching positions of the left and right hand relative to the sternum are shown as a top view (x,y) along with, for example, the reaching areas for comparing the left and right arm. Extracting relevant assessment measures and how to visualize them is, and remains an ongoing process within INTERACTION.

6 Implementation

We finished and tested the complete system architecture, from sensors to web portal, with a full body configuration. This includes all system components with over 14 Xsens MTw sensors. Prior to this, the systems architecture was constructed by first using one sensor, then extending it to three sensors with a combination of a basic upper body biomechanical model and finally to a total of 14 sensors with a full body biomechanical model. The gateway software, web-portal software and biomechanical model were developed in parallel and merged together in May 2014.

To measure with the INTERACTION system, the Xsens Awinda base station is connected to a laptop which runs the gateway software. A pre-determined sensor configuration (for example, full body with gloves and shoes), subject ID and the type of measurement have to be set using the gateway software options menu. The mode to initialize the system then becomes available. Initializing the system includes waking up the sensors (they are in sleep mode when not used and can be woken up by a slow turning motion), and waiting until all sensors are synchronized with the Awinda base station. We use an anatomical print with sensor locations upon which sensors can be placed during the initialization phase. The clinician then knows which sensor is assigned to which body segment and can later place the sensor boxes in the correct textile suit pockets. After all sensors are synchronized, the software automatically goes into calibration mode and waits for the user input to measure. In this phase all MTw sensor boxes are placed in the e-textile suit pockets and additional sensors like strain, goniometers and force sensors are connected. The subject is now instructed to stand in an N-pose (standing up straight with arms alongside the body) for 20 s for a calibration measurement. When a calibration is successfully performed, the software continues to the specified measurement mode and when the user and subject are ready, the measurement can be started by the clinician by pressing "start measurement".

A number of test were done on healthy subjects with the complete sensing system, prior to the start of patient measurements, in the lab and in a home environment in April and May 2014. An example of a healthy subject performing a 10 - m walking test is shown in Fig. 10. A 3D visual reconstruction is shown on top and the left knee joint angle is plotted for the different axes according to the ISB standards. The walking test started by sitting in a chair, then standing up, walking 10 m in a normal pace, turning around, waiting, walking back and finally sitting down again. During sitting, the knee joint angle is at about 90 degrees flexion, and when walking it oscillates from 5 to 75 degrees for this particular subject. The activity recognition schemes are successfully implemented and healthy subject measurement data was successfully selected for the specified activities for further analysis. Upper extremity clinical assessment measures were shown on the web-portal based on the test data.

Several training days were organised in Enschede and Zurich to train clinicians in using the INTERACTION sensor system. The training included how to use the hardware and software of the complete sensing system in several measurement scenarios and practicalities such as how to wash the specialized e-textile suits. In Fig. 11, clinicians are measuring with the complete sensing system. The sensing suit is worn under regular clothing. Support protocols were created so that technical experts are available during patient measurements and one complete set of sensors was built as backup in case of sensor failure. Measurement protocols were made for patient measurements in the clinic and at home.

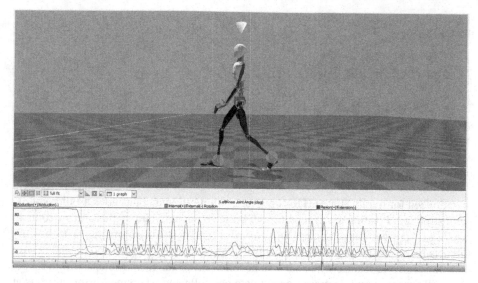

Fig. 10. INTERACTION 3D full body reconstruction during a walking test. The left knee joint is plotted over time for abb/adduction, internal/external rotation and flexion/extension.

Fig. 11. Clinicians measuring with the INTERACTION system. The sensing suit is worn under regular clothing.

A number of lab and in-home tests were done by wearing an on-body tablet pc with an Xsens Awinda USB dongle to overcome connection issues with the Xsens Awinda base station when walking out of range. These tests were successful and further options are explored like how patients can safely wear a small tablet on-body (in for example a pouch attached to a belt) and optimizing the wireless connection by the placement of the USB dongle.

7 Conclusion and Future Work

The INTERACTION project aims to develop and validate an unobtrusive and modular system for objectively monitoring of upper and lower extremity motor function in stroke patients during daily-life activities. The system's complete architecture was developed according to the requirements identified at the beginning of the project. The architecture, including all its components, have been tested for up to 14 MTw sensors for a full body configuration. The biomechanical model, including the Xsens MoCap Engine (XME), is optimized for the INTERACTION sensor configuration and provides a full body 3D reconstruction. Position and orientation of body segments, joint angles and anatomical distances were computed successfully from several test measurements with healthy subjects in the lab and at home. The gateway software is fully tested and extended with an auto-reconnect feature when sensors become out of range and return within range. This will ensure that more measurement data are collected when measuring subjects at home, which is prone to out of range issues. Furthermore, the portal's MVC structure has been designed to be extensible and provides a flexible coding environment for engineers by the inclusion of a Matlab-Java bridge for back-end data processing algorithms. The XME, and the algorithms for activity recognition and for computing clinical assessment measures are included within the back-end data processors.

An option was investigated for the out of range issues, namely by measuring with an on-body tablet. This option was successfully tested in the lab and at home with a full body configuration and is currently being optimized for use. The next step is to validate the sensor suit by comparing it with Vicon [14] as an optical reference measurement system. The activity recognition algorithms for upper and lower extremity measures need to be validated as well and the specificities need to be determined. Patients measurement are starting in June 2014 at both the Roessingh Research and Development clinic in the Netherlands and the neurorehabilitation clinic Cereneo in Switzerland as well as at the patients home. At home measurements will provide additional challenges as external factors such as unpredictable magnetic distortions and movement of clothing will have larger influences on sensor data than in a controlled environment.

In this project, we have identified an extensive list of potential clinical assessment measures. The list of clinical assessment measures given in this paper is an example of what the INTERACTION system will deliver. We are now in the process, together with clinicians and engineers, to make a final selection of these measures to be implemented by the system.

References

1. Berg, K.O., Wood-Danphinee, S., Williams, J.T.: Measuring balance in the elderly: preliminary development of an instrument. Physiotherapy 41(6), 304–311 (1989)

2. European Data Format: A simple and flexible format for exchange and storage of multichannel biological and physical signals (2014). http://www.edfplus.info/ 05 June 2014
3. Highsoft Solutions AS: Highcharts JS, interactive JavaScript charts for your website (2014). http://www.highcharts.com/ 05 June 2014
4. INTERACTION: Official INTERACTION project website (2014). http://www.interaction4 stroke.eu 30 September 2013
5. Interlink Electronics Inc.: (2014). http://www.interlinkelectronics.com 05 June 2014
6. Liferay, Inc: Liferay delivers open source enterprise solutions for portals, publishing, content, and collaboration. (2013). http://www.liferay.com/ 30 September 2013
7. Matlab: The MathWorks Inc., Natick, Massachusetts, United States (2014)
8. MatlabControl: Matlabcontrol API for JAVA. (2013) http://code.google.com/p/matlabcontrol/ 30 September 2013
9. Pawar, P., Jones, V., Van Beijnum, B.J.F., Hermens, H.: A framework for the comparison of mobile patient monitoring systems. J. Biomed. Inform. **45**(3), 544–556 (2012)
10. Roetenberg, D., Luinge, H., Slycke, P., Xsens MVN: Full 6DOF Human Motion Tracking Using Miniature Inertial Sensors. Xsens Motion Technologies BV. Technical Report (2009)
11. Sumner, B., Mancuso, C., Paradiso, R.: Performances evaluation of textile electrodes for EMG remote measurements. In: Proceedings of the 35th Annual International Conference of the IEEE Engineering in Medicine and Biology Society. pp. 6510–6513 (2013)
12. Tekscan, Inc.: FlexiForce® (2014). http://www.tekscan.com/flexiforce.html 05 June 2014
13. Twitter Bootstrap: Sleek, intuitive, and powerful front-end framework for faster and easier web development (2014). http://getbootstrap.com/ 05 June 2014
14. Vicon Motion Systems Ltd. Oxford, England (2014)
15. Xsens Technologies B.V: MTW Development KIT Lite (2014). http://www.xsens.com/en/ mtw-dk-lite 05 June 2014

Author Index

Printed in the United States
By Bookmasters